COOPERATIVE EFFECTS

COOPERATIVE EFFECTS

PROGRESS IN SYNERGETICS

Lectures given at a Summerschool at Erice / Sicily
May 1974

Edited by

H. HAKEN

Institute for Theoretical Physics
University of Stuttgart, Germany

1974

NORTH-HOLLAND PUBLISHING COMPANY – AMSTERDAM • OXFORD
AMERICAN ELSEVIER PUBLISHING COMPANY, INC. – NEW YORK

Library of Congress Catalog Card Number 74 84209
ISBN North-Holland 0 7204 0309 x
ISBN American Elsevier 0 444 10778 9

Publishers:
NORTH-HOLLAND PUBLISHING COMPANY - AMSTERDAM
NORTH-HOLLAND PUBLISHING COMPANY, LTD. - OXFORD

Sole distributors for the U.S.A. and Canada:
AMERICAN ELSEVIER PUBLISHING COMPANY, INC.
52 VANDERBILT AVENUE
NEW YORK, N.Y. 10017

PRINTED IN THE NETHERLANDS

PREFACE

In quite different disciplines, such as physics, chemistry, biology and sociology, we observe the phenomenon that out of chaotic, disordered or structureless states there arise completely new states which have well defined spatial structures or which behave in a well regulated manner. The present book gives numerous examples of such disorder-order transitions and provides the reader with the concepts and mathematical tools to cope with these phenomena. One of the big surprises which came to scientists working in these fields was the fact that there are striking analogies in the behaviour of complex systems though these systems may be composed of completely different subsystems (atoms, molecules, animals, humans etc.). Without any doubt a new rapidly developing field has come into existence namely that of studying and comparing cooperative effects in different disciplines. One of its main goals is to find the common roots of the specific behaviour exhibited by systems composed of interacting subsystems. The analogies found in these systems go far beyond those which are now well known in phase transition theory of systems in thermal equilibrium.

An important impetus (but certainly not the only one) for the development of this new field came, without doubt, from laser physics. It was this reason which led Professor Arecchi and Dr. Roess to incorporate a course on "Cooperative effects in multi-component systems" into the summer-school on Quantum Electronics which is held annually at the Centro Majorana at Erice, a center under the directorship of Professor Zichichi.

The present book contains the lectures held at this course. The course directors were Professor Bonifacio and the undersigned. The school- and course-directors took great care in selecting topics and lectures in such a way that a coherent and consistent account of this new field was reached. The authors made all attempts to write their articles in a manner so that they can be understood by students and non-experts. We are sure that this way the book will appeal to a broad audience which is interested in establishing ties between different disciplines. During this course particular emphasis was given to the problem of superradiance, which is of great current interest in quantum optics and which is an excellent example for selforganization.

I should like to use this occasion to thank the director of the scientific center of Erice, Professor Zichichi, for making these courses possible, to thank Professor Arecchi and Dr. Roess for their initiative and support and to thank my colleague, Professor Bonifacio, for his very efficient cooperation. It is a particular pleasure for me to thank my secretary, Mrs. U. Funke, for her tireless and valuable assistance in editing this book.

H. Haken

LIST OF CONTENTS

Cooperative Phenomena, H. Haken, ed.

SYNERGETICS+: BASIC CONCEPTS AND MATHEMATICAL TOOLS

Hermann Haken
Institut für theoretische Physik
der Universität Stuttgart
Stuttgart, Germany

The reader of this book might be puzzled by the variety of its subjects which range from physics over chemistry and other fields to sociology. Thus the obvious question arises: what have all these different topics in common and is it reasonable to try to correlate them? It is the purpose of this introductory chapter to answer such questions, at least in a preliminary fashion.
Let us start with some rather obvious remarks: All systems we are dealing with have in common that they are composed of many subsystems (components) of one or a few kinds (compare table 1 on the next page). Furthermore very often the subsystems do not behave independently of each other, but act in a well regulated manner which in some cases one might even call purposeful . Such systems may show pronounced transitions from disordered states to ordered states or from one ordered state to a different one. Usually these ordering phenomena show up on a macroscopic level. A number of examples are given in the first column of table 2. The second column indicates the ordering phenomenon. Such ordering phenomena have been investigated in great detail in physics with respect to systems in thermal equilibrium, e.g. the ferromagnet and the superconductor. A great deal of work has been devoted in recent years to explain the pronounced similarities of the phase transitions of quite different systems. Here we just mention the theory of scaling laws or the more recent Wilson techniques like the renormalization group.++ The main purpose of this book is, however, to go far beyond this scheme. In recent years it became more and more obvious that phase transition-like phenomena are not the privilege of physical systems in thermal equilibrium but may be found in a wide variety of problems scattered over quite different disciplines. So we will see that a number of the ordering phenomena listed in table 2 are very similar to each other and are governed by only very few principles. This is quite surprising because we are dealing with systems which are composed of quite different subsystems and whose interaction with each other may be of quite a different nature. Nevertheless at a certain level of description which one might call "macroscopic" or "phenomenological", striking similarities become apparent.

+ The word "Synergetics" is constructed from Greek words indicating "joint efforts". It has been chosen as name of a new discipline which aims at the investigation and the comparison of cooperative effects in different fields of research.

++ Compare the contribution by Kadanoff where further references are given.

Table 1

EXAMPLES OF SYSTEMS

Discipline	System	Subsystem
physics	ferromagnet	spins
	superconductor	electrons + phonons
	tunnel diode	electrons
	laser	light + atoms
	super radiant light source	light + atoms
	optical parametric oscillator	light
	fluids	atoms (molecules)
	liquid crystals	molecules
astrophysics	stars	elementary particles
mechanical engineering	thin shells	"atomic" continuum
chemistry	chemical reactions	molecules
biology	brain	neurons
	physiological clocks	molecules
evolution	biological molecules	molecules
ecology	forest	trees
	eco-system	animals + plants
population dynamics	population	animals, humans
sociology	society	individuals

Table 2

EXAMPLES OF ORDER PARAMETERS

System	Ordering phenomenon	Order parameter
ferromagnet	macroscopic magnetization	magnetization
superconductor	superconductivity (macroscopic wave function)	pairwave function
tunnel diode	e.g. bistability	capacitance charge
laser	coherence	electric field strength
super-radiant source	coherent polarization	atomic polarization
optical parametric oscillator	coherence	electric field strength
fluids	e.g. Bénard cells	amplitudes of mode (wave) configurations
liquid crystals	e.g. alignment of molecules	quadrupole-moment of molecules
thin shells (of buildings or containers)	buckling cells	amplitudes of mode configurations
stars	spatial structures, pulsations	density of particles+
chemical reactions	spatial structures, oscillations	density of molecules+
brain (neuron networks)	"coherent action"	firing rate of nerve cells+
physiological clocks	oscillations	density of molecules+
forest	selection of trees	density of trees+
populations	e.g. oscillations	density of animals+
sociology	social groups, formation of political opinion	number of individuals+

+of a given kind

Before describing some general principles which will allow us to deal with these systems we briefly sketch the main features of the above mentioned systems.

In the ferromagnet the subsystems are the elementary magnets (the spins) which above a certain critical temperature (the Curie temperature) point into random directions whereas below that temperature they get aligned giving rise to a macroscopic magnetization.

Fig. a

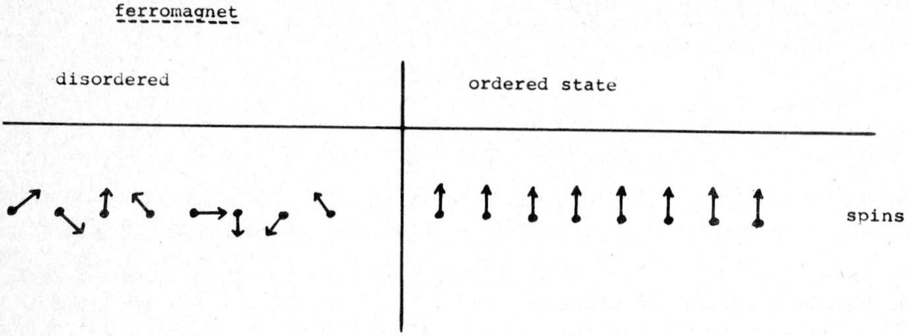

ferromagnet

disordered

ordered state

spins

In the superconductor the electrons are uncorrelated above a certain critical temperature but below that temperature they form pairs. These electron pairs form a state described by **a macroscopic wave** function which finally leads to an infinitely high conductivity and a perfect diamagnetism (in the ideal case)

In a circuit containing a tunnel diode one may find certain transitions between different states of the current depending on the battery voltage. This allows for the construction of bistable elements and logical elements of computers.

When a laser is pumped only weakly it acts as a lamp in which the atoms emit light tracks independently of each other giving rise to completely incoherent light with coherence length of one meter or less. Above a certain threshold of the pump laser light becomes extremely coherent which is caused by an internal ordering of the atomic dipole moments.

The optical parametric oscillator consists essentially of a crystal whose polarization depends nonlinearly on the fieldstrengths. Thus e.g. an incoming coherent light wave may be split into two other waves, the so-called signal and idler. When the incoming field amplitude is small the signal and idler waves are incoherent but they become completely coherent beyond a certain threshold.

Fig. b

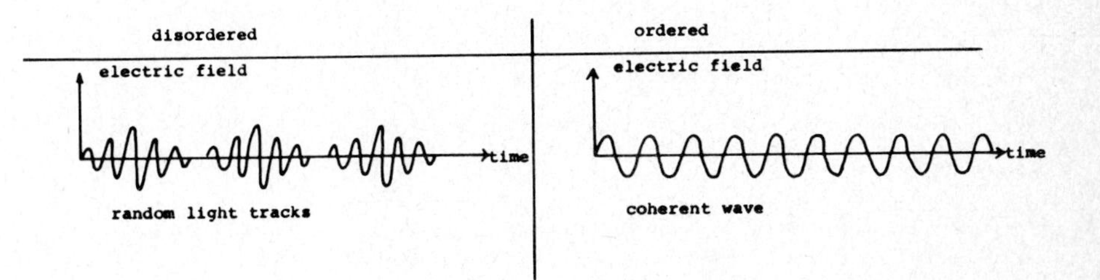

In fluids a number of ordering phenomena is known. We just mention the Bénard instability. In it a fluid layer heated from below remains quiescent if the temperature gradient is small enough, but it shows a convection pattern on a macroscopic scale if the temperature gradient exceeds a certain critical value

Fig. c

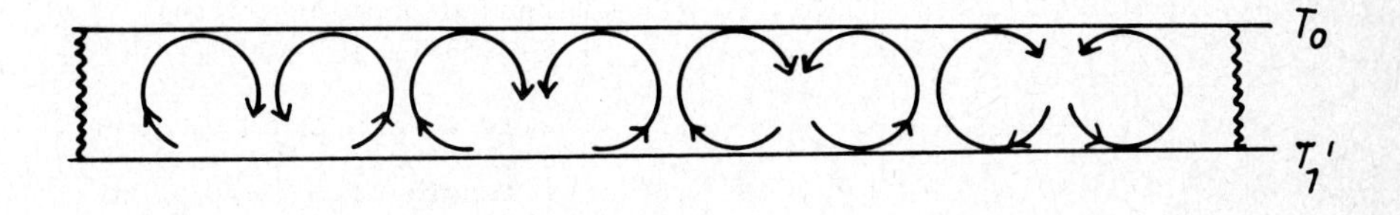

The formation of Bénard cells by convection. The fluid layer is kept externally at two different temperatures.

Fig.d

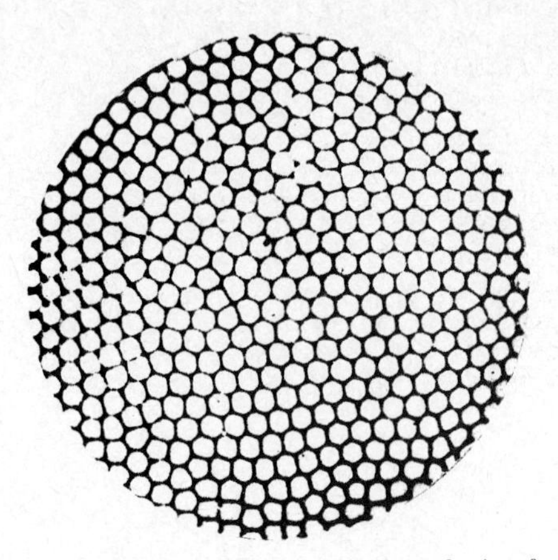

Bénard cells in spermaceti. A reproduction of one of Bénard's original photographs.

Liquid crystals consist of long molecules which get aligned and are even put into layers under certain conditions.

Fig. e

A liquid crystal (schematic)

We conclude this list of examples taken from physics by one of high energy physics. Here in some theories it is assumed that the strongly interacting elementary particles (neutrons, protons, mesons, i.e. the so-called hadrons) may be considered as elementary excitations in analogy to superfluid helium.

In astrophysics it is well known that stars may show spatial structures or pulsations. Again at certain internal parameters of the stars transitions from one state of internal structure to another state of internal structure are found.

In the field of engineering, modern constructions try to use thin shells. Here the following situation occurs. Without external load the shells show, of course, a smooth surface. Under symmetric load suddenly periodic structures of deformation (buckling cells) show up which have an amazing analogy e.g. to Bénard cells.

In chemistry there occur certain reactions in which the reacting molecules and their products form spatial structures or show oscillations. E.G. some of these substances show a periodic change of colour from red to blue and vice versa.

Fig. f

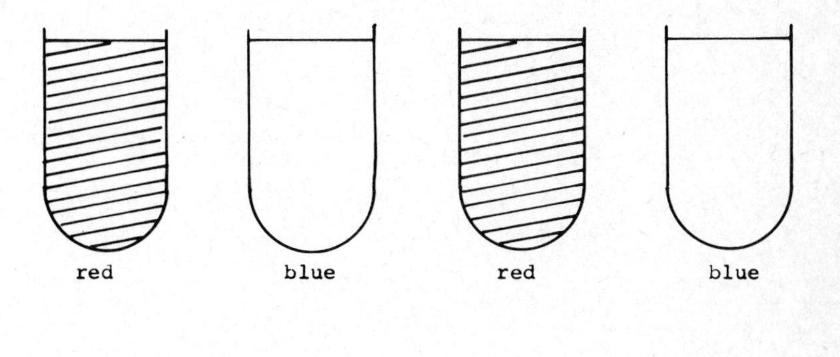

The brain may be considered as a neuron network which shows coherent actions under different external excitations. In particular, information is stored in a delocalized manner.

In biology many examples of physiological clocks are known (think of the heart) which show oscillations quite similar to those of chemical reactions. To quote another ordering phenomena in biology we mention the transition of the spherical shape of the blood cells to a disk shape which may be explained as differently ordered forms of lipid bilayers.

In ecology an example for the transition from one ordered state to another ordered state is provided by the forest. Here it is well known that forests of one kind of trees are replaced by forests of another kind of trees depending e.g. on the climate.

Population dynamics provides further numerous examples for certain ordered states. E.g. in Lotka-Volterra cycle two kinds of fishes, the predator and the prey fishes "interact" with each other. If the

Fig. g

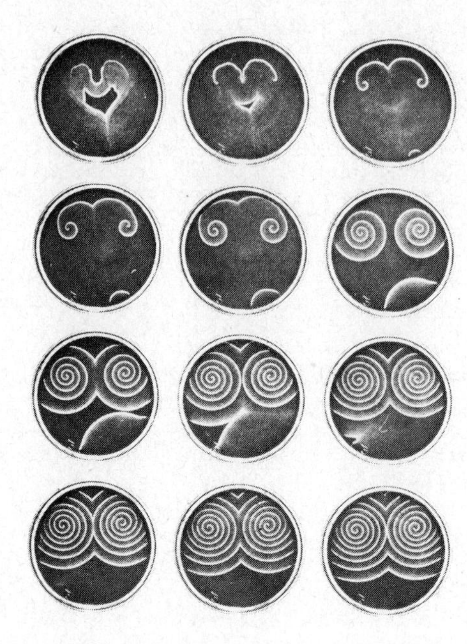

Example of the temporal and spatial development of a chemical reaction (after A.T.Winfree) Scientific American Vol. 23o p.83 (1974)

predator population increases, too many prey fishes are eaten so that the food supply for the predator fishes decreases. Their number drops which allows for the prey fish number to increase again which gives rise to an oscillatory behaviour of the number of predator and prey fishes.

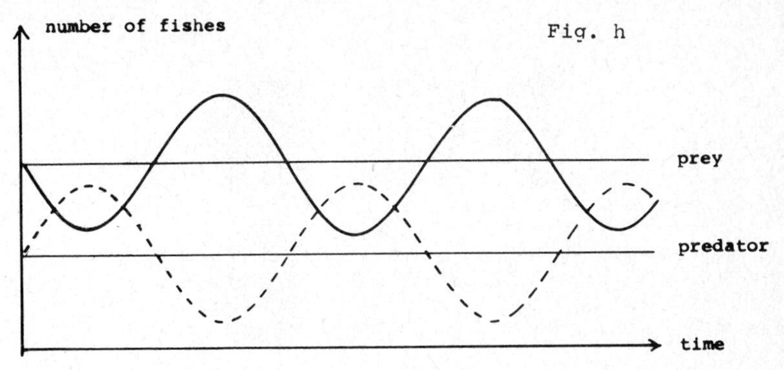

The Lotka-Volterra cycle (qualitatively)

Another example of ordering processes occurs in evolution of biological molecules where an internal selection mechanism which by the way is completely analogous to the selection of laser modes, leads to a selection of the "fittest".

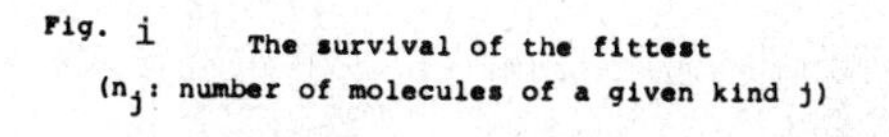

Fig. i The survival of the fittest (n_j: number of molecules of a given kind j)

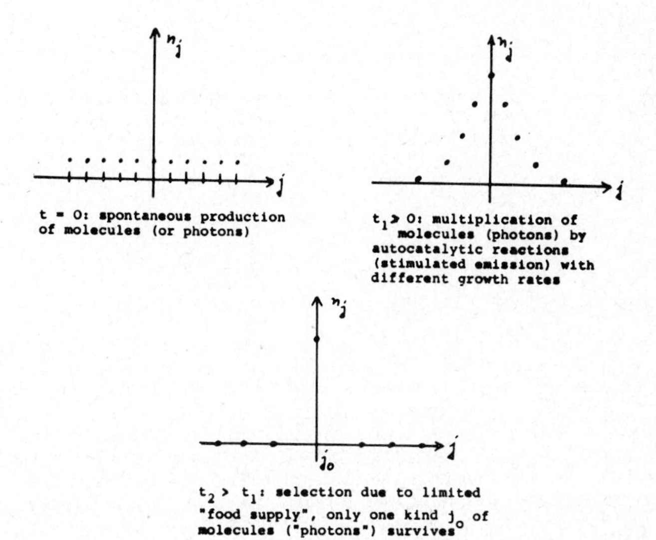

t = 0: spontaneous production of molecules (or photons)

$t_1 > 0$: multiplication of molecules (photons) by autocatalytic reactions (stimulated emission) with different growth rates

$t_2 > t_1$: selection due to limited "food supply", only one kind j_o of molecules ("photons") survives

We conclude this list which could be extended far beyond the present scheme with an example from sociology. Here we observe the formation and growth of political groups and sometimes transitions from one political situation to another one which may be described by phase transitions.

We hope that the above list (and the many examples provided in the present book) convince the reader that there are indeed numerous examples in nature where pronounced transitions from a disordered state to an ordered state occur. What is less obvious at this stage of our presentation, however, is the fact that these ordering phenomena may be described by means of a few principles and exhibit amazing similarities. It would go far beyond the scope of this article to elaborate all these analogies but we hope that the reader will get a feeling how closely these problems are connected with each other.

In the next part we first want to give a few general ideas and then exhibit by means of two examples taken from laser physics and from fluid mechanics how order is brought about and how analogies may show up. In the end part of our chapter we will then briefly sketch several mathematical tools.

The order parameter concept

We now want to discuss the basic concepts which allow us to deal with all these different systems. We have found that the concept of the order parameter, originally introduced by Landau to describe systems in thermal equilibrium, is an excellent tool for an adequate description of many kinds of systems (compare table 2). Let us recall the concept of the order parameter by reminding the reader of the ferromagnet. Here the magnetization plays the role of the order parameter. It fulfills the following purposes:

1) It describes the order on a macroscopic level (it is zero in the disordered state and acquires a maximum value in the completely ordered state).
2) It replaces the direct interaction between the spins (which are brought about e.g. by the Coulomb exchange interaction). So on the one hand it is constructed from all spins and on the other hand it gives orders to the spins.

Before we present an example how to deal with order parameters we briefly describe the general procedure to establish adequate equations for the order parameters. One way starts from first principles, i.e. from the microscopic equations for the subsystems. Describing the subsystems j by "coordinates" or "parameters" q_j depending on the time t, a typical form of such equations may be:

$$\dot{q}_j = K_j(q_1,\ldots,q_N) \quad (1)$$

We then look for macroscopic quantities which may serve as order parameters. Usually there exists a hierarchy of time constants for the relaxation times. The order parameters are usually those quantities which have the longest relaxation times. Due to this hierarchy the subsystems or suitable modes of the subsystems obey the orders of the order parameter immediately and can therefore be eliminated "adiabatically". This leaves us with a set of equations for the order parameters alone which describe the behaviour of a system on a macroscopic level. Usually the microworld of the subsystems leaves its trace in the equations of the order parameters not only by deterministic forces but also by fluctuating forces. So the basic

equations will be in general of a stochastic nature. As will be demonstrated by an example in this chapter and many others in this book, the order parameter equations govern in particular the disorder-order transition. Near such transitions, the order-parameter equations of quite different systems may become quite similar or even identical which explains the analogous behaviour of different systems. Before passing over to explicit examples, we make the following comment. While the order parameter concept has proven to be applicable to amazingly large classes of problems, we should also be aware of its limitations. We mention a few of them:
1) In the "conventional" phase-transitions at thermal equilibrium, refined experimental techniques may reveal a behaviour of the systems which requires more sophisticated theoretical methods for explanation.
2) There may be systems which cannot be described by order parameters at all.
Thus future work should be directed towards more general classifications.

We now proceed, however, within the frame of the order parameter concept and present an example from laser physics which exhibits a good deal of characteristic features. Many other contributions to this book could serve the same purpose, however. We hope to phrase our example in a way, which is also understandable to non-experts of that particular field.

Example: The single mode laser

In the laser we have to treat the interaction of the lightfield with atoms. We describe the lightfield by its electric field strength which we decompose into modes

$$E(x,t) = \sum_{\lambda} \left(b_{\lambda} e^{ik_{\lambda}x} c_{\lambda} + b_{\lambda}^{+} e^{-ik_{\lambda}x} c_{\lambda} \right) \tag{2}$$

We assume field propagation only in one dimension x and running waves. k_{λ} is the wave vector or wave number, c_{λ} are suitable normalization constants which we don't consider explicitly here. What we are mainly interested in are the field amplitudes b_{λ}, b_{λ}^{+}. We may treat b_{λ}, b_{λ}^{+} as classical time dependent amplitudes though the experts will notice that b_{λ}, b_{λ}^{+} are, more strictly speaking, quantum mechanical photon operators. In the following we consider only a single mode and drop the index λ . We want to investigate the mode amplitude b as a function of time and its statistical properties. We distinguish the laser atoms by an index μ and denote the dipole moments of the atom μ by $\alpha_{\mu}^{+},\alpha_{\mu}$. The expert again will recognize immediately that the α's are, strinctly speaking, quantum mechanical flip operators. Finally we introduce the inversion, i.e. the difference of the population numbers of the upper and lower atomic level $\sigma_{\mu} = 1/2(N_2-N_1)_{\mu}$. Then in laser theory it is shown that the field amplitudes b, b^{+} and the atomic quantities $\alpha_{\mu}^{+},\alpha_{\mu},\sigma_{\mu}$ obey the following equations of motion [+]

[+] A block diagram of these equations is given in the figure on the next page. The other figure shows the feedback loop described by these equations.

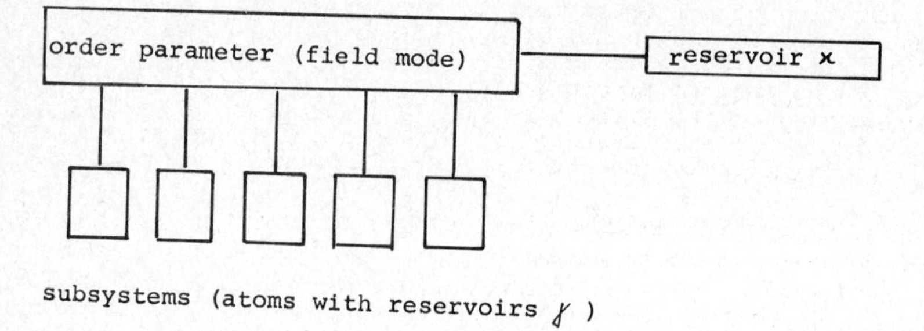

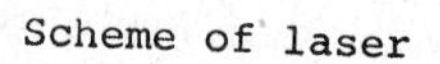
Scheme of laser

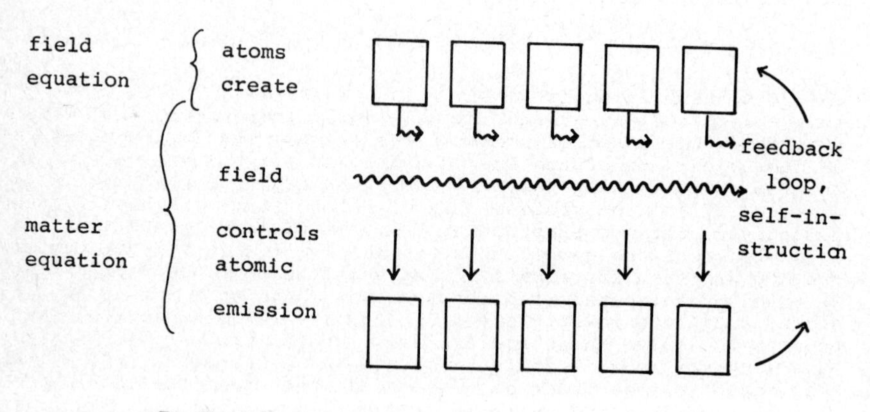

Feedback loop of the laser

a) the field equation

$$\dot{b}^+ = i\omega b^+ + ig \sum_\mu \alpha_\mu - \kappa b^+ + F^+(t) \qquad (3)$$

The left hand side describes the temporal change (indicated by the dot) of b^+. The first term on the right hand side describes the oscillation of the field mode if not interacting with the atoms. The second term describes the driving force of the atomic dipole moments on the field mode where g is the coupling constant which is essentially the optical dipole matrix element. The third term describes the damping of the field e.g. due to absorption of the field by the end mirrors of the laser and $F^+(t)$ describes fluctuations which are always inseparably connected with the damping.

b) the matter equations

b1) equations for the atomic dipole moments

$$\dot{\alpha}_\mu^+ = i\nu\alpha_\mu^+ - i2g\, b^+\sigma_\mu - \gamma\alpha_\mu^+ + F_\mu^+(t) \qquad (4)$$

In eq. (4) the first term of the right hand side describes the undamped oscillation of the atoms, the second term the action of the field on the atom. This term will play a crucial role in the later considerations because it is nonlinear. The third term $\sim \alpha_\mu^+$ describes the damping of the atomic oscillating dipole moments whereas the last term describes fluctuations of the surroundings of the atoms causing all the time "pushes" on their electrons.

b2) equations for the inversion

$$\dot{\sigma}_\mu = 2ig\,(b\alpha_\mu^+ - b^+\alpha_\mu) + \frac{1}{T}(\sigma_o - \sigma_\mu) + \Gamma_{\sigma,\mu}(t) \qquad (5)$$

The first term in eq. (5) describes how the inversion changes due to the interaction with the atomic dipole moment. The second term describes the impact of the external pumping and decay processes on the relaxation of the atomic inversion to its mean value d_o and the last term finally describes the fluctuating forces connected with the relaxation process. The equations (3) - (5) have all the same structure. The first terms describe the coherent motion, the following term the damping and the last term the fluctuations. As one may see from the equations (3) - (5) our system of equations is rather complicated if one imagines that there may be about 10^{14} atoms within the laser. Thus at a first sight it appears rather hopeless to solve these equations which are obviously nonlinear and have furthermore stochastic forces. We want to show, however, that the concept of the order parameter and some additional concepts will help us to cut down the whole problem to a very simple equation which covers the whole dynamics of the system. For simplicity we assume exact resonance, i.e.

$$\nu = \omega \qquad (6)$$

(though this is not a vital assumption) and we further do away the mean oscillatory behaviour by the substitutions

$$b^+ = B^+ e^{i\omega t} \; , \; \alpha_\mu^{\;+} = \tilde{\alpha}_\mu^+ e^{i\nu t} \tag{7}$$

$$F^+(t) = \tilde{F}^+ e^{i\omega t} \; , \; \Gamma_\mu^{\;+}(t) = \tilde{\Gamma}_\mu(t) e^{i\nu t} \tag{8}$$

With the transformations (7), (8) we obtain a new system of equations which reads after dropping the tilde

$$\dot{B}^+ = -\kappa B^+ + ig \sum_\mu \alpha_\mu^{\;+} + F^+(t) \tag{9}$$

$$\dot{\alpha}_\mu^{\;+} = -\gamma \alpha_\mu^+ - i2g B^+ \sigma_\mu + \Gamma_\mu^+(t) \tag{10}$$

$$\dot{\sigma}_\mu^{\;+} = \frac{1}{T} (d_o - \sigma_\mu) + ig(B \alpha_\mu^+ - B^+ \alpha_\mu) + \Gamma_{\sigma\mu}(t) \tag{11}$$

We now describe the procedure how one usually deals with such a system. First of all one assumes that the fluctuations are small which means that we neglect in a first step F^+, Γ^+. The next step is to look for stationary solutions. Because for a small inversion σ_o there is no laser light emission we expect that the field amplitude B^+ vanishes. One then convinces oneself that (9) to (11) is fulfilled by $B^+ = 0$, $\alpha_\mu^{\;+} = 0$, $\sigma_\mu = \sigma_o$. The next step in a usual analysis is to check the stability. To do this one linearizes a system around its stationary values. To show the essential features we over-simplify the treatment somewhat by keeping $\sigma_\mu = \sigma_o$ fixed (as is done very often in this context actually) and confine our further considerations to the equations (9) and (10) which then read

$$\dot{B}^+ = -\kappa B^+ + ig \sum_\mu \alpha_\mu^+ \tag{12}$$

$$\dot{\alpha}_\mu^+ = -\gamma \alpha_\mu^+ - 2 ig B^+ \sigma_o \tag{13}$$

Because B^+ interacts with the atomic system only by the sum

$$S^+ = \sum_\mu \alpha_\mu^{\;+} \tag{14}$$

we introduce this macroscopic polarization as a new variable. Summing up eq. (13) over μ and putting

$$D_o = 2N \sigma_o \tag{15}$$

we then obtain instead of (12) and (13) the equations

$$\dot{B}^+ = -\kappa B^+ + ig S^+ \tag{16}$$

$$\dot{S}^+ = -\gamma S^+ - ig B^+ D_o \tag{17}$$

We now study the stability of the stationary solutions $B^+ = 0$, $S^+ = 0$ by making the hypothesis

$$B^+ = e^{\Omega t} B_o^+ \; , \; S^+ = e^{\Omega t} S_o^+ \tag{18}$$

We confine ourselves to the case $\gamma \gg \kappa$. In that case the eigenvalues Ω read

$$\Omega_1 \approx -\kappa + \frac{g^2 D_o}{\gamma} \tag{19}$$

$$\Omega_2 \approx -\gamma - \frac{g^2 D_o}{\gamma}$$

From (19) we conclude that the system becomes unstable if the total inversion D_o exceeds a critical value

$$D_c = \frac{\kappa\gamma}{g^2} \tag{21}$$

To see what kind of modes get unstable we form the ratios

$$\text{for} \qquad \Omega_1 \; : \; R_1 = \left| \left(\frac{S_o^+}{B_o^+} \right)_1 \right| \approx \frac{g\, D_o}{\gamma} \approx \frac{\kappa}{g} \tag{22}$$

$$\text{for} \qquad \Omega_2 \; : \; R_2 = \left| \left(\frac{S_o^+}{B_o^+} \right)_2 \right| \approx \frac{g\, D_o \gamma}{g^2\, D_o} \approx \frac{\gamma}{g} \tag{23}$$

Apparently (22) is much smaller than (23) so that losely spoken the mode which gets unstable is field-like :

$$B^+ = e^{\Omega t} B_o^+ \,, \quad S^+ \approx e^{\Omega t}\, i\, \frac{\kappa}{g}\, B_o^+ \tag{24}$$

We thus see an important feature which will be observed in many other multicomponent systems. If one changes certain experimental parameters, in our case the inversion, there occur certain unstable modes which grow exponentially. We now want to demonstrate that this exponential growth is limited due of the nonlinearities of the system. To deal with the nonlinearity we go back to the total set of equations (9) to (11) applying now a second typical approximation scheme. Because we have just seen that one mode (namely the field mode) gets undamped whereas the other mode (the atom-like mode) is still strongly damped, we now eliminate the strongly damped mode adiabatically which means mathematically spoken that we neglect the time derivatives compared to the damping constants γ and $1/T$ in (1o) and (11), respectively. The next steps are only of technical nature. From (1o) we obtain

$$\alpha_\mu^+ = \frac{1}{\gamma} \left(-2\, ig\, B^+ \alpha_\mu + \Gamma_\mu{}^+ \right) \tag{25}$$

from (11)

$$\sigma_\mu = \sigma_o + ig\, T (B\, \alpha_\mu^+ - B^+ \alpha_\mu) + \text{fluctuations} \tag{26}$$

or by use of (25)

$$\sigma_\mu = \sigma_o - 4g^2 T \frac{1}{\gamma} B^+ B \sigma_\mu \quad + \text{fluctuations} \tag{27}$$

Because σ_μ is supposed still to be of the vicinity of σ_o and B^+B also to be still rather small we replace on the right hand side of (27) σ_μ by σ_o. Thus (27) acquires the form

$$\sigma_\mu \approx \sigma_o - C B^+ B \sigma_o \quad + \text{fluctuations} \tag{28}$$

with
$$C = 4 g^2 T \frac{1}{\gamma} \tag{29}$$

From (25) - (28) we may now eliminate σ_μ which yields

$$\alpha_\mu^+ = - \frac{1}{\gamma} 2 ig B^+ \sigma_o (1 - CB^+B) + \frac{1}{\gamma} \Gamma_\mu^+ + \text{further fluctuations} \tag{3o}$$

Thus we are able to express α^+ which occurs in (9) completely by the field modes. Inserting now (3o) into (9) we obtain our fundamental equation for the order parameter which is of the form

$$\dot{B}^+ = - \alpha B^+ - \beta B^+ B B^+ + F^+_{tot}(t) \tag{31}$$

with the abbreviations

$$\alpha = \kappa - \frac{2D_o g^2}{\gamma} \tag{32}$$

and

$$\beta = \frac{2 g^2 D_c}{\gamma} C \tag{33}$$

The behaviour of the system on a macroscopic level is now completely described by (31). In our present case we are insofar particularly fortunate because the property of the order parameter B^+ can be measured directly experimentally. We will discuss these properties lateron and demonstrate that our example allows us to discuss a number of well known features of many phase transitions or more generally spoken for transitions from disorder to order.

A second example: the convection instability[+]

Before we enter the discussion of the order parameter equation (31), we demonstrate by means of a second example from a completely different discipline the usefulness of the approach described in the foregoing. We consider a horizontal fluid layer heated from below. If the temperature gradient (or, in dimensionless units, the Rayleigh number R) is small, the fluid remains quiescent. If, however, $R \geq R_c$, convection sets in, giving rise to regular cells (see e.g. fig. c).

+) Readers who are only interested in a quick survey may immediately pass on to the next chapter starting with equation (34).

We briefly sketch the treatment of this problem using exactly the same steps as before.

a) start from equations of motion (with fluctuations)
b) seek stationary solutions of suitably linearized equations
c) check stability
d) find unstable solution
e) eliminate stable solutions adiabatically to find nonlinear equations for the amplitudes of the formerly unstable solutions

The physical quantities are the velocity field $\underset{\sim}{u}(\underset{\sim}{x}) = (u_1, u_2, u_3)$, the density ρ and the temperature T, all taken as functions of space and time.

a) The basic equations read in dimensionless units and after putting $T = T_o - \Theta z$ (z:coordinate in vertical direction)

1) continuity equation

$$\frac{\partial u_i}{\partial x_j} = 0$$

2) momentum equation

$$\frac{\partial u_i}{\partial t} = - u_j \frac{\partial u_i}{\partial x_j} - \frac{\partial \bar{\omega}}{x_i} + P \Theta \lambda_i + P\Delta u_i + F_i^{(u)} ,$$

$$\lambda = (0,0,1)$$

3) heat conduction equation (Θ is the dimensionless temperature after the above substitution)

$$\frac{\partial \Theta}{\partial t} = - u_j \frac{\partial \Theta}{\partial x_j} + R u_z + \Delta\Theta + F^{(\Theta)}$$

P and R are the Prandtl and Rayleigh number, respectively. $\bar{\omega}$ is defined by

$$\bar{\omega} = P/\rho_o + gz - \frac{1}{2} \beta\alpha z^2$$

b) the stationary solutions are given by

$$\underset{\sim}{u} = 0, \Theta = 0$$

c) d) checking the stability reveals that for $R \to R_c$ a certain set of modes gets unstable. They are denoted by $v^{(o)}$ in the table 3 on the next page, where $v^{(o)}$ stands for $\underset{\sim}{u}$ and T.

e) the elimination procedure and its results are sketched in that table.

Table 3: The Bénard problem: stabilization via unstable modes (schematic)

Hydrodynamic equations have the form

v: vector $(\underset{\sim}{u}, T, ..)$,

$\dot{v} = Lv + vNv + F$ (a) N,L contain differential operators; F fluctuating forces

solution of linearized equations

$$\dot{v}^{(o)} = Lv^{(o)} \longrightarrow e^{-\gamma_n t} \, v^{(o)}_{n,k}(\underset{\sim}{r}) \qquad \text{(b)}$$

index n distinguishes discrete modes, k is continuous vector

expansion of v into complete set of modes

$$v(\underset{\sim}{r}) = \sum_{k,n} v^{(o)}_{n,k}(\underset{\sim}{r}) \, A_{n,k}(t) \qquad \text{(c)} \qquad \gamma_o \rightarrow 0 \qquad \gamma_n > 0 \quad n > 1$$

insertion of (c) into (a) yields equations for amplitudes A

unstable mode obeys the equation

$$\dot{A}_o = L^{(o)} A_o + \sum_{n>1} A_o N^{(on)} A_n + F_o \qquad (+A_o N^{(oo)} A_o)$$

(the index k is dropped for simplicity)

stable mode obeys the equation

$$\dot{A}_n = L^{(n)} A_n + A_o N^{(no)} A_o + \ldots + F_n$$

these modes are eliminated adiabatically to yield effective equation for A_o

$$\boxed{\dot{A}_o = L^{(o)} A_o + b \, A_o \, N^{(on)} \, A_o \, N^{(no)} \, A_o + F_o \; (+A_o N^{(oo)} \, A_o)}$$

Table 4

Analogies between the laser threshold and the Bénard instability

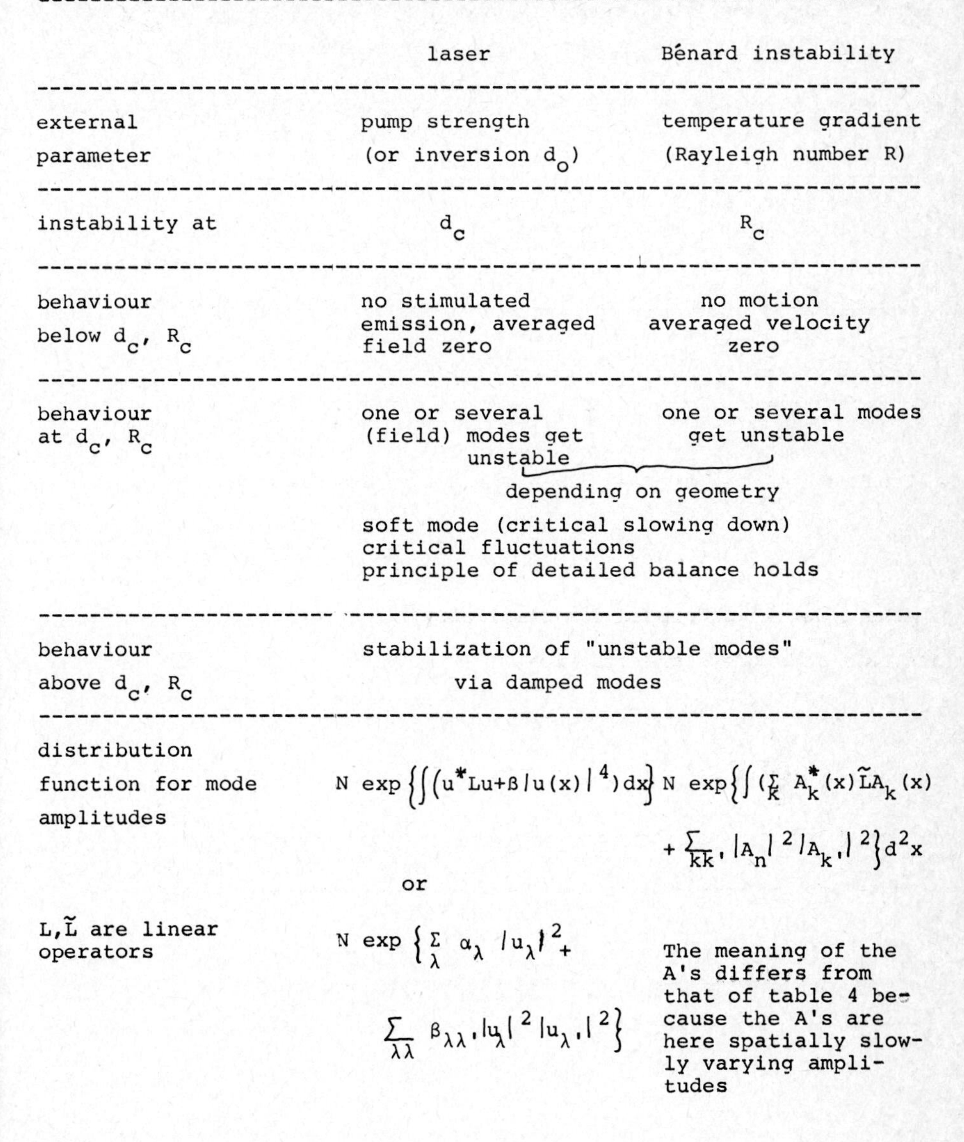

	laser	Bénard instability
external parameter	pump strength (or inversion d_o)	temperature gradient (Rayleigh number R)
instability at	d_c	R_c
behaviour below d_c, R_c	no stimulated emission, averaged field zero	no motion averaged velocity zero
behaviour at d_c, R_c	one or several (field) modes get unstable	one or several modes get unstable
	depending on geometry	
	soft mode (critical slowing down) critical fluctuations principle of detailed balance holds	
behaviour above d_c, R_c	stabilization of "unstable modes" via damped modes	
distribution function for mode amplitudes $L, \tilde{L}$ are linear operators	$N \exp\left\{\int\left(u^* L u + \beta \lvert u(x) \rvert^4\right) dx\right\}$ or $N \exp\left\{\sum_\lambda \alpha_\lambda \lvert u_\lambda \rvert^2 + \sum_{\lambda\lambda'} \beta_{\lambda\lambda'} \lvert u_\lambda \rvert^2 \lvert u_{\lambda'} \rvert^2\right\}$	$N \exp\left\{\int\left(\sum_k A_k^*(x) \tilde{L} A_k(x) + \sum_{kk'} \lvert A_n \rvert^2 \lvert A_{k'} \rvert^2\right\} d^2x$ The meaning of the A's differs from that of table 4 because the A's are here spatially slowly varying amplitudes

Further examples of a similar kind of problems are provided by instabilities of liquid crystals, where one may again use equations of motion including fluctuating forces.

Discussion of a typical order parameter equation

Because in this introductory chapter we want to show only some of the main features we replace in eq.(31) the complex variable B^+ by the real coordinate q and we discuss the equation

$$\dot{q} = -\alpha q - \beta q^3 + F(t) \tag{34}$$

a) stationary states

We neglect the fluctuating force F(t) and obtain as stationary solutions ($\dot{q} = 0$):

$$\alpha > 0 : \quad q_s = 0 \tag{35a}$$

$$\alpha < 0 : \quad q_s = \pm (|\alpha|/\beta)^{1/2} \tag{35b}$$

The further solution (35a) is still stationary but unstable. To investigate the local stability one linearizes

$$q = q_s + \delta q$$

and obtains for $\alpha > 0$

$$\delta\dot{q} = -\alpha\delta q \tag{36}$$

which has the solution

$$\delta q = q_o e^{-\alpha t} \tag{37}$$

Fig. j

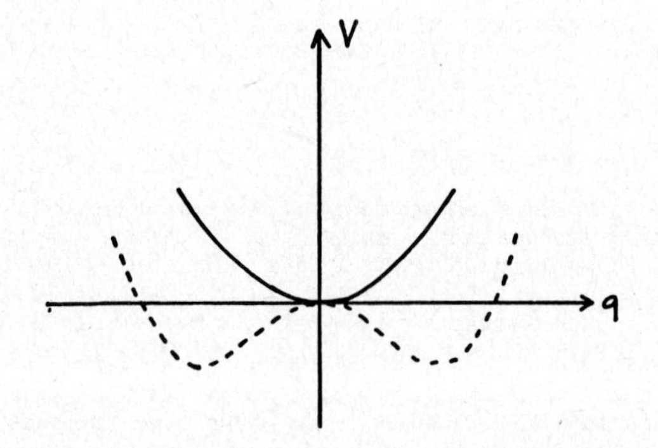

Potential of fictitious particle (eq.(43)

——— $\alpha > 0$

– – – – $\alpha > 0$

Apparently the relaxation time $1/\alpha$ increases if we let the parameter α (compare also (32)) go to zero. This phenomenon is known as critical slowing down or occurrence of a soft mode. In a similar manner we find for $\alpha < 0$ by means of (35) and (34) after linearization the equation

$$\delta\dot{q} = -2|\alpha|\delta q \tag{38}$$

which possesses the solution

$$\delta q = q_o e^{-2|\alpha|t} \tag{39}$$

We now consider the role of the fluctuations F(t) in the linearized theory. We assume that the fluctuating forces are δ-correlated (Markovian)

$$\langle F(t)\, F(t')\rangle = C\, \delta(t-t') \tag{4o}$$

In the linearized theory we then obtain for $\alpha > 0$ after a short calculation

$$\langle \delta q(t)\ \delta q(t')\rangle = \frac{C}{2\alpha} e^{-\alpha(t-t')} \tag{41}$$

In particular we obtain for $\alpha \rightarrow 0$ a divergency of the fluctuations of the coordinate (critical fluctuations). It should be noted that the divergencey

$$\langle (\delta q)^2\rangle \propto \frac{1}{\alpha} \longrightarrow \infty \quad \text{for} \quad \alpha \rightarrow 0 \tag{42}$$

indicates the breakdown of the linearization procedure near a critical point. Instead one has to solve the nonlinear equation (34). Before we describe methods of solution we discuss eq.(34) by means of a somewhat different interpretation. Adding to the left hand side of (34) a term $m \cdot \ddot{q}$ where m is the mass of a particle we may interpret (34) as the overdamped motion of a particle. The first terms on the right hand side possess the potential

$$V = \frac{\alpha}{2} q^2 + \frac{\beta}{4} q^4 \tag{43}$$

which is plotted in the figure of the preceding page.

It is one of the simplest examples to demonstrate how the symmetry may be broken when an instability occurs (symmetry breaking instability). We obtain two new equivalent stable states (35b) instead of the stable state $q = 0$ for $\alpha > 0$. There exist now numerous examples of such instabilities in systems far from thermal equilibrium and even in non-physical systems.

We now briefly return to the example of the laser because it allows us to demonstrate how the behaviour of the order parameter characterizes the macroscopic behaviour of the system. q is now again to be replaced by the field amplitude b. We recognize that below the critical inversion the stable state is $q = 0$, the average field vanishes and the field is only supported by fluctuations (spontaneous emission noise). This is the usual emission of a lamp. If we go beyond the laser threshold suddenly a non-vanishing amplitude q_s shows up. The stability of the field amplitude guarantees that light is now completely coherent. This coherence is only disturbed by small fluctuations caused by F(t). Multiplying eq.(31) by B and adding to it the complex conjugate we find an equation for the photon numbers

$$\dot{n} = -\ 2\alpha n - 2\beta n^2 \qquad \text{(+ fluctuations)} \tag{44}$$

or, in a different form

$$\dot{n} = an \tag{44a}$$

where

$$a = -2\alpha - 2\beta n$$

Order parameter equations in other systems

Equations of the type (44a) occur also in quite other disciplines e.g. in population dynamics. Here this equation describes the exponential growth of a population be it plants, animals or humans. The growth factor a consists of the difference of gain and losses. The gain depends on external sources e.g. the food supply. If there is a limited food supply we may assume that a decreases with increasing population so that one may put for instance

$$a = 2\alpha - 2\beta n \tag{45}$$

which leads back to the eq.(44). Eq.(44) can be generalized to many species leading to equations of the form

$$\dot{n}_\lambda = (g_\lambda - \sum_{\lambda'} \alpha_{\lambda\lambda'} n_{\lambda'})n_\lambda \qquad \text{(+ fluctuations)} \tag{46}$$

These equations govern the mode selection of the laser or the population dynamics of competing species or selection of molecules of the evolution process. Indeed it had been suggested some time ago by us that laser type equations may be very suitable for the description of biological processes because laser equations describe the selection of few modes due to the filter action of g_λ in eq.(46). Thus it is not surprising that equations proposed by Eigen for the evolution process are basically identical with our laser equations. They even contain stochastic production rates of molecules in the same sense as the laser is fed by originally spontaneous emitted photons. The main difference in the interpretation of the two processes consists in the following. The number of available laser modes is fixed all the time, but new types of molecules are created in the evolution process so that one has an open system also with respect to the available species. As the reader will notice in the subsequent lectures there are many more examples of this or related types in which a sudden change of the system occurs if external parameters are altered (in our present example the g_λ's are changed).

Mathematical tools

We conclude this introductory chapter by going over to explain some of the main mathematical tools.
Equations of the type (34) have long been known in physics as Langevin equations which we write in the form

$$\dot{q} = g(q) + F(t) \tag{47}$$

where g may be in general a nonlinear function of q. While the solution of (47) is rather simple in the linear domain it represents a formidable task in the nonlinear region. Therefore we want to discuss some other methods of solving (47). If we had solved (47) we would have obtained $q(t)$ as a stochastic function. Equivalently we could ask for the probability to find a given value of q at a given time t. Let us first consider the stationary state. We then ask for the probability $f(q)dq$ to find q in the region $q...q+dq$. Let us further write this probability distribution function in the form

$$f(q) = N \exp \phi(q) \tag{48}$$

As is well known in equilibrium thermodynamics, $\phi(q)$ is here given e.g. by the entropy or the free energy. The consideration of the probability distribution function (48) has turned out to be of considerable use in non-equilibrium systems as well, because $\phi(q)$ governs the stability, the fluctuations and the dynamics of the system.

a) We may call such states of the system stable for which $\phi(q)$ possesses an absolute maximum. We may call such states metastable for which ϕ possesses a relative maximum.

b) the fluctuations
By expanding $\phi(q)$ around its stable points we find the probability to obtain a fluctuation of the size δq by

$$p \propto \exp\left\{\left(\frac{\partial^2\phi}{\partial q^2}\right)_{q_s} \frac{1}{2}(\delta q)^2\right\} \tag{49}$$

c) dynamics
Finally we may show in several concrete cases that the system develops in time in such a way that it goes from one initial state q_a to a final state q_f so that $\phi(q_f) > \phi(q_a)$. In view of the aspects a) - c) it is an important task to construct $\phi(q)$. This may be done in several ways which we first discuss for the special case of our simple example of a single coordinate but which may be readily generalized to more coordinates.

α) A phenomenological approach (compare also the table on the next page)

The Landau theory of phase transition originally devised for systems in thermal equilibrium has turned out to be an extremely useful tool in dealing with non-equilibrium systems. While the Landau theory starts by the assumption that ϕ is the entropy or the free energy etc. and considers it as a function of the order parameter q we now quite generally assume f in the form (48). Provided (48) is a unique function, $\phi(q)$ must possess the same symmetry as the original problem. We then expand (48) into a power series of q so that thesymmetry of the original problem is conserved. Because our example (34) possesses an inversion symmetry ($q \rightarrow -q$), $\phi(q)$ must have the form

$$\phi(q) = aq^2 - bq^4 + ... \tag{5o}$$

where we have chosen b as positive so that $f(q)$ is a normalizable function. For systems in thermal equilibrium these steps appear quite natural. In systems far from thermal equilibrium or still more in non-physical systems, there exists no free energy in that sense. Thus the forms (48) with e.g. (5o) are here a nontrivial result and found only by explicit examples. (Landauer: tunnel diodes, Laser: Risken, interpretation by Landau theory: Haken, Graham and Haken, De Giorgio and Scully).

Landau Theory of Second Order Phase Transitions

$f = \mathcal{N} \exp(-F/kT)$, F: free energy, $F(T,\varphi) = F_o(T) + \alpha(T)\varphi^2 + \beta\varphi^4$

$\alpha(T) = a(T - T_c)$

state	disordered	ordered
temperature	$T > T_c$	$T < T_c$
"external" parameter	$\alpha > 0$	$\alpha < 0$
most probable order parameter φ_o: $f = max!$; $F = min!$	$\lvert\varphi_o\rvert^2 = 0$	$\lvert\varphi_o\rvert^2 = -\frac{\alpha}{2\beta}$
entropy $S = -\frac{\partial F(T,\varphi_o)}{\partial T}$	$S_o \; (= -\frac{\partial F_o}{\partial T})$	$S_o + \frac{a}{2\beta}(T - T_c)$
	continuous at $T = T_c$	$(\alpha = 0)$
specific heat $c = T(\frac{\partial S}{\partial T})$	$T \frac{\partial S_o}{\partial T}$	$T\frac{\partial S_o}{\partial T} + T\frac{a^2}{2\beta}$
	discontinuous at $T = T_c$	$(\alpha = 0)$

ß) Methods to derive $\phi(q)$ or $f(q)$ from first principles
Another way to establish $f(q)$ is to derive first an equation for it. Without going into all details we just discuss several of such typical equations. One important is the so-called master equation which reads

$$\dot{f}(q,t) = \int w(q,q')f(q',t)dq' - f(q,t)\int w(q'q)dq' \tag{51}$$

In it we consider the temporal change of f which is caused by transitions from other states q' to q and vice versa. If q' is a discrete variable we may write instead of (51) the master equation in the form

$$\dot{f}(m,t) = \sum_n w(m,n)f(n,t) - f(m,t) \sum_n w(n,m) \tag{52}$$

In (51) and (52) the transition rates w may be obtained from (47) but we don't want to discuss this procedure here any further. We want to show instead how (51) and (52) can be solved. Several classes of solutions are now known. We mention the most important ones:

a) The system obeys detailed balance
This means that the rate of transition from state m to n is equal to the reverse rate or written mathematically

$$w(n,m)\ f(m) = w(m,n)f(n) \tag{53}$$

We assume that the stationary solution is determined uniquely and choose a sequence m_λ which connects m_o with m. One then readily deduces from (53) by dividing (53) by $w(m,n)$ and multiplying subsequent equations (λ = o, 1, 2,...)

$$f(m) = f(m_o) \prod_{\lambda=1} \frac{w(m_\lambda, m_{\lambda-1})}{w(m_{\lambda-1}, m_\lambda)} \tag{54}$$

Using the exponential function we may write (54) in the form

$$f(m) = f(m_o) \exp V \tag{55}$$

with

$$V = \sum_{\lambda=1} \ln \{w(m_\lambda, m_{\lambda-1})/w(m_{\lambda-1}, m_\lambda)\} \tag{56}$$

Thus in the case of detailed balance the master equation can be always solved explicitly by summations or in the continuous case (51) by quadratures.

Distribution Functions and Landau Theory of Phasetransitions I

$f = N \exp \phi$

System	Order parameter	$-\phi =$	Symmetry
superconductor	pair-wavefunction		
zero-dimensional		$\frac{F}{kT} = a\lvert\psi\rvert^2 + b\lvert\psi\rvert^4$	$\psi \longrightarrow e^{i\varphi}\psi$
one-dimensional		$= \int \{ a\lvert\psi\rvert^2 + b\lvert\psi\rvert^4 + \lvert(i\frac{d}{dx} - \frac{2e}{c}A)\psi\rvert^2 \} dx$	$\psi \longrightarrow e^{i\varphi}\psi$
liquid crystal e.g. Smectic $A \leftrightarrow$ nematic transition	complex amplitude of density, ψ	$\iint \{ a\lvert\psi\rvert^2 + b\lvert\psi\rvert^4 + (\nabla + i\delta\tilde{n})\psi^* : \frac{1}{2M} : (\nabla - i\delta\tilde{n})\psi \} dx^2$	$\psi \longrightarrow e^{i\varphi}\psi$
liquid convection instability	amplitude of unstable mode, A	$\frac{2}{Q}\tilde{\phi}\,;\ \tilde{\phi} =$	$A \longrightarrow e^{i\varphi}A$
one-dimensional		$\int \{ a\lvert A\rvert^2 + b\lvert A\rvert^4 + cA^*(\frac{\partial}{\partial x} - \frac{i}{\sqrt{2\pi}}\frac{\partial^2}{\partial y^2})^2 A \} dx$	+ translation + inversion
two- " (including hexagons)	amplitudes of unstable modes, A_k	$\iint \{ \sum_k a\lvert A_k\rvert^2 + \sum_{k,k'} b_{kk'}\lvert A_k\rvert^2\lvert A_{k'}\rvert^2$	translation rotation
		$+ \sum_k cA_k^* (\frac{\partial}{\partial x_{(k)}} - \frac{i}{\sqrt{2\pi}}\frac{\partial^2}{\partial y^2_{(k)}})^2 A_k$	
		$+ \sum_{k,k',k''} (d A_k A_{k'} A_{k''} \delta_{k+k'+k'',0} + c.c.) \} dx$	no inversion

Distribution functions and Landau Theory of Phasetransitions II

$$f = N \exp \phi, \quad \phi = \frac{2}{Q} \tilde{\Phi}$$

System	order parameter	$-\tilde{\phi} =$	Symmetry
laser	mode amplitude		
single mode	u	$a\lvert u\rvert^2 + b\lvert u\rvert^4$	$u \longrightarrow u\, e^{i\varphi}$
single mode with nonlinear absorption	"	$a\lvert u\rvert^2 + b\lvert u\rvert^4 + c\lvert u\rvert^6$	"
multi-mode no phase relations	mode amplitudes u_λ	$\Sigma a_\lambda \lvert u_\lambda\rvert^2 + \sum_{\lambda\lambda'} b_{\lambda\lambda'} \lvert u_\lambda\rvert^2 \lvert u_{\lambda'}\rvert^2$	$u_\lambda \longrightarrow u_\lambda e^{i\varphi_\lambda}$
with phase relations	"	$\Sigma a_\lambda \lvert u_\lambda\rvert^2 + \sum_{\lambda\lambda'\lambda''\lambda'''} b_{\lambda\lambda'\lambda''\lambda'''} u_\lambda^* u_{\lambda'}^* u_{\lambda''} u_{\lambda'''}$	$u_\lambda \longrightarrow u_\lambda e^{i\varphi}$
mode continuum	$u(x)$	$\int \{a\lvert u\rvert^2 + b\lvert u\rvert^4 + c\lvert (iv \frac{d}{dx} - v) u\rvert^2\} dx$	$u \longrightarrow e^{i\varphi} u$
parametric oscillator			
with coherent pump field	mode amplitudes u_λ	$\alpha(\lvert u_1\rvert^2 + \lvert u_2\rvert^2) + \beta u_1^* u_2^* + \text{c.c.} + \gamma \lvert u_1\rvert^2 \lvert u_2\rvert^2$	$u_\lambda \longrightarrow u_\lambda e^{i\varphi_\lambda}$; $\beta \longrightarrow \beta\, e^{i(\varphi_1 + \varphi_2)}$
with incoherent pump field		$\alpha\lvert u_1\rvert^2 + \beta\lvert u_2\rvert^2 + \gamma(\lvert u_1\rvert^2 - \lvert u_2\rvert^2)^2$	$u_\lambda \longrightarrow u_\lambda e^{i\varphi_\lambda}$

b) Scaling

In a number of cases it is possible to introduce smallness parameters ε e.g. the inverse volume ($\varepsilon = 1/v$). We then put

$$\eta = \varepsilon q \tag{57}$$

assume

$$\hat{w}(q,r) = \frac{1}{\varepsilon}\tilde{w}(\eta,r) \tag{58}$$

and make the replacement

$$f(q,t) \longrightarrow f(\eta,t) \tag{59}$$

(51) then transforms into

$$\varepsilon(\partial/\partial t)\, f(\eta,t) = -\int \tilde{w}(\eta,r)dr\, f(\eta,t) + \int \tilde{w}(\eta - \varepsilon r,r)dr\, f(\eta - \varepsilon r,t) \tag{6o}$$

If we expand the right hand side into powers of the expansion parameters ε we obtain the Kramers-Moyal expansion

$$\frac{\partial}{\partial t} f(n,t) = \sum_{n=1}^{\infty} \frac{(-)^n}{n!} \varepsilon^{n-1} \left(\frac{\partial}{\partial \eta}\right)^n c_n(\eta)\, f(\eta,t) \tag{61}$$

where the coefficients are defined by

$$c_n(\eta) = \int r^n\, \tilde{w}\,(\eta,r)dr \tag{62}$$

(61) is also called the generalized Fokker-Planck equation. If we retain in (61) only the first two terms we obtain the usual Fokker-Planck equation

$$\frac{\partial}{\partial t} f(\eta,t) = -\frac{\partial}{\partial \eta}(c_1(\eta)f(\eta,t)) + \frac{1}{2}\frac{\partial^2}{\partial \eta^2}(c_2(\eta,f(\eta,t)) \tag{63}$$

Without going into all the details we just mention that some discussion is going on in the literature under which circumstances one may replace the master equation by the Fokker-Planck equation (63). It should also be noted that in some systems the expansion parameter may have a different meaning. E.g. in the laser ε is the inverse square root of the number of laser atoms. In many cases it is important to divide the motion of the system into an average motion and residual fluctuations. It has become possible at least in several explicit examples to distinguish the mean motion and the fluctuations with respect to their scaling, that means with respect to a certain smallness parameter which in most cases is the inverse of the volume. This case has been treated recently in detail by Mori, who succeeds in classifying resulting Fokker-Planck equations. Here we mention only a well known example in which the mean motion is volume independent but the fluctuations scale with the inverse of the volume. In this case we put

$$\eta = y(t) + \varepsilon^{1/2}\,\xi \tag{64}$$

and the Fokker-Planck equation (61) may be transformed, following van Kampen and Kubo et al into

$$\frac{\partial}{\partial t} \bar{f}(\xi,t) = - \frac{\partial}{\partial \xi} (c_1'(y)\ \xi\ \bar{f}(\xi,t)) + \frac{1}{2} \frac{\partial^2}{\partial \xi^2} (c_2(y)\ \bar{f}\ (\xi,t)) \tag{65}$$

where y obeys the equation

$$\dot{y}(t) = c_1(y) \tag{66}$$

The solution of (65) can be found explicitly in the form

$$\bar{f} = N \exp \left\{ - \frac{1}{2\varepsilon} \sigma^{-1} (\xi - \xi(t))^2 \right\} \tag{67}$$

where σ obeys the equation

$$\dot{\sigma}(t) = 2\, c_1'\ (y)\ \ \sigma(t) + c_2(y) \tag{68}$$

The whole procedure can be extended to many coordinates and forms the basis of recent work by Tomita, who calculates chemical oscillations and the accompanying fluctuations.

c) We mention finally the method which generalizes the Boltzmann distribution function. It starts from the Fokker-Planck equation and assumes that the proper system is only weakly coupled to reservoirs. Without coupling to the reservoirs the proper system possesses a number of constants of motion

$$h_1(q),\ h_2(q), \ldots,\ h_M(q) \tag{69}$$

(in general q is to be replaced, of course, by a set of coordinates). Then without coupling to the reservoirs any function of the form

$$f(h_1, \ldots, h_M) \tag{7o}$$

is a possible distribution function. By coupling to the reservoirs the degeneracy may be lifted and a unique function may be found. This method has been proved rather powerful and has been applied to lasers and nonlinear optics. As an example of classical physics we consider several particles which are strongly interacting with each other each of them being coupled to a reservoir at a temperature T_j with a friction force having the friction constant κ_j. The distribution function is then given by

$$f = N \exp\ (-\beta_{eff} H),\ \beta_{eff} = \frac{\kappa_1 + \kappa_2 + \ldots + \kappa_N}{\kappa_1 T_1 + \kappa_2 T_2 + \ldots + \kappa_N T_N} \tag{71}$$

where now the effective temperature depends on the strength of the coupling of each particle to its reservoir.

So far we have discussed the systems mainly in terms of their distribution function f(q). We now briefly discuss if this is all what one can do or should do. First of all the stationary solution is certainly not sufficient if one wants to calculate correlation functions e.g. of the form $\langle q(t)q(t')\rangle$ where the brackets denote the statistical average In this case even the simple equation (34) or its associate Fokker-Planck equation require computer calculations which have been done in this case by Risken and Vollmer or Hempstead and Lax. But even if we look for the stationary state, functions of the form (48) may disguise important features of the systems. This is, of course the case if the system shows oscillatory behaviour. No trace is left of these oscillations in (48) perhaps except for a strong correlation of certain variables. In the believe of the present author it is here where important further steps must be done, particularly in the close vicinity of the instability and including fluctuations.

Some quantum mechanical approaches

So far we have discussed our problems in the frame of classical physics. We now discuss some tools of quantum mechanics. It has turned out that Langevin equations e.g. of the type (47) can also be established in quantum mechanics where q and the fluctuating force F are quantum mechanical operators. Such equations have been used e.g. in laser theory. Because it is not possible to define a distribution function of an operator at least in the usual sense one has used instead the density matrix and the density matrix equation. The most important problems arise when one couples the proper system to reservoirs representing the external world. One then eliminates the coordinates of the reservoirs and is left with certain reduced density matrix equations. Our space does not allow us to discuss these procedures in detail. We just mention as an interesting example one which describes the coupling of the field mode with operators b^+, b to a reservoir which is at thermal equilibrium. This reservoir causes a damping with the friction constant κ and the mean photon number n. The corresponding density matrix equation derived by Weidlich and Haake then reads

$$\frac{d\rho}{dt} = \kappa\bar{n} \left\{ \left[b^+\rho, b \right] + \left[b^+, \rho b \right] \right\} + \kappa(\bar{n}+1) \left\{ \left[b\rho, b^+ \right] + \left[b, \rho b^+ \right] \right\} \tag{72}$$

Quantum classical correspondence

In recent years it has become possible to transform density matrix equations into generalized Fokker-Planck equations or the ordinary Fokker-Planck equation. This scheme, now called "quantum classical correspondence", permits to calculate all quantum mechanically relevant quantities by pure c-number procedures according to the following scheme:

quantum mechanical quantity	classical quantity
1) density matrix	(quasi)distribution function
2) operator	variable or/and differentiation
3) density matrix equation	stochastic equation (generalized or ordinary Fokker-Planck equation)
4) expectation values (traces)	expectation values (averages over distribution function)
5) time-ordered correlation functions of operators	correlation functions of classical quantities

The crux of the problem consists in finding a transformation of the density matrix into a classical distribution function or, more precisely spoken, into a quasi-distribution function (which in principle can also acquire negative values). We mention just two examples. In the case of Bose operators b^+, b this may be achieved e.g. by the Glauber-Sudarshan P-representation

$$P(u,u^*) = \pi^{-2} \iint e^{-i\beta u} e^{-i\beta^* u^*} \mathrm{tr}(e^{i\beta^* b^+} e^{i\beta b} \rho) d^2\beta \qquad (73)$$

The recipe is as follows: Multiply the density matrix ρ by exponential functions containing b^+, b take the trace and Fourier analyze the result. We then obtain a classical function of u and u^* corresponding to the operators b^+, b. An analogous representation has been also found for arbitrary quantum systems described by projection operators P_{ik}. The function reads as follows

$$f(v) = N \int \exp\left(-\sum_{ik} v_{ik} x_{ik}\right) \mathrm{tr}\left(\prod_{ik} \exp(x_{ik} P_{ik})\, \rho\right) d\{x\} \qquad (74)$$

We mention as the most simple example a single two-level atom in which case one puts

$$\left.\begin{aligned} &P_{21} = \alpha^+, \quad P_{12} = \alpha, \quad P_{11} + P_{22} = 1 \\ &P_{22} - P_{11} = 2\sigma \end{aligned}\right\} \qquad (75)$$

This may be considered as a transcription of the coherent state representation inherent in (73) to a coherent atomic state representation. A somewhat similar approach has been recently developed by Arecchi et al and will be discussed in detail in his chapter.

Concluding remarks

We hope that the reader has obtained the feeling that a great many different systems occurring in quite different disciplines can be treated by the same concepts and rather similar mathematical methods and furthermore that at a macroscopic level of description there exist fascinating analogies in behaviour of such systems. This has two main consequences. First of all one may now more easily understand the behaviour of complex systems and secondly by establishing some analogies one can draw conclusions from one system which is already well understood to other less understood ones. So in a certain sense some systems may serve as sort of analogue computer for other systems. We feel that we are at present just at the beginning of such a research and we hope that this book will stimulate in particular young research workers to go ahead along these lines.

References

We refer the reader for detailed references to the individual contributions to this volume. Here we give only a few key references or those which cannot be found elsewhere in this volume.

Further references on synergetics may be found in "Synergetics" ed. H.Haken, B.G.Teubner, Stuttgart 1973, and in H.Haken "Cooperative Phenomena in Systems far from Thermal Equilibrium and in non-physical Systems" Rev.Mod.Physics, to be published.

Phase transition of blood cells: Helfrich, FU Berlin, private communication

Laser theory

H.Haken, "Laser Theory" Vol. XXV/2c of the Encyclopedia of Physics Springer, New York, (197o)

M.Sargent, M.O.Scully and W.Lamb, "Laser Physics" to be published 1974

M.Lax, "Statistical Physics, Phase Transitions and Superfluidity" eds. M. Chrétien, E.P.Gross and S. Deser, Gordon & Breach, New York, 1968

W. Louisell, "Quantum Statistical Properties of Radiation" Wiley and Sons, New York, 1973

Convection instability

H.Haken, Physics Letters **46A**, 3, 143 (1973) and unpublished material

R.Graham, Phys.Rev.Letters **31**, 1479 (1973)

A procedure similar to that of table 3 (without fluctuations, however) was first developed by T. Tsuzuki and Y.Kuramoto, preprint March 1974

For the Landau theory of liquid crystals see the references in the lecture by de Gennes.

For details on scaling theories see

N.G. van Kampen, Canad. J.Phys. **39**, 551 (1961)

R. Kubo, K. Matsuo, K. Kitahara, J. Stat.Phys. **9**, 51 (1973)

H. Mori, H. Fujisaba, H. Shigematsu, Progr.Theor.Physics **51**, 1o9 (1974)

H. Mori, to be published

Generalized Botzmann distribution function

H.Haken, Z.Physik **263**, 267 (1973)

Cooperative Phenomena, H. Haken, ed.

ATOMIC COHERENT STATES IN QUANTUM OPTICS

F.T. Arecchi

University of Pavia and C.I.S.E., Milano, Italy

1. Introduction

The central problem of Quantum Optics (laser theory, superradiance, resonant propagation, etc.) is the description of the interaction between N atoms and an electromagnetic field confined in a cavity of finite volume. A suitable model Hamiltonian for this problem is the following one ($\hbar = 1$)

$$H = \sum_k \omega_k a_k^+ a_k + \frac{\omega_0}{2} \sum_{i=1}^{N} S_{3(i)} + \sum_{k,i} g_k \left(a_k S_i^+ e^{i \underline{k} \cdot \underline{x}_i} + a_k^+ S_i^- e^{-i \underline{k} \cdot \underline{x}_i} \right),$$

where a_k, a_k^+ are Bose operators describing the k-th field mode and $S_i^{\pm}$, S_{3i} are Pauli operators describing the atom located at position $\underline{x}_i$ as a two level system.

We can introduce the collective operators [1]

$$J_k^{\pm} = \sum_i S_i^{\pm} e^{\pm i \underline{k} \cdot \underline{x}_i}$$

$$J_z = \sum_i S_{3i}$$

They obey the commutation rule:

$$[J_k^+, J_{k'}^-] = \sum_i S_{3i} \, e^{i(\underline{k}-\underline{k}') \cdot \underline{x}_i}$$

This reduces to $\delta_{kk'}$, J_z in the following particular cases which are of extreme physical importance:

i) point laser (cavity volume $\ll \lambda^3$)
ii) single mode laser (travelling wave)
iii) travelling wave field in an amplifying or absorbing medium.

In such cases the above operators obey standard angular momentum commutation rules. The associated Heisenberg equations of motion become (leaving out for simplicity the k index)

$$\dot{a} = -i\omega a - i g J^-$$

$$\dot{J}^- = -i\omega_0 J^- + i g a J_z$$

$$\dot{J}_z = -i g (a J^+ - a^+ J^-) ,$$

plus similar equations for a^+ and J^+. This set of five equations is not closed. For instance, to solve the second equation, we must know the evolution of the binary operator aJ_z, whose equation of motion will imply ternary operators, and so on.

In the self-consistent approximation (SCA), or semi-classical approach, we introduce the approximation

$$\langle a\, J^{+} \rangle \approx \langle a \rangle \, \langle J^{+} \rangle \, ,$$

that is, we consider only the interaction among the mean fields. The three equations for $J^{\pm}$, J_z can be summarized in the vector equation

$$\langle \dot{\underline{J}} \rangle = \underline{\Omega} \times \langle \underline{J} \rangle \, ,$$

where $\underline{\Omega} \equiv (g \langle a \rangle, 0, \omega_o)$. This is a Bloch [2] equation for an angular momentum $\underline{J}$ precessing around a classical field $\langle a \rangle$.

The field equation becomes

$$\langle \dot{a} \rangle = - i\omega \langle a \rangle - ig \langle J^{-} \rangle \, .$$

This is the equation for a field acted upon by a classical current $\langle \underline{J} \rangle$. Starting with a field in the vacuum state, it leads to a particular field state with $\langle a \rangle \neq 0$ which was introduced by Bloch and Nordsieck to deal with the "infrared catastrophe", [3] then formalized by Schwinger [4], and later used by Glauber in Quantum Optics, [5] under the name of coherent state.

We will similarly call a coherent atomic state, or Bloch state, that state with zero induced dipole ($\langle J^{-} \rangle \neq 0$) generated from the ground state by a classical field.

We want to show that these states, besides being generated by classical sources, give expectation values of quantum operators whose limits, for large excitations, are the classical value (see Table I). For a more detailed treatment see Ref. 6.

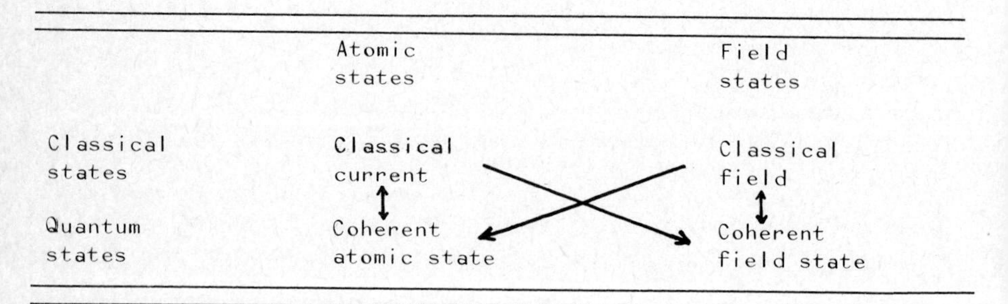

	Atomic states	Field states
Classical states	Classical current	Classical field
	↕	↕
Quantum states	Coherent atomic state	Coherent field state

Table I. Classical excitation and coherent states. The single arrows indicate the direction of production of coherent states starting from classical states. The double arrows indicate states connected by the correspondence principle.

2. Description of the free field

A. The Harmonic Oscillator States

In order to point out with maximum clarity the analogies between the free-field description and the free-atom description we start by listing here, in simple terms, the properties of the single harmonic oscillator. The equation numbering in this section and in Sec.3 is done in parallel.

The single harmonic oscillator is described by its canonically conjugated coordinates q, p with the commutation relation

$$[q,p] = i\hbar \tag{2.1}$$

One forms the usual lowering and raising operators

$$a = (2\hbar\omega m)^{-1/2} (\omega mq + ip) \tag{2.2a}$$

$$a^{+} = (2\hbar\omega m)^{-1/2} (\omega mq - ip) \tag{2.2b}$$

Where $\omega m > 0$ is characteristic of the oscillator. These operators satisfy

$$[a,a^{+}] = 1 \tag{2.3a}$$

from which one obtains

$$[a,a^{+} a] = a \tag{2.3b}$$

$$[a^{+},a^{+} a] = -a^{+} \tag{2.3c}$$

The harmonic oscillator states, or Fock states, are the eigenstates of

$$N = a^{+}a \tag{2.4}$$

and are given by [9]

$$|n\rangle = \frac{(a^{+})^{n}}{\sqrt{n!}} |0\rangle \quad (n= 0,1,2...) \tag{2.5}$$

with eigenvalue n. The vacuum state $|0\rangle$ is the harmonic oscillator ground state defined by

$$a|0\rangle = 0 \tag{2.6}$$

B. Coherent States of the Field.

Let us consider the translation operator which produces a shift ξ in q and η in p:

$$T_{\alpha} = \exp -(i/\hbar)(\xi p-\eta q) = \exp (\alpha a^{+} - \alpha^{*}a) \tag{2.7}$$

where

$$\alpha = (2\hbar\omega m)^{-1/2} (\omega m\xi + i\eta) \tag{2.7b}$$

A coherent state $|\alpha\rangle$ is obtained by translation of the ground state [4,5]

$$|\alpha\rangle \equiv T_{\alpha}|0\rangle \tag{2.8}$$

We shall name these states Glauber states, as they have been used extensively by Glauber in quantum optics [5]. Since

$$T_\alpha a T_\alpha^{-1} = a - \alpha, \tag{2.9}$$

the state $|\alpha\rangle$ satisfies the eigenvalue equation

$$(a - \alpha)|\alpha\rangle = 0 \tag{2.10}$$

Using the Baker-Campbell-Hausdorf theorem [7] or Feynman disentangling techniques [8] the translation operator can be written in the following forms:

$$T_\alpha = e^{|\alpha|^2/2} e^{-\alpha^* a} e^{\alpha a^\dagger} = e^{-|\alpha|^2/2} e^{\alpha a^\dagger} e^{-\alpha^* a} \tag{2.11}$$

The second of these forms, which is known as the normally ordered form, gives immediately the expansion of $|\alpha\rangle$ in terms of Fock states

$$|\alpha\rangle = T_\alpha |0\rangle = e^{-|\alpha|^2/2} e^{\alpha a^\dagger} |0\rangle \tag{2.12}$$

from which, expanding the exponential and using (2.5), one obtains

$$\langle n|\alpha\rangle = e^{-|\alpha|^2/2} \frac{\alpha^n}{\sqrt{n!}} \tag{2.13}$$

The scalar product of Glauber states can be obtained either from (2.12), using the disentangling theorem (2.11), or from (2.13), using the completeness property of Fock states, $\sum |n\rangle\langle n| = 1$. One gets

$$\langle\alpha|\beta\rangle = \exp\left[-(1/2)(|\alpha|^2 - 2\alpha^*\beta + |\beta|^2)\right] \tag{2.14a}$$

from which one obtains

$$|\langle\alpha|\beta\rangle|^2 = \exp(-|\alpha - \beta|^2) \tag{2.14b}$$

The coherent states are minimum uncertainty packets. For three observables A,B,C, which obey a commutation relation $[A,B] = iC$, it is easy to show [9] that $\langle A^2\rangle\langle B^2\rangle \geq \langle C\rangle^2/4$. In particular, with $A = q-\xi$, $B = p-\eta$ and $C = \hbar$ one has

$$\langle(q-\xi)^2\rangle\langle(p-\eta)^2\rangle \geq \hbar^2/4 \tag{2.15}$$

for any state. It is easy to show [5] that the equality sign holds for the coherent state $|\alpha\rangle$, where α is related to ξ and η by (2.7b). This establishes the minimum uncertainty property.

C. The Coherent States as a Basis

We now consider the completeness properties of the coherent states. Using (2.13), and the completeness of Fock states $\sum_n |n\rangle\langle n| = 1$, one obtains straightforwardly

$$\int \frac{d^2\alpha}{\pi} |\alpha\rangle\langle\alpha| = 1 \tag{2.16}$$

The expansion of an arbitrary state in Glauber states follows

$$|c\rangle \equiv \sum_n c_n|n\rangle = \int \frac{d^2\alpha}{\pi} \sum_n c_n|\alpha\rangle\langle\alpha|n\rangle$$

$$= \int \frac{d^2\alpha}{\pi} \exp\left[-(1/2)|\alpha|^2\right] f(\alpha^*)|\alpha\rangle \tag{2.17a}$$

where

$$f(\alpha^*) \equiv \sum_n c_n \frac{(\alpha^*)^n}{\sqrt{n!}} = e^{-|\alpha|^2/2}\langle\alpha|c\rangle \tag{2.17b}$$

Using (2.5) one sees that $|c\rangle$ can also be written as

$$|c\rangle = f(a^\dagger)|0\rangle \tag{2.18}$$

where $f(a^\dagger)$ is defined by its expansion (2.17b). The scalar product of any two states $|c'\rangle$ and $|c\rangle$ is obtained from (2.16) and (2.17b)

$$\langle c'|c\rangle = \int \frac{d^2\alpha}{\pi} \langle c'|\alpha\rangle\langle\alpha|c\rangle$$

$$= \int \frac{d^2\alpha}{\pi} e^{-|\alpha|^2} \left[f'(\alpha^*)\right]^* f(\alpha^*) \tag{2.19}$$

In view of the completeness relations, operators F acting on this Hilbert space can be expanded as

$$F = \sum_{m,n} |m\rangle\langle m|F|n\rangle\langle n| \tag{2.20a}$$

or

$$F = \iint \frac{d^2\alpha\, d^2\beta}{\pi^2} |\beta\rangle\langle\beta|F|\alpha\rangle\langle\alpha| \tag{2.20b}$$

Due to the overcompleteness of the $|\alpha\rangle$ states, the expansion (2.20b) is in general not unique. This expansion is especially useful if it can be written in the diagonal form

$$F = \int d^2\alpha \; f(\alpha)|\alpha\rangle\langle\alpha| \tag{2.20c}$$

This will be further discussed for the case of the density matrix.

D. Statistical Operator for the Field

Up to now we have considered pure quantum states. Since a field in thermal equilibrium with matter at ordinary temperatures is essentially in the ground state ($\hbar\omega \gg k_B T$), this is an adequate description for any field obtained from thermal equilibrium in response to a classical current. However, the field radiated by an incoherently pumped medium is a statistical mixture described by a statistical operator ρ, which we assume normalized to unity,

$$\mathrm{Tr}\,\rho = 1 \tag{2.21}$$

With the help of this operator, the statistical average of any observable $F(a,a^\dagger)$ is obtained as

$$\langle F\rangle = \mathrm{Tr}\,\rho F \tag{2.22}$$

Of particular interest are statistical ensembles described by a statistical operator which is diagonal in the Glauber representation [5]

$$\rho = \int P(\alpha)\,|\alpha\rangle\langle\alpha|\,d^2\alpha \tag{2.23}$$

where the normalization (2.21) requires

$$\int P(\alpha)\,d^2\alpha = 1 \tag{2.24}$$

The statistical average of an observable F is then given by an average over the diagonal elements $\langle\alpha|F|\alpha\rangle$:

$$\langle F\rangle = \int P(\alpha)\langle\alpha|F|\alpha\rangle\,d^2\alpha \tag{2.25}$$

The weight function $P(\alpha)$ has thus the properties of a distribution function in α-space, except that it is not necessarily positive.

Let us define a set of operators $\hat{X}(\lambda)$ such that their expectation values for coherent states

$$b^{\alpha}(\lambda) = \langle\alpha|\hat{X}(\lambda)|\alpha\rangle \tag{2.26}$$

form a basis in the function space of functions of α. If the statistical ensemble has a diagonal representation (2.23), then the statistical averages of the operators form a kind of "characteristic function" of $P(\alpha)$:

$$X(\lambda) \equiv \langle\hat{X}(\lambda)\rangle = \int d^2\alpha\,P(\alpha)b^{\alpha}(\lambda) \tag{2.27}$$

The weight function $P(\alpha)$ can be expressed in terms of $X(\lambda)$ with the help of the reciprocal basis $\bar{b}^{\lambda}(\alpha)$,

$$P(\alpha) = \int d^2\lambda\,X(\lambda)\,\bar{b}^{\lambda}(\alpha) \tag{2.28}$$

A convenient basis is the Fourier basis

$$b^{\alpha}(\lambda) = \exp(\lambda\alpha^* - \lambda^*\alpha) \tag{2.29}$$

$$\bar{b}^{\lambda}(\alpha) = (2\pi)^{-2}\exp(-\lambda\alpha^* + \lambda^*\alpha) \tag{2.29b}$$

which is generated by the normally ordered operators

$$\hat{X}_N(\lambda) = \exp(\lambda a^\dagger)\,\exp(-\lambda^* a) \tag{2.29c}$$

The question of the existence of the P representation is a complicated one [5,9]. Using the Fourier basis (2.29) it can be shown however that the mere existence of the inverse transformation (2.28) guarantees that the resulting function $P(\alpha)$ can be used to calculate the statistical average of any moment $\langle a^{\dagger m}a^n\rangle$ as if $P(\alpha)$ was the weight function defined in (2.23). This is due to the fact that the characteristic function $X_N(\lambda)$ plays the role of a generating function for $\langle a^{\dagger m} a^n \rangle$:

$$\langle a^{\dagger m} a^n \rangle = \left(\frac{\partial}{\partial \lambda}\right)^n \left(-\frac{\partial}{\partial \lambda^*}\right)^m \chi_N(\lambda)\Big|_{\lambda=0}$$

From which, by derivation of (2.27), one obtains

$$\langle a^{\dagger m} a^n \rangle = \int d^2\alpha \; P(\alpha) \; \langle \alpha | a^{\dagger m} a^n | \alpha \rangle$$

which is a particular case of (2.25) and proves the above statement. One could moreover introduce, in addition to (2.29c), symmetrically ordered $\hat{\chi}_S(\lambda)$, and antinormally ordered $\hat{\chi}_A(\lambda)$, exponential operators [5,7] . The Fourier transform of their statistical averages are the Wigner distribution, and the matrix element $\langle \alpha | \rho | \alpha \rangle / \pi$, respectively. We will not develop these aspects further as the corresponding expression for atomic coherent states are rather involved, and of no clear use as yet.

3. Description of the free atoms.

A. The Angular Momentum States

Angular momentum operators can be defined which act on the N-atom Hilbert space. In particular we can consider a subspace of degenerate eigenstates of J^2 with eigenvalues $J(J+1)$. Since J^2 commutes with J_x, J_y, J_z, these operators only connect states within the same subspace. In general, J^2 and J_z do not form a complete set of commuting observables. As explained in Ref.6, such a complete set is formed by adding to J^2 and J_z some operators of the permutation group of N objects P_N. These operators play with respect to P_N the same role that J^2 and J_z have with respect to the three-dimensional rotation group. We shall assume that the subspace considered here has also been made invariant under these permutation operations, but for simplicity we shall omit for the time being to indicate this in the labelling of the states. The subspace we are dealing with is identical to a constant angular momentum Hilbert space. The Dicke states, which are the analog of the Fock states (2.5), and the Bloch states, which correspond to the Glauber state (2.8), are most easily defined within such a subspace. The equation numbering is in parallel with that of Sec.2. From the angular momentum operators J_x and J_y, which satisfy the commutation relation

$$[J_x, J_y] = i J_z \tag{3.1}$$

the lowering and raising operators are formed

$$J_- = J_x - iJ_y \tag{3.2a}$$

$$J_+ = J_x + iJ_y \tag{3.2b}$$

which obey

$$[J_-, J_+] = -2J_z \tag{3.3a}$$

$$[J_-, J_z] = J_- \tag{3.3b}$$

$$[J_+, J_z] = -J_+ \tag{3.3c}$$

The Dicke states, which are simply the usual angular momentum states, are defined as the aigenstates of

$$J_z = \frac{1}{2}(J_+J_- - J_-J_+) \tag{3.4}$$

They are given by [10]

$$|M\rangle = \frac{1}{(M+J)!}\begin{pmatrix} 2J \\ M+J \end{pmatrix}^{-1/2} J_+^{M+J}|-J\rangle \tag{3.5}$$

$$(M = -J, -J+1, \ldots, J)$$

with eigenvalue M. They span the space of angular momentum quantum number J. The ground state $|-J\rangle$ is defined by

$$J_-|-J\rangle = 0 \tag{3.6}$$

B. Coherent Atomic States

Let us consider the rotation operator which produces a rotation through an angle θ about an axis $\hat{n} \equiv (\sin\varphi, \cos\varphi, 0)$:

$$R_{\theta,\varphi} = e^{-i\theta J_n} = \exp\left[-i\theta(J_x \sin\varphi - J_y \cos\varphi)\right]$$
$$= \exp(\zeta J_+ - \zeta^* J_-) \tag{3.7a}$$

where

$$\zeta = \frac{\theta}{2} e^{-i\varphi} \tag{3.7b}$$

A coherent atomic state, or Bloch state, $|\theta,\varphi\rangle$ is obtained by rotation of the ground state $|-J\rangle$:

$$|\theta,\varphi\rangle \equiv R_{\theta,\varphi}|-J\rangle \tag{3.8}$$

Furthermore

$$R_{\theta,\varphi} J_n R^{-1}_{\theta,\varphi} = J_n$$

$$R_{\theta,\varphi} J_k R^{-1}_{\theta,\varphi} = J_k \cos\theta + J_z \sin\theta$$

$$R_{\theta,\varphi} J_z R^{-1}_{\theta,\varphi} = -J_k \cos\theta + J_z \sin\theta$$

where

$$J_n = J_x \sin\varphi - J_y \cos\varphi$$

$$J_k = J_x \cos\varphi + J_y \sin\varphi$$

which gives

$$J_+ = (J_k - i\, J_n)\, e^{i\varphi}$$

$$J_- = (J_k + i\, J_n)\, e^{-i\varphi}$$

Using these relations one obtains

$$R_{\theta,\varphi} J_- R^{-1}_{\theta,\varphi} = e^{-i\varphi}\left[J_-\, e^{i\varphi}\cos^2(\theta/2) - J_+\, e^{-i\varphi}\sin^2(\theta/2) + J_z \sin\theta\right] \quad (3.9a)$$

and similar relations for J_+ and J_z :

$$R_{\theta,\varphi} J_+ R^{-1}_{\theta,\varphi} = e^{i\varphi}\left[J_+ e^{-i\varphi}\cos^2(\theta/2) - J_-\, e^{-i\varphi}\sin^2(\theta/2) + J_z \sin\theta\right] \quad (3.9b)$$

$$R_{\theta,\varphi} J_z R^{-1}_{\theta,\varphi} = J_z \cos\theta - J_-\, e^{i\varphi}\sin(\theta/2)\cos(\theta/2) - J_+\, e^{-i\varphi} \sin(\theta/2)\cos(\theta/2) \quad (3.9c)$$

From (3.9a), and definition (3.8), one obtains the eigenvalue equation

$$\left[J_-\, e^{i\varphi}\cos^2(\theta/2) - J_+\, e^{-i\varphi}\sin^2(\theta/2) + J_z \sin\theta\right]|\theta,\varphi\rangle = 0 \quad (3.10a)$$

This equation, together with

$$J^2|\theta,\varphi\rangle = J(J+1)|\theta,\varphi\rangle \quad (3.10b)$$

specifies uniquely the Bloch state $|\theta,\varphi\rangle$. Note that the harmonic oscillator analog of (3.10b) would have been the trivial relation $(a^+ - \alpha^*)(a-\alpha)|\alpha\rangle=0$.
Other forms of the eigenvalue equation can be obtained using the relation

$$R_{\theta,\varphi} J_z R^{-1}_{\theta,\varphi}|\theta,\varphi\rangle = -J|\theta,\varphi\rangle$$

and (3.9c). The resulting equation can be combined with (3.10a) to eliminate one of the operators J_z, J+, or J-, giving

$$\left[J_-\, e^{i\varphi}\cos^2(\theta/2) + J_+\, e^{-i\varphi}\sin^2(\theta/2)\right]|\theta,\varphi\rangle = J\sin\theta|\theta,\varphi\rangle \quad (3.10c)$$

$$\left[J_-\, e^{i\varphi}\cos(\theta/2) + J_z \sin(\theta/2)\right]|\theta,\varphi\rangle = J\sin(\theta/2)|\theta,\varphi\rangle \quad (3.10d)$$

$$\left[J_+\, e^{-i\varphi}\sin(\theta/2) - J_z\cos(\theta/2)\right]|\theta,\varphi\rangle = J\cos(\theta/2)|\theta,\varphi\rangle \quad (3.10e)$$

These additional relations are not independent of (3.10a) and (3.10b) One notes that these eigenvalue equations are more complicated than their counterpart (2.10). In particular they involve at least two of the three operators J_-, J_+, J_z. This feature is required by the more complicated commutation relation (3.1) which applies here.
Using the disentangling theorem for angular momentum operators [Ref.6], the rotation $R_{\vartheta,\varphi}$ given by (3.7a) becomes

$$R_{\vartheta,\varphi} = \exp(-\tau^* J_-)\exp\left[-\ln(1+|\tau|^2)J_z\right]\exp(\tau J_+)$$
$$= \exp(\tau J_+)\exp\left[\ln(1+|\tau|^2)J_z\right]\exp(-\tau^* J_-) \qquad (3.11a)$$

where

$$\tau \equiv e^{-i\varphi}\tan(\vartheta/2) \qquad (3.11b)$$

Let us point out that these expressions are singular for $\vartheta=\pi$, i.e., for the uppermost state. We may have to exclude from some of the following considerations the states contained within an infinitesimally small circle around $\vartheta=\pi$. The validity of expressions such as (3.13) for $\vartheta=\pi$ is usually not affected and can be checked directly. The last form of (3.11a), which we call the normally ordered form, gives immediately the expansion of $|\vartheta,\varphi\rangle$ in terms of Dicke states:

$$|\vartheta,\varphi\rangle = R_{\vartheta,\varphi}|-J\rangle = \left[1/(1+|\tau|^2)\right]^J e^{\tau J_+}|-J\rangle \qquad (3.12)$$

from which, expanding the exponential and using (3.5) one obtains

$$\langle M|\vartheta,\varphi\rangle = \binom{2J}{M+J}^{1/2}\frac{\tau^{M+J}}{(1+|\tau|^2)^J} = \binom{2J}{M+J}^{1/2}\sin^{J+M}(\vartheta/2)\cos^{J-M}(\vartheta/2)\,e^{-i(J+M)\varphi} \qquad (3.13)$$

Since the Dicke states form a basis for a well-known irreducible representation of the rotation group, these results could have been derived using the appropriate Wigner $\mathcal{D}^{(J)}$ matrix [10]. The same remark applies to Eqs. (2.12) and (2.13): these could have been obtained without using the Baker-Campbell-Hausdorf formula, from the transformation properties of an irreducible representation of the group of operations T_α.

The overlap of two Bloch states is obtained either from (3.12), using the disentangling theorem for exponential angular momentum operators, or from (3.13), using the completeness property of Dicke states $\sum_M |M\rangle\langle M| = 1$. One obtains

$$\langle\vartheta,\varphi|\vartheta',\varphi'\rangle = \left(\frac{(1+\tau^*\tau')^2}{(1+|\tau|^2)(1+|\tau'|^2)}\right)^J$$
$$= e^{iJ(\varphi-\varphi')}\left(\cos\frac{\vartheta-\vartheta'}{2}\cos\frac{\varphi-\varphi'}{2} - i\cos\frac{\vartheta+\vartheta'}{2}\sin\frac{\varphi-\varphi'}{2}\right)^{2J}$$

(3.14a)

from which one obtains

$$|\langle\theta,\varphi|\theta',\varphi'\rangle|^2 = \cos^{4J}(\Theta/2) \tag{3.14b}$$

where τ is given by (3.11b), τ' is given by the same equation written with the primed quantities, and Θ is the angle between the (θ,φ) and (θ',φ') directions, as given by

$$\cos\Theta = \cos\theta\cos\theta' + \sin\theta\sin\theta'\cos(\varphi-\varphi')$$

The Bloch states form minimum uncertainty packets. The uncertainty relation can be defined in terms of the set of rotated operators $(J_\xi, J_\eta, J_\zeta) = R_{\theta,\varphi}(J_x, J_y, J_z)R^{-1}$. These three observables obey a commutation relation of the type $[A,B] = iC$ with $A = J_\xi$, $B = J_\eta$, $C = J_\zeta$, from which they have the uncertainty property

$$\langle J_\xi^2\rangle\langle J_\eta^2\rangle \geq \frac{1}{4}\langle J_\zeta\rangle^2 \tag{3.15}$$

for any states. It is easy to show that the equality sign holds for the Bloch state $|\theta,\varphi\rangle$, which is therefore a minimum uncertainty state.

C. The Bloch States as a Basis

Let us now consider the completeness properties of the Bloch states. Using (3.13), and the completeness of Dicke states $\sum_M |M\rangle\langle M| = 1$, one obtains

$$(2J+1)\int\frac{d\Omega}{4\pi}|\theta,\varphi\rangle\langle\theta,\varphi|$$

$$=(2J+1)\int\frac{d\Omega}{4\pi}\sum_{M,M'}\binom{2J}{M+J}^{\frac{1}{2}}\binom{2J}{M'+J}^{\frac{1}{2}}e^{i(M'-M)\varphi}$$

$$\times\left(\cos\frac{\theta}{2}\right)^{2J-M-M'}\left(\sin\frac{\theta}{2}\right)^{2J+M+M'}|M\rangle\langle M'|$$

$$=(2J+1)\int_0^\pi d\theta\,\frac{\sin\theta}{2}\sum_M\binom{2J}{M+J}\left(\cos\frac{\theta}{2}\right)^{2J-2M}\left(\sin\frac{\theta}{2}\right)^{2J+2M}|M\rangle\langle M|$$

$$=\sum_M|M\rangle\langle M| = 1 \tag{3.16}$$

The expansion of an arbitrary state in Bloch states follows:

$$|c\rangle = \sum_M c_M|M\rangle = (2J+1)\int\frac{d\Omega}{4\pi}\sum_M c_M|\theta,\varphi\rangle\langle\theta,\varphi|M\rangle$$

$$= (2J+1)\int\frac{d\Omega}{4\pi}\,\frac{f(\tau^*)}{(1+|\tau|^2)^J}|\theta,\varphi\rangle \tag{3.17a}$$

where

$$f(\tau^*) \equiv \sum_M C_M \begin{pmatrix} 2J \\ J+M \end{pmatrix}^{1/2} (\tau^*)^{J+M} = (1 + |\tau|^2)^J \langle \theta, \varphi | c \rangle \qquad (3.17b)$$

Using (3.5) one sees that $|c\rangle$ can also be written as

$$|c\rangle = f\left(\frac{1}{J + 1 - J_z} J_+\right) | -J\rangle \qquad (3.18)$$

The amplitude function $f(\tau^*)$ is, by its definition (3.17b), a polynomial of degree 2J. However any function which has a Maclaurin expansion can be taken as a suitable amplitude function in (3.17a) or (3.18). Indeed the powers of τ^* higher than 2J gives zero contribution in (3.17a) and (3.18). The coefficients C_M are then obtained from the first (2J + 1) terms of the Maclaurin series, using (3.17b).

The scalar product of two states characterized by their amplitude function is, from (3.16) and (3.17b)

$$\begin{aligned} \langle c' | c\rangle &= (2J+1) \int \frac{d\Omega}{4\pi} \langle c' | \theta, \varphi\rangle\langle \theta, \varphi | c\rangle \\ &= (2J + 1) \int \frac{d\Omega}{4\pi} \frac{1}{(1 + |\tau|^2)^{2J}} \left[f'(\tau^*)\right]^* f(\tau^*) \qquad (3.19) \end{aligned}$$

Since (3.17b) was used to derive this equation, its validity is restricted to amplitude functions which are polynomials of degree 2J.

In view of the completeness relations, operators G acting on this Hilbert space can be expanded as

$$G = \sum_{M,M'} |M\rangle\langle M| G | M'\rangle\langle M'| \qquad (3.20a)$$

or

$$G = \frac{(2J + 1)^2}{(4\pi)^2} \iint d\Omega\, d\Omega' \, |\theta, \varphi\rangle\langle \theta, \varphi | G | \theta', \varphi'\rangle\langle \theta', \varphi'| \qquad (3.20b)$$

However, G is completely defined by the $(2J + 1)^2$ matrix elements $\langle M|G|M'\rangle$. It results that, except for pathological cases, an operator can always be written in the diagonal form

$$G = \int d\Omega \, g(\theta, \varphi) |\theta, \varphi\rangle\langle \theta, \varphi| \qquad (3.20c)$$

where $g(\theta, \varphi)$ is given by a series expansion

$$g(\theta, \varphi) = \sum_{\ell, m} G_{\ell, m} \; Y_\ell^m (\theta, \varphi)$$

In accordance with Appendix IV of Ref.6 only the $(2J + 1)^2$ first terms of this sum contribute to (3.20c). These are the terms for which $0\leq \ell \leq 2J$. The corresponding coefficients $G_{\ell,m}$ can be expressed as a function of the matrix element $\langle M|G|M'\rangle$.

D. Statistical Operators for the Atoms

In order to describe an incoherently pumped system of atoms we introduce a statistical operator ϱ with the properties

$$\mathrm{Tr}\,\varrho = 1 \tag{3.21}$$

$$\langle G\rangle = \mathrm{Tr}\,\varrho\, G \tag{3.22}$$

As before, the considerations are restricted to states belonging to a single constant angular momentum subspace, and therefore the statistical operator described here does not allow for the most general mixing of atomic states. Of particular interest is the expression of ϱ in a diagonal Bloch representation

$$\varrho = \int P(\vartheta,\varphi)\,|\vartheta,\varphi\rangle\langle\vartheta,\varphi|\,d\Omega \tag{3.23}$$

with the normalization

$$\int P(\vartheta,\varphi)\;d\Omega = 1 \tag{3.24}$$

The statistical average of an observable G is then given by

$$\langle G\rangle = \int P(\vartheta,\varphi)\langle\vartheta,\varphi|G|\vartheta,\varphi\rangle \tag{3.25}$$

The weight function $P(\vartheta,\varphi)$ has thus the properties of a distribution function on the unit sphere, except that it is not necessarily positive.

Let us define a set of operators $\hat{X}_\lambda$ such that their expectation values for Bloch states

$$b_\lambda(\vartheta,\varphi) = \langle\vartheta,\varphi|\hat{X}_\lambda|\vartheta,\varphi\rangle \tag{3.26}$$

form a basis in the space of functions on the unit sphere. Since in this space a discrete basis can be chosen, the parameter λ can be restricted to discrete values $\lambda = 1,2,\ldots$. For a statistical ensemble described by (3.23), the statistical averages of the operators $\hat{X}_\lambda$ form a set of "characteristic coefficients" of $P(\vartheta,\varphi)$

$$X_\lambda \equiv \mathrm{Tr}\,\varrho\,\hat{X}_\lambda = \int d\Omega\, P(\vartheta,\varphi)\, b_\lambda(\vartheta,\varphi) \tag{3.27}$$

The weight function can be expressed as a series with the help of the reciprocal basis $\bar{b}^\lambda(\vartheta,\varphi)$,

$$P(\vartheta,\varphi) = \sum_\lambda X_\lambda\, \bar{b}^\lambda(\vartheta,\varphi) \tag{3.28}$$

A convenient basis is given by the spherical harmonics

$$b_\lambda(\vartheta,\varphi) = Y_\ell^m(\vartheta,\varphi) \qquad (\lambda \equiv \ell, m) \tag{3.29a}$$

$$\bar{b}^{\lambda}(\theta,\varphi) = Y_{\ell}^{-m}(\theta,\varphi) \qquad (3.29b)$$

which are generated by the spherical harmonic operators [11]

$$\hat{X}_{\lambda} = \mathcal{Y}_{\ell}^{m}(\vec{J}) \qquad (3.29c)$$

The already mentioned fact that a diagonal representation always exists in the atomic case also corresponds to the fact that for a given J only the $(2J + 1)^2$ operators $\mathcal{Y}_{\ell}^{m}$ with $\ell \leq 2J$ are different from zero. The finite dimensionality of the basis is required, since ρ is completely determined by its $(2J + 1)^2$ matrix elements $\langle M|\rho|M'\rangle$ in the Dicke representation.

Other differences with the field case should also be noted. First, the spherical harmonic operators are usually written in a fully symmetrized form, whereas the operators (2.29c) are normally ordered. This is only a formal difficulty as it should be possible to write normally ordered and antinormally ordered multipole operators with properties similar to the $\mathcal{Y}_{\ell}^{m}(\vec{J})$. A second, and more fundamental, difference is that the expectation values X_{λ} are not generating functions for products of the type $\langle J_+^m J_z^n J_-^p \rangle$, in view of the discreteness of the set. This does not cause much difficulty, as generating functions can be defined from exponential operators whose expectation values can be calculated with the help of the disentangling theorem. It is tempting to use for the $\hat{X}_{\lambda}$'s of Eq. (3.29c) these exponential operators themselves. Though the parallel with the field case then seems more transparent, the use of the discrete set $\hat{X}_{\lambda}$ may be of more fundamental significance as it takes into account symmetry properties of the states.

A final comment should be made about the difficulty of dealing with creation and annihilation operations in a finite Hilbert space. The existence of two terminal states, $|J\rangle$ and $|-J\rangle$, requires the presence of a third operator with the properties of J_z, and prevents the writing of an eigenvalue equation in terms of one compound operator alone. For instance, the comparison of (3.18) and (2.18) suggests that $J_-(J + 1 - J_z)^{-1}$ could be a "good" annihilation operator. Using (3.13) one finds immediately

$$J_-(J + 1 - J_z)^{-1}|\theta,\varphi\rangle = \tau|\theta,\varphi\rangle - \tau \sin^{2J}(\theta/2)\, e^{-2iJ\varphi}|J\rangle$$

which for small θ and large J is almost an eigenvalue equation: so the application of the operator reproduces $|\theta,\varphi\rangle$ except for the uppermost Dicke state. There is no doubt that a theory could be developed in terms of more complicated annihilation and creation operators of such type, but the advantages are not clear.

4. Contraction, or the relation between atomic states and field states.

The extreme similarity between the treatments of Secs.3 and 2 suggests a close connection between atomic and field states. This connection is made here through a process known as group contraction. The time evolution of a single two-level atom is governed by a 2 x 2 unitary transformation matrix. The commutation relations for the generators of the group U(2) are rewritten here:

$$\begin{aligned} [J_z, J_\pm] &= \pm J_\pm \\ [J_+, J_-] &= 2 J_z \\ [\vec{J}, J_0] &= 0 \end{aligned} \qquad (4.1)$$

where J_0 in the third relation is essentially the identity. An arbitrary 2 x 2 unitary transformation matrix is given by

$$U(2) = \exp(i \sum_\mu \lambda_\mu J_\mu) \qquad (4.2)$$

where the summation is over all four indices and the λ_μ's are c-number parameters which characterize the group operation.

If another set of generators h_+, h_-, h_z, h_o, is related to J_+, J_-, J_z, J_o by a nonsingular transformation $A_{\nu\mu}$

$$h_\nu = \sum_\mu A_{\nu\mu} J_\mu \qquad (4.3a)$$

then the group operation (4.2) may be written

$$\exp(i\sum_\mu \lambda_\mu J_\mu) = \exp(i\sum_\nu \alpha_\nu h_\nu) \qquad (4.4)$$

with

$$\alpha_\nu = \sum \lambda_\mu (A^{-1})_{\mu\nu} \qquad (4.3b)$$

We select the following transformation A, which depends on a real parameter c:

$$\begin{bmatrix} h_+ \\ h_- \\ h_z \\ h_0 \end{bmatrix} = \begin{bmatrix} c & 0 & 0 & 0 \\ 0 & c & 0 & 0 \\ 0 & 0 & 1 & 1/2c^2 \\ 0 & 0 & 0 & 1 \end{bmatrix} \begin{bmatrix} J_+ \\ J_- \\ J_z \\ J_0 \end{bmatrix} \qquad (4.5)$$

It is easily verified that the h_ν's satisfy the commutation relations

$$\begin{aligned} [h_z, h_\pm] &= \pm h_\pm \\ [h_+, h_-] &= 2c^2 h_z - h_0 \end{aligned} \qquad (4.6)$$

$$\left[\vec{h},\ h_o\right] = 0 \qquad (4.6)$$

In the limit $c \to 0$ the transformation A becomes singular and A^{-1} fails to exist. Nevertheless, the commutation relations (4.6) are well defined, and in fact identical to the commutation relations (2.6) under the identification

$$\lim_{c\to 0} h_z = n = a^{\dagger}a; \quad \lim_{c\to 0} h_+ = a^{\dagger}; \quad \lim_{c\to 0} h_- = a \qquad (4.7)$$

Although the inverse A^{-1} (4.3b) does not exist as $c \to 0$, the parameter α_ν may approach a well-defined limit if we demand all the parameters λ_μ to shrink ("contract") to zero in the limit $c \to 0$, in such a way that the following ratios are well defined:

$$\begin{aligned} \lim i\lambda_+/c &= \lim e^{-i\varphi}\theta/2c = \alpha \\ \lim i\lambda_-/c &= \lim \left(-e^{i\varphi}\theta/2c\right) = -\alpha^* \\ \lim \lambda_z/c &= \lim c\,\lambda_z/c^2 = 0 \end{aligned} \qquad (4.8)$$

Within any $(2J + 1)$-dimensional representation of the group U(2) the eigenvalue of the diagonal operator h_z is

$$h_z \left| J,M \right\rangle = (J_z + 1/2c^2) \left| J,M \right\rangle = \left| J,M \right\rangle (M + 1/c^2) \qquad (4.9)$$

We demand this have a definite limit as $c \to 0$. Physically, for both Fock and Dicke states we progress upward from the ground or vacuum state. It is convenient to demand that the (energy) eigenvalue in (4.9) be zero for the ground state $M = -J$

$$\lim_{c\to 0} (-J + 1/2c^2) = 0 \qquad (4.10)$$

In the limit $c \to 0$, $2Jc^2 = 1$, the unitary irreducible representation $D^J\left[U(2)\right]$ goes over into the unitary irreducible representation for the contracted group with generators (4.6). In simple words, the contraction procedure amounts to letting the radius of the Bloch sphere tend to infinity as $1/c^2$, while considering smaller and smaller rotations on the sphere. The motion on the sphere then becomes identical to the motion on the bottom tangent plane which goes over into the phase plane of the harmonic oscillator.

This procedure for contracting groups, commutation relations, and representations will now be used to show the similarity between Dicke and Fock states. We define

$$\left|\infty, n\right\rangle = \lim_{c\to 0} \left| J,M \right\rangle \qquad (J + M = n \text{ fixed}) \qquad (4.11)$$

Then

$$a^\dagger a|\infty,n\rangle = \lim (J_z + 1/c^2)\,|J,M\rangle$$
$$= \lim |J,M\rangle\left[J+M + (-J + 1/2c^2)\right] \qquad (4.12a)$$
$$= n|\infty,n\rangle$$

The computations for $a^\dagger$ and a are handled in an entirely analogous way:

$$a^\dagger|\infty,n\rangle = \lim h_+|J,M\rangle = \sqrt{n+1}\,|\infty,n+1\rangle \qquad (4.12b)$$
$$a|\infty,n\rangle = \lim h_-|J,M\rangle = \sqrt{n}\,|\infty,n-1\rangle \qquad (4.12c)$$

These equations provide a straightforward connection between Dicke and Fock state. The operators h_+, h_- and h_z contract to $a^\dagger$, a, and $a^\dagger a$ with the proper commutation relations (2.3), and with the proper matrix elements between contracted Dicke states as shown in (4.12). The contracted Dicke states (4.11) can thus be identified with the Fock states, and we conclude that every property of Dicke and Bloch states listed in Sec. 3 must contract to a corresponding property of Fock and Glauber states listed in Sec.2. The contraction procedure is summarized in Table II.

Table II. Rules for Contraction of The Angular Momentum Algebra to The Harmonic Oscillator Algebra.
[The limit of the angular momentum quantities (1st line) for $c \to 0$ are the corresponding harmonic oscillator quantities (2nd line).]

Operators	Coordinates	Eigen-values	Eigen-states	Coherent states
Angular momentum				
cJ_+, cJ_-, $J_z + \frac{1}{2c^2}$	$\frac{\theta}{2c}e^{-i\varphi}$	$2c^2J$, $J + M$	$\|J,M\rangle$ (Dicke)	$\|\theta,\varphi\rangle$ (Bloch)
Harmonic oscillator				
$a^\dagger$, a, $a^\dagger a$	α	1, n	$\|\infty,n\rangle$ (Fock)	$\|\alpha\rangle$ (Glauber)

We demonstrate this correspondence in some particular cases:
Example 1: Just as the angular momentum eigenstates $|J,M\rangle$ are obtained from the ground state $|J,-J\rangle$ by (J+M) successive applications of the shift-up operator J_+, the Fock state $|n\rangle$ is obtained by n successive applications of $a^\dagger$. By contraction of (3.5) we get

$$|\infty,n\rangle = \lim \frac{(J_+)^n}{\left[2J!n!/(2J-n)!\right]^{1/2}}\,|J,-J\rangle$$

$$= \lim \frac{(cJ_+)^n}{\left[(2Jc^2)^n n!\right]^{1/2}} |J,-J\rangle$$

$$= \frac{(a^\dagger)^n}{\sqrt{n!}} |\infty, 0\rangle \qquad (4.13)$$

which is nothing but (3.5)

Example: Let us now contract Bloch states to Glauber states, using equation (4.12) and Table I:

$$|\alpha\rangle = \lim |\theta, \varphi\rangle = \lim \left(\frac{1}{1 + |\tau|^2}\right)^J e^{\tau J_+} |-J\rangle$$
$$= \lim \left[1 - 2c^2(\alpha\alpha^*/2)\right]^{1/2c^2} e^{\alpha a^\dagger} |0\rangle$$
$$= e^{-\alpha\alpha^*/2} e^{\alpha a^\dagger} |0\rangle \qquad (4.14)$$

which is nothing but (2.12).

We leave it to the reader to verify that every equation of Sec.3 goes over to the corresponding equation of Sec.2 under contraction. This is true in particular of the disentangling theorem (3.11) whose contracted limit is the Baker-Campbell-Hausdorf formula (2.11). In general all properties related to angular momentum have as a counterpart.a harmonic oscillator property. Thus, the total angular momentum contracts to the harmonic oscillator Hamiltonian, with the spherical harmonics (and their properties) contracting to the harmonic oscillator eigenfunctions (and corresponding properties).

REFERENCES

1. R.H.Dicke, Phys.Rev.93, 99 (1954).
2. F.Bloch, Phys.Rev. 70, 460 (1946).
3. F.Bloch and A.Nordsieck, Phys.Rev. 52, 54 (1937).
4. J.Schwinger, Phys.Rev. 91, 728 (1953).
5. R.J.Glauber, Phys.Rev.131, 2766 (1963); R.J.Glauber, in Quantum Optics and Electronics, edited by C.De Witt et al. (Gordon and Breach, New York,1965).
6. F.T. Arecchi, E.Courtens,R.Gilmore, and H.Thomas, Phys.Rev. A 6, 2211 (1972).
7. H.Haken, in Encyclopedia of Physics edited by S.Flügge (Springer-Verlag, Berlin,1970), Vol.XXV, Chap.2.
8. R.P.Feynman, Phys.Rev. 84, 108 (1951).
9. R.J. Glauber in Physics of Quantum Electronics, edited by P.L. Kelley, B.Lax and P.E. Tannenwald (Columbia, New York,1966); J.R. Klauder and E.C.G. Sudarshan, Fundamentals of Quantum Optics (Benjamin, New York, 1968).
10. E.P.Wigner, Group Theory and Its Application to the Quantum Mechanics of Atomic Spectra (Academic, New York, 1959).
11. E.Callen and H.Callen, Phys.Rev. 129, 578 (1963).

Cooperative Phenomena, H. Haken, ed.

EXPERIMENTAL INVESTIGATIONS OF COHERENT RESONANT PHENOMENA

Eric Courtens
IBM Zurich Research Laboratory,
8803 Rüschlikon, Switzerland.

The Bloch-Maxwell theory is briefly reviewed, and phenomena associated with coherent resonant interaction of two-level systems with electromagnetic waves are categorized. The attention is then focussed on three particular experimental realizations made at the IBM Research Laboratories: 1) an experiment on adiabatic rapid passage; 2) a group of experiments which constitute the optical analog of E.P.R.; 3) a group of propagation experiments performed in resonant vapors slightly away from the resonance line, in the so-called adiabatic following regime.

The present description is restricted to the essential points and the reader should consult the original publications for details.

I. THEORETICAL INTRODUCTION [1]

a. The single two-level atom

For a single two-level atom:

i) a pure state can be represented by two complex numbers $\underline{a}$ and $\underline{b}$ with $aa^* + bb^* = 1$;

ii) the three Pauli operators plus the identity form a complete set of operators;

iii) the "spin-up" and "spin-down" operators, σ_+ and σ_-, respectively, are those that connect the ground and the excited states;

iv) the dipole operator can therefore be expressed in terms of

$$\sigma_{\pm} = (\sigma_1 \pm i\sigma_2)/2 \quad , \tag{1}$$

as

$$\vec{\mu} = \vec{p}\,\sigma_+ + \text{h.c.} \quad , \tag{2}$$

where $\vec{p}$ is a complex vector that specifies the polarization properties of the particular dipole transition;

v) the hamiltonian taking into account the interaction with a (real) classical field E is (in the dipole approximation)

$$\mathcal{H} = (1/2)\,\hbar\Omega\sigma_3 - \vec{\mu}.\vec{E} \quad , \tag{3}$$

where Ω is the "resonant" frequency;

vi) the statistical properties of a mixed state is represented as usual by a density matrix

$$\rho = \overline{|\psi><\psi|} \tag{4}$$

b. An assembly of two-level atoms

The interest is not in a single system but rather in a large assembly. Cooperative effects occur due to the feedback mechanism of the common radiation field **E**. Except for this field the systems are non-interacting. The density matrix of the assembly ρ_{tot} is assumed to be the direct product of single system density matrices:

$$\rho_{tot} = \rho_1 \otimes \rho_2 \otimes \dots \otimes \rho_n \otimes \dots \quad . \tag{5}$$

Such states are called uncorrelated [2]. They have a nice property: "An uncorrelated state acted upon by a (non-stochastic) classical field remains uncorrelated". The proof of this theorem is left as an exercise. Initial states are often uncorrelated (in particular the ground state is uncorrelated), and the applied field often behaves rather classically. It results that the restriction (5) is not too severe for the description of a number of experiments. In particular it seems appropriate for the experiments presented here.

This being the case, for any single-atom operator O_n one can define an operator density

$$O_{tot} \equiv \sum_n O_n \, \delta(\vec{r} - \vec{r}_n) \tag{6}$$

with the expectation value

$$\langle O_{tot} \rangle = Tr(\rho_{tot} O_{tot}) = \sum_n Tr_n(\rho_n O_n) \, \delta(\vec{r} - \vec{r}_n) \quad . \tag{7a}$$

Defining a continuous function ρ taking the value ρ_n at $\vec{r} = \vec{r}_n$, (7a) becomes

$$\langle O_{tot} \rangle = N(\vec{r}) \, Tr(\rho O) \tag{7b}$$

with

$$N(\vec{r}) \equiv \sum_n \delta(\vec{r} - \vec{r}_n) \quad . \tag{8}$$

In particular one can define a space-dependent spin vector

$$\vec{S} \equiv \langle \sigma_1, \sigma_2, \sigma_3 \rangle \quad , \tag{9}$$

whose equation of motion tells us about the system evolution.

c. The equation of motion

It is derived from the usual density matrix equation $i\hbar \, \partial\rho/\partial t = [\mathcal{H}, \rho]$ and the proof is left as an exercise. More instructive is the analogy with the Bloch equation [3] describing the precession of a magnetic moment $\vec{M}$ in a magnetic field $\vec{H}$:

$$\frac{\partial \vec{M}}{\partial t} = \gamma \vec{M} \times \vec{H} - \underline{\underline{\Gamma}} \cdot (\vec{M} - \vec{M}_o) \quad . \tag{10a}$$

Here γ is the gyromagnetic ratio and $\underline{\underline{\Gamma}}$ is a (phenomenological) relaxation tensor describing the relaxation of $\vec{M}$ towards its equilibrium value $\vec{M}_o$. Similarly one obtains

$$\frac{\partial \vec{S}}{\partial t} = \frac{2\mu}{\hbar} \vec{S} \times \vec{E}_{eff} - \underline{\underline{\Gamma}} \cdot (\vec{S} - \vec{S}_o) \quad . \tag{10b}$$

The exact definition of the various vectors in (10b) depends on the polarization state of the E.M. field. For simplicity we assume a right circularly polarized (plane-wave) field

$$\mathbf{E} = \hat{x} E_x + \hat{y} E_y = Re\left[(\hat{x} - i\hat{y})(E_x + iE_y)\right] = Re\{(\hat{x} - i\hat{y}) \, E(z,t) \exp\left[i\omega t - ikz + i\phi(t,z)\right]\} \tag{11}$$

propagating in the $\hat{z}$ direction, and a right-polarized transition

$$\vec{p} \equiv (\hat{x} - i\hat{y})\mu \tag{12}$$

in (2). In this case $\vec{S}$ has the definition given in (9) and

$$\vec{E}_{eff} = \vec{E} - \hat{z}\frac{\hbar}{2\mu}\Omega \quad , \tag{13}$$

that is, the effective field in the real field (in the x-y plane) plus a fictitious $\hat{z}$ component that guarantees the atomic precession at frequency Ω in the absence of E. The meaning of the various components of $\vec{S}$ rather clear, and has been elaborated elsewhere. Note that:

i) Equation (10b) also applies to other polarization states with minor changes in (9) and (13).

ii) The definition of $\phi(t,z)$ in (11) is arbitrary as it depends on the choice of ω and k. Usually ω is taken as some suitable central frequency, and k is related to ω by the low-level phase velocity in the medium (sometimes neglecting the two-level systems). Note that $E(z,t)$ is real.

(iii) The expectation value of the dipole density is

$$\langle\vec{\mu}_{tot}\rangle = N\mu[\mathrm{Re}\,(\hat{x} - i\hat{y})(S_x + iS_y)] \tag{14}$$

which is the source term entering Maxwell's equation.

iv) If there is a distribution of atomic frequencies Ω (inhomogeneous broadening) one writes $\vec{S}(\Omega,r,t)$, and (14) contains an integration over the frequency distribution.

d. Rotating frames

As written, (10b) applies to the laboratory frame, which means that $\vec{S}$ and $\vec{E}_{eff}$ precess rapidly at a frequency of the order of Ω. It is convenient to rewrite (10b) in a coordinate frame rotating around $\hat{z}$. Let $\varphi(t)$ be the angle which determines the position of the rotating axes x'y' with respect to the laboratory axes xy (Fig.1). One finds that (10b) still applies in this new frame with new vectors

$$\vec{S}' \equiv R\,\vec{S} \tag{15a}$$

$$\vec{E}'_{eff} \equiv R\,\vec{E}_{eff} + z\frac{\hbar}{2\mu}\frac{\partial\varphi}{\partial t} \tag{15b}$$

where R is the rotation operation. One can take $\varphi = \omega t - kz$ in which case $(E'_{eff})_z \propto \omega - \Omega$, or $\varphi = \Omega t - kz$ in which case there is no z component left. One very convenient choice is $\varphi = \omega t - kz + \phi$ in which case the x'y' component of E'_{eff} has a stationary direction. With an appropriate choice of phase one selects then $(E'_{eff})_y = 0$.

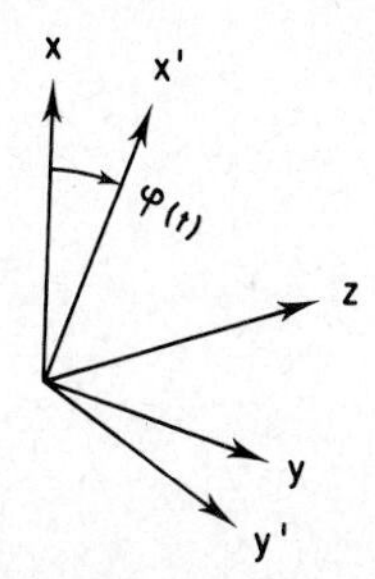

Fig.1. The laboratory frame (xyz) and the rotating frame (x'y'z).

Very convenient definitions are then

$$\mathcal{E} \equiv \frac{2\mu}{\hbar} R E \equiv \frac{2\mu}{\hbar} E(z,t) \text{ , (Rabi precession frequency) ;} \tag{16a}$$

$$\gamma \equiv \Omega - \omega - \frac{\partial\phi}{\partial t} \text{ , (instantaneous frequency offset) .} \tag{16b}$$

The equation of motion is simply

$$\frac{\partial \vec{S}'}{\partial t} = \vec{S}' \times (\hat{x}\mathcal{E} - \hat{z}\gamma) - \underline{\underline{\Gamma}} \cdot (\vec{S}' - \vec{S}'_o) \quad . \tag{17}$$

The three components of $\vec{S}'$ can be labelled in the following way:

$$S' \equiv \hat{x}u - \hat{y}v + \hat{z}w \tag{18}$$

(The minus sign is to conform with previous notation in [1]). The equations for these components are (with phenomenological relaxation terms in parenthesis)

$$\frac{\partial u}{\partial t} = \gamma v + (-u/T_2') \tag{19a}$$

$$\frac{\partial v}{\partial t} = -\gamma u - \mathcal{E}w + (-v/T_2') \tag{19b}$$

$$\frac{\partial w}{\partial t} = \mathcal{E}v + [(w - w_o)/T_1] \tag{19c}$$

e. Coupling to the wave equation

The equation for plane waves is

$$\frac{\partial^2 \vec{E}}{\partial z^2} - \frac{n^2}{c^2} \frac{\partial^2 \vec{E}}{\partial t^2} = \frac{4\pi}{c^2} \frac{\partial^2 \vec{P}}{\partial t^2} \quad , \tag{20}$$

where $\vec{P}$ is given by (14) , or, in the previous notation

$$\vec{P} = N\mu \, \mathrm{Re}[(\hat{x} - i\hat{y})(u - iv) \exp(i\omega t - ikz + i\phi)] \quad , \tag{21}$$

(plus an integration over atomic frequencies if there is a distribution). Making the "slowly varying envelope approximation" one gets

$$\left(\frac{\partial}{\partial z} + \frac{n}{c}\frac{\partial}{\partial t}\right)\mathcal{E} = -\frac{4\pi\omega N\mu^2}{nc\hbar} v \quad , \tag{22a}$$

$$\mathcal{E}\left(\frac{\partial}{\partial z} + \frac{n}{c}\frac{\partial}{\partial t}\right)\phi = -\frac{4\pi\omega N\mu^2}{nc\hbar} u \quad . \tag{22b}$$

As is well known, the amplitude evolution is caused by the out-of-phase component of the polarization and the phase evolution by the in-phase component. Dividing (22b) by $\mathcal{E}$ and taking the time derivative on both sides one gets

$$\left(\frac{\partial}{\partial z} + \frac{n}{c}\frac{\partial}{\partial t}\right)\gamma = \frac{4\pi\omega N\mu^2}{nc\hbar} \frac{\partial}{\partial t} \left(\frac{u}{\mathcal{E}}\right) \quad . \tag{22c}$$

This step is not as trivial as it seems: ϕ propagates with the phase velocity and γ with the group velocity. The phase velocity can be derived from (22b), and the group velocity from (22c), and these can be very different close to a resonance line. It is then more convenient to use the couple of variables $(\mathcal{E},\gamma)$ and the corresponding equations (22a), (22c), as these variables propagate with the same velocity.

f. Local effects and propagation effects

Equations (19) and (22) constitute a coupled system. This is at the root of the cooperativity in this field-atom interaction: the field drives the atoms which produce a polarization which modifies the field. *Local effects* are those for the description of which this chain does not have to be closed self-consistently. One uses the applied field in (19), derives a polarization, and uses it in (22) to calculate reradiation. This is all right for "small" systems (the relevant length is the so-called "cooperation length" [4]). In other words small systems are those for which the reaction of the system onto the field can be treated as a perturbation. The number of phenomena that can be observed under these conditions is quite large. Firstly, one can in principle produce all the optical analogs of well-known NMR and EPR effects such as adiabatic rapid passage, nutation, echoes, free-induction decay, etc. Secondly, some new effects can be seen thanks to the far greater transition probabilities in the optical range. Examples of this are the coherent Raman beats observed by Brewer and co-workers (see Section II). The local effects can mainly be used to:
1) Investigate relaxation processes;
2) Prepare systems in their excited states (adiabatic rapid passage).

A second category consists of the "propagation effects", for which the reaction of the two-level systems onto the field is so severe that Eqs.(19) and (22) have to be solved simultaneously. This is usually quite an involved problem and very few analytic solutions are known. Steady-state solutions usually involve solitary waves, and transients are usually accompanied by very strong pulse reshaping. At, or extremely close to, resonance a phenomenon known as self-induced transparency occurs [5]. The medium becomes quasi-transparent to pulses of definite area, in particular those for which $\int \mathcal{E} dt$ is 2π, and of definite shape. These pulses propagate very slowly through the medium. Further away from resonance, considerable non-linearity can still be observed. In particular self-modulation and optical shock formation have been seen and analyzed in detail as described in Section III. The propagation effects are:
1) Very interesting for their own sake. In this situation the cooperative interaction of atoms and field is at its strongest;
2) Potentially useful for modulation.

In closing, two comments are in order having to do with the relation to superradiance [6] and to the laser itself. With regard to superradiance one has to watch out not to confuse the quantum concepts of spontaneous and stimulated emission with the classical dipole radiation. The classical theory allows emission both in the presence or in the absence of field. It is incorrect to state that classical emission is all stimulated. There is however a small part of the spontaneous emission process which cannot be described classically, and there are various *ad hoc* ways by which this can be patched up. In this sense the semiclassical theory described above can to a great extent be used to describe superradiance. Echo pulses (see Section II) are proportional to N^2. Slowly propagating 2π pulses of self-induced transparency consist mostly of material excitation with very little electromagnetic content, and they can rightly be thought of as superradiant packets grinding their way through the material. Finally, the cooperation length mentioned above is also the maximum sample length that can emit proportionally to N^2. Incidentally, a very interesting problem is that of the emission from a very large sample (of size much larger than the cooperation length) which is initially placed all in the upper state. Such emission should be stochastic, with a characteristic time related to the cooperation time [4].

With respect to the laser, one should just note that Eqs.(19) and (22) with minor modifications are also adequate for an elementary description of the phenomenon. Experimentally, the laser turned out to be a well-suited device for refined measurements of field statistics. The description of those constitutes a field of its own and requires much more elaborated techniques, such as those described in the lectures of Prof. Haken.

II. EXPERIMENTS ON LOCAL EFFECTS

Two types of "local" coherent resonant optical experiments have been performed recently at the IBM Research laboratories. The first kind is the optical analog of adiabatic rapid passage. The second kind is a series of experiments where the level structure is modified non-adiabatically by Stark switching allowing the observation of nutation, free-induction decay, echoes, as well as some novel phenomena.

a. Adiabatic rapid passage

It was observed by M.M.T. Loy [7] in ammonia using the resonance of the Stark-tuned ($\nu_2 = 0 \to 1$, $J = 5$, $K = 5$, $M = \pm 5$) line with the R6 line of the CO_2 laser. Equation (17) describes the precession of $\vec{S}'$ around the effective field. If the direction of the effective field varies *sufficiently slowly* the angle it makes with the spin vector will remain constant. How much slow is slow,results from the proof of the adiabatic theorem [8]. With $\theta = \arctan(\mathscr{E}/\gamma)$, (see Fig.2), the adiabatic conditions are:

1) that $\left|\frac{d\theta}{dt}\right| < \Delta = \sqrt{\mathscr{E}^2 + \gamma^2}$ (the precession frequency) ,

2) that $\frac{d\theta}{dt}$ does not contain strong Fourier components at Δ .

Keeping $\mathscr{E}$ approximately constant, γ is changed from a large value in one direction to a large value in the other direction. The effective field sweeps the x-z plane (Fig.2), carrying along the vector $\vec{S}'$. If the system is initially in the ground state, it finds itself in the upper state after one passage (independently of the initial direction of γ). The adiabatic condition puts an upper limit to the rate of change of γ ; relaxation imposes a lower bound. Hence the expression *adiabatic rapid* passage.

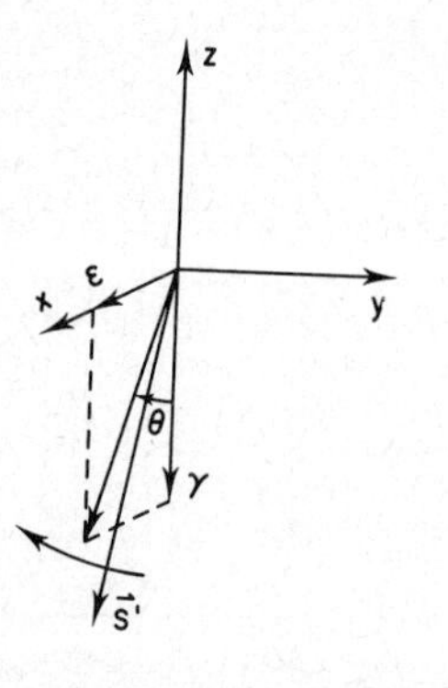

Fig.2. Position of the various vectors in adiabatic rapid passage.

Experimentally a "long" Q-switched CO_2 laser pulse is focused between the plates of a Stark cell. A D.C. voltage is applied to the cell in order to have the appropriate line close to resonance with the laser (about half a GHz away). Near the peak of this laser pulse an additional Stark voltage pulse is applied. During the adiabatic rise, an absorption signal is observed on the main laser beam. For the duration of the voltage pulse the excited system is allowed to relax by T_1 ("longitudinal") processes. At the end of the pulse a second passage occurs which probes the new state of the system. If the pulse duration was short enough for the system to have remained in the excited state, an emission signal is seen. On the other hand if the pulse is long compared to the relaxation time a new absorption signal is seen (Fig.3). The experimental difficulties are:

i) to avoid dielectric breakdown of the cell;
ii) to obtain clean electrical pulses of controllable rise time;
iii) to have sufficient laser power to satisfy easily the adiabatic condition.

The experiment was used to measure the longitudinal relaxation time T_1 in the excited state of ammonia. For the interpretation of the results one should note that:

i) the transit time of the molecules across the laser beam is a trivial cause of decay which must be accounted for;
ii) the refilling of the ground state by rotational relaxation within the ground states contributes as well as the emptying of the excited state by the same and other processes.

The detailed results are found in Ref. [7].

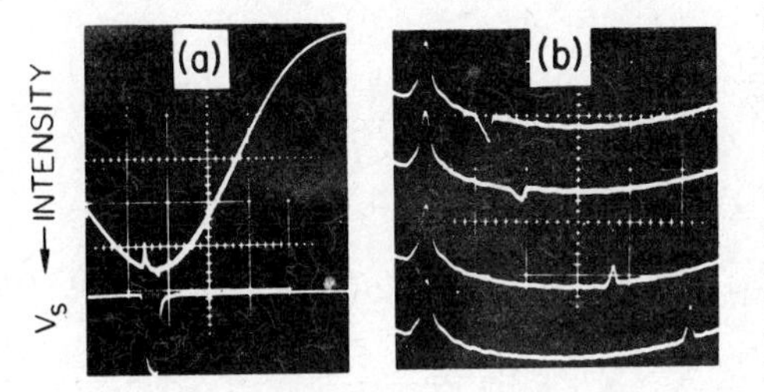

Fig.3.a) Dual-beam oscilloscope traces showing the Stark pulse (lower beam) and the laser pulse (upper beam) after the cell. Absorption appears as a positive signal riding on top of the slowly-varying pulse. Time scale: 1 μsec/div.

b) The relevant part of the laser pulse on expended time scale (0.2 μsec/div), and for different Stark pulse durations. (From Phys. Rev. Letters, Ref. [7].)

b. Stark switching experiments

This refers to a series of experiments by R. Brewer and coworkers [9-13] which are the optical analog of NMR and EPR. The experimental arrangement is sketched in Fig.4. In contrast to the case of adiabatic rapid passage, a strong laser power is not usually required, but the laser frequency should be quite stable: the use of a CW laser is thus indicated. The Stark pulses move velocity packets of a Doppler broadened profile in and out of resonance, as indicated in Fig.5. Upon application of the field those molecules with velocity v are shifted out of resonance, those with velocity v' are shifted in resonance, and vice versa upon termination of the pulse. The Stark pulses are applied non-adiabatically. Typical systems used to-date are:

i) $C^{13}H_3F$ (90% enriched) whose ($\nu_3 = 0\rightarrow1$, $J = 4\rightarrow5$, $K = 3$, $M\rightarrow M\pm1$) lines are resonant with the P(32) line of the CO_2 laser;

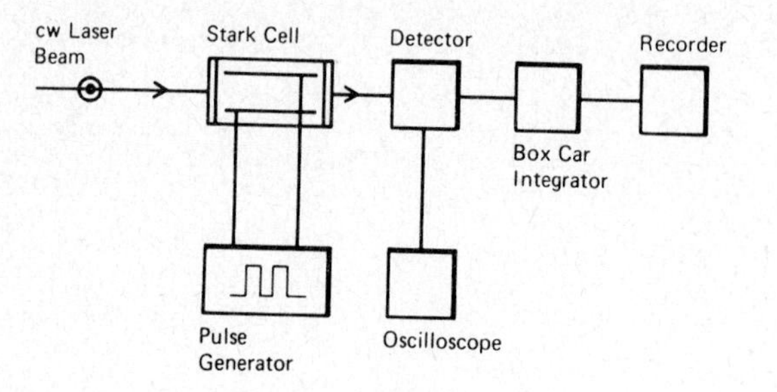

Fig.4. Monitoring technique for observing optical transients with one or more Stark pulses. (From Phys. Rev. Letters [9].)

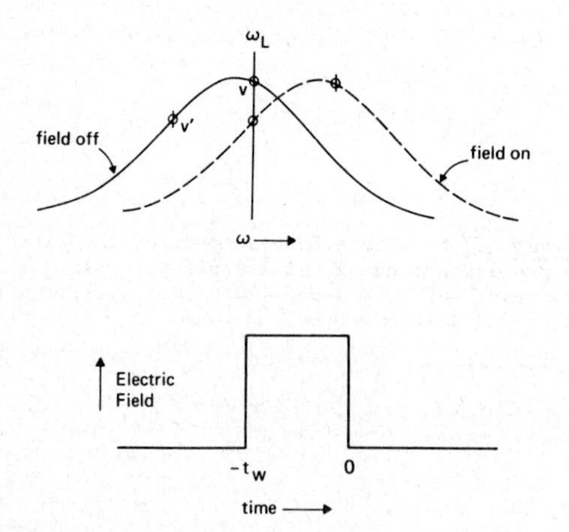

Fig.5. Switching behavior of a non-degenerate Doppler line under the influence of a Stark pulse of duration t_w . The laser frequency is ω_L . (From Phys. Rev. [11] .)

ii) NH_2D which has several coincidences with CO_2 and offers the advantage of having non-degenerated levels (except for the remaining M degeneracy, which does not affect the transition dipole moment). The particular coincidences used are of the type (ν_2 = 0→1 , J = 4→5 , M→M±1).

Phenomena so far observed are:

1) Optical nutation [9]: Upon sudden application of the field, the spin vector starts to precess around the new effective field direction. This motion causes successive absorption and emission which is seen as a modulation at frequency Δ on the transmitted beam (Fig.6). This signal decays with a relaxation time which is a mean of T_1 and T_2 . If several different dipole matrix elements are involved there will be an interference effect.

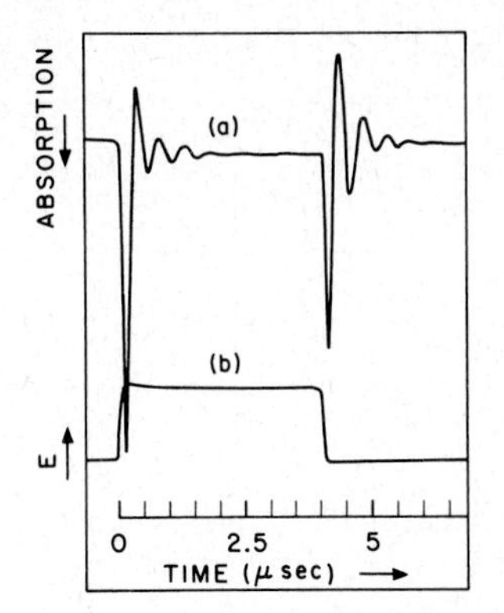

Fig.6. Nutation in $C^{13}H_3F$ at ~5 mT pressure with a Stark pulse of 35 V/cm. (From Phys. Rev. Letters [9].)

2) Two-pulse nutation [13]: First, a $\pi/2$ pulse places the spin vector horizontally. (For a given laser power, a $\pi/2$ pulse is obtained by adjusting the Stark pulse length.) The spin vectors corresponding to different homogeneous packets then fan out in the horizontal plane due to inhomogeneous broadening. In addition, if there is T_1 relaxation, there is an overall decay. Secondly, a step function applied after a given delay, causes nutation. The nutation amplitude monitors the T_1 decay. Indeed, the nutation of a uniformly populated *plane* of spins gives no absorption or emission signal (the sum of all spins is zero).

3) Free-induction decay [11] : Upon sudden removal of the field, excited spin vectors start to precess around $\hat{z}$ at frequency γ in the rotating frame. In the laboratory frame this is a precession at their eigenfrequency Ω (which is Stark shifted). It produces an emission at Ω . This emission beats with the laser, and the beat frequency is the Stark shift, (Fig.7). Again, if several different groups of molecules are involved, with different Stark shifts, interference effects can occur which wash out the beat. Inhomogeneous broadening also has this effect. It should be noticed that upon application of a step function (Fig.5) there will be nutation for the packet at v' *together* with free-induction decay for that at v . Upon removal of the field the packet at v nutates and that at v' emits by free-induction decay.

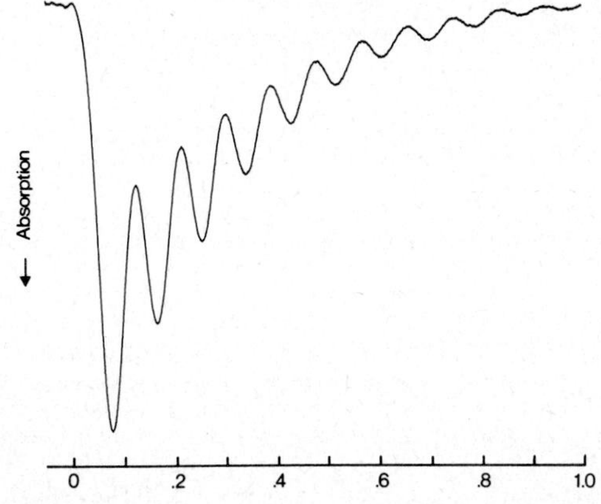

Fig.7. Free-induction decay in NH_2D at 10.6μm. The signal appears as an heterodyne beat on top of a slowly-varying background which is the result of the velocity group v' undergoing nutation. (From Phys.Rev. [11] .)

4) Photon echo [9]: This is a three-step process, identical to spin echo [14] i) at time $t = 0$ a $\pi/2$ pulse places the spin vectors in the horizontal plane where they fan out by inhomogeneous broadening; ii) at time $t = \tau$, a π pulse flips around the pancake of fanned-out spins; iii) due to the same inhomogeneous broadening and because they have been flipped, the fanned-out spin rephase at $t = 2\tau$. These rephased spins emit an echo pulse. This echo pulse is emitted at a Stark- shifted frequency which beats with the laser giving the heterodyne signal of Fig.8. The echo amplitude measures T_2.

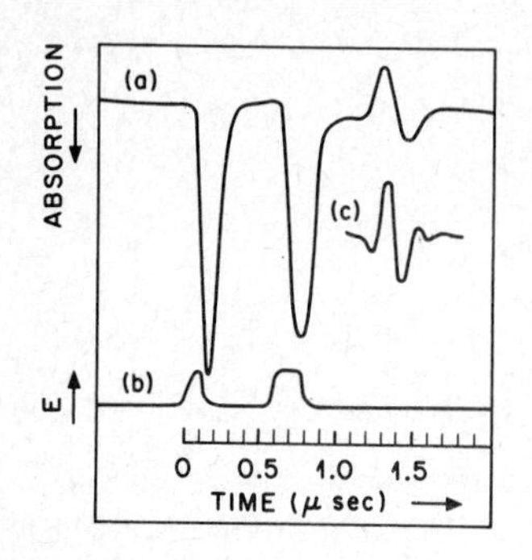

Fig.8. Photon echo in NH_2D at ~5mT pressure. (a) The optical response to $1/2\pi$ and π pulses followed by the echo beat signal, (b) The Stark pulses (E = 35 V/cm). (c) Echo beat signal for E = 60 V/cm. In this case the beat frequency is about twice that of (a). (From Phys.Rev. Letters [9].)

5) Multiple echoes [13]: The rephasing process described above can be repeated at times 3τ, 5τ ..... leading to echoes at 4τ, 6τ ..... (Fig.9).

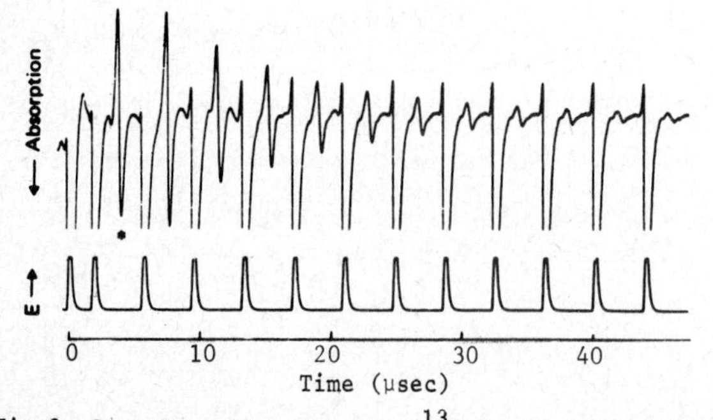

Fig.9. Carr-Purcell echoes in $C^{13}H_3F$ at 0.3mT, with 40 V/cm Stark pulses and 80 V/cm Stark bias. (From Phys. Rev. Letters [13].)

6) Raman beats [10,12]: This new effect is simplest to grasp for a three-level system (Fig.10) which is degenerate in absence of Stark field. Levels 1 and 2 (M±1) are connected to 3(M) (laser polarization perpendicular to Stark field) but not connected to each other. The presence of the CW laser produces an off-diagonal density matrix component ρ_{12}. Upon application of the Stark field which lifts the degeneracy both Stokes (10b) and Anti-Stokes (10c) are produced parametrically due to the simultaneous presence of this ρ_{12} and of the laser field. The emission is shifted by the Stark shift ω_{12} and therefore beats with the laser at that frequency. It is clear that inhomogeneous broadening will affect the emitted frequency only by $\omega_{12}v/c$, which is negligible. The decay of the signal is essentially due to homogeneous relaxation of ρ_{12}, plus a weak contribution related to power broadening

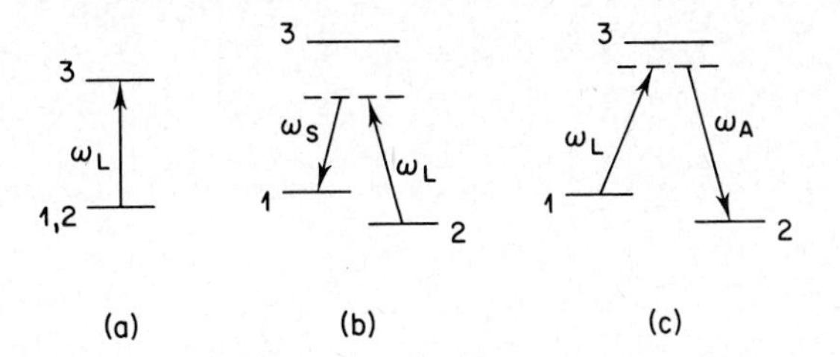

Fig.10. Level diagram for coherent Raman beats: (a) in the absence of Stark field. (b) The Stokes emission ω_S and, (c) the anti-Stokes emission ω_A , in the presence of the field.

of 1-3 and 2-3. In real systems(Fig.11) the level structure is generally more complicated. In the particular case shown the transition 2-3 (Fig.12) happens to remain resonant with the laser frequency so that the emission signal contains beats of ω_{12} and ω_{34} with the laser frequency.

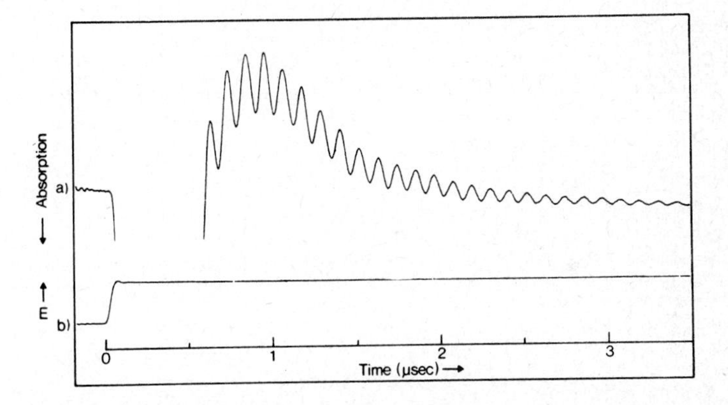

Fig.11. (a) Coherent beat signal in $C^{13}H_3F$ at ~3mTorr following the step-junction Stark field in (b). The beat signal is riding on a stronger optical nutation signal whose downward spike goes off-scale. The beat actually contains two frequencies at ~6 and 8 MHz. (From Phys. Rev. Letters [10] .)

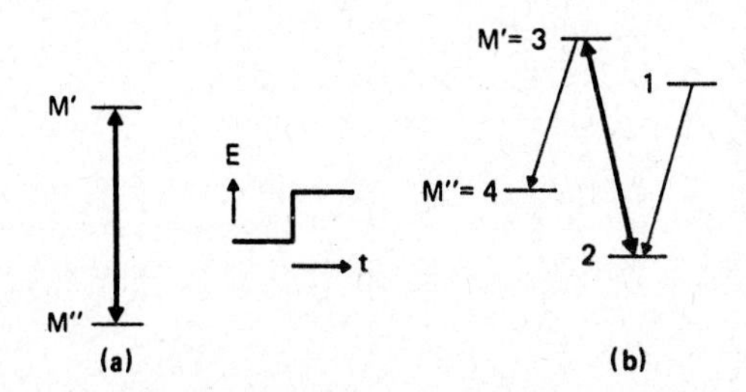

Fig.12. Energy-level diagram for the beat of Fig.11, before (a), and after (b) the Stark step. The laser is resonant with all transitions in (a) but only with the 2-3 transition in (b), giving rise to the emissions shown.

Mostly two kinds of information are derived from these experiments:
i) precise Stark shifts, in free-induction decay signals and Raman beats;
ii) relaxation times, in free-induction decay, two-pulse nutation, echoes, etc.
In favorable cases one can distinguish between relaxation due to hard collisions and due to velocity-changing collisions [13]. If the r.m.s. velocity change Δu in these collisions is small (it was found to be 1/2% of the thermal velocity in methyl fluoride) one can adjust the Stark pulse length τ such that $\omega\Delta u/c$ be either large or small compared to τ^{-1} which is the bandwidth excited by the pulse. In the first case ($\omega\Delta u/c > \tau^{-1}$) velocity-changing collisions will place the molecules outside the relevant bandwidth, thus contributing to decay of, say, photon echo. In the opposite case only hard collision will produce decay.

III. EXPERIMENTS ON PROPAGATION EFFECTS

A series of experiments has been performed on propagation in an atomic vapor (Rb) at a frequency slightly away from resonance (a few tenths of a cm^{-1}) [15-19]. The in-phase polarization component is then *usually* much larger than the out-of-phase component. Except for extremely steep transients one can write

$$\left|\frac{\partial v}{\partial t}\right| << |\gamma u| \tag{23}$$

in (19b). This is the so-called "adiabatic following" regime [19], in which the spin vector follows closely the effective field. In this regime, and close to resonance,

i) the medium is extremely dispersive;
ii) non-linearities set in rapidly. They can in first-order be thought of as the familiar "non-linear index" $n_{eff} = n_o + n_2E^2$ but a more exact description is usually required.

It should be noted that n_2 is positive above resonance and negative below. The steepening effects described below *are independent of this sign*. However a positive n_2 leads to self-focusing, an effect on the transverse beam dimension which easily becomes catastrophic [18]. Propagation studies are better performed with n_2 negative, as defocusing is easily controllable [19]. The steepening effects are due to two causes [16]:

i) the nonlinear group velocity: it becomes higher as the system is in a more excited state;
ii) the self-phase modulation which, in a simple-minded first-order description can be thought of as being due to the non-linearity of the index. The instantaneous phase is $\omega t - kz = \omega t - n_o\omega z/c - n_2\omega zE^2/c$; the instantaneous frequency is the time derivative of this phase: $\omega - n_2 z\frac{\omega}{c}\,\partial E^2/\partial t$. Therefore the instantaneous frequency is always pushed towards resonance on a pulse front ($\partial E^2/\partial t > 0$).

a. Experimental arrangement

It is sketched in Fig.13. The reader should refer to the literature for details Note only that in order to obtain all data required for the interpretation a large number of simultaneous measurements must be performed:

i) the monochromaticity of the incoming pulse must be checked (Fabry-Perot #1),
ii) its position with respect to a reference lamp must be measured (Fabry-Perot #2),
iii) the input pulse shape must be monitored (photo-diode #1),
iv) similarly for the output pulse shape (photo-diode #2),
v) the magnetic field which splits the Zeeman components of the Rb $^2S_{1/2}\rightarrow{}^2P_{1/2}$ resonance (a few thousand gausses) must be synchronously measured,
vi) the transversal beam pattern of the output, and sometimes of the input, must be checked.

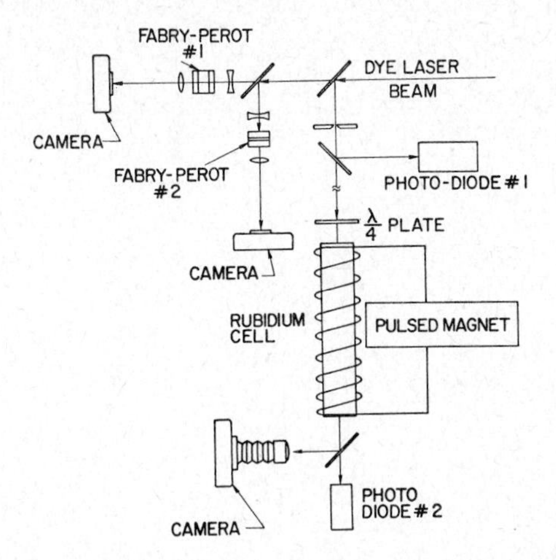

Fig.13. Schematic diagram of the experiment. The rubidium cell is 1 meter long and the pulsed magnetic field is typically 8.5 kG. The effective density of interacting atoms (1/2 the actual density in view of the ground-state splitting) is around $10^{13}cm^{-3}$ at $120^{o}C$. The dye laser emits a *transform limited* pulse. It is pumped by a ruby laser. (From Phys. Rev. [15].)

Without the $\lambda/4$ plate the two circularly polarized components of the linearly-polarized input pulse separate in the Rb cell as shown in Fig.14. The σ^- part is nearly resonant and propagates slowly, whereas the σ^+ component is further away (by Zeeman splitting) and is fast.

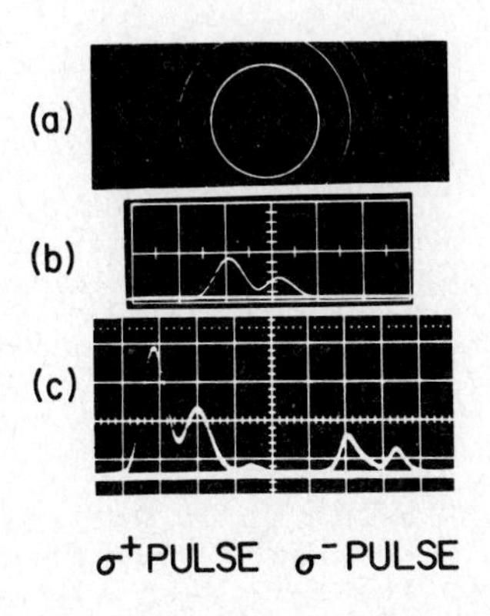

Fig.14. Fabry-Perot interferograms (a), input pulse (b), and output pulse (c) for linearly-polarized light. The interferometer has a finesse of ~100 and a spectral range of 0.33 cm^{-1}. The measured linewidth is less than 0.005 cm^{-1}. The sweep speed in (b) is 5nsec/div, the peak power is 7W, and the laser frequency is 0.25cm^{-1} below the center of the σ^- hyperfine components and 1.30 cm^{-1} below that of the σ^+ components. In (c) the pulse separation is 26nsec, corresponding to $c/v_g = 9$ for the σ^- pulse. (From Phys. Rev. [15].) The intensity of the input pulse corresponds to $\mathcal{E} \ll \gamma$.

b. Theory

Using (23) in (19b) (neglecting relaxation) one gets

$$u = -\frac{w}{\gamma} \quad ; \tag{24a}$$

(19a) is

$$v = \frac{1}{\gamma}\frac{\partial u}{\partial t} \quad ; \tag{24b}$$

and (19c) can then be integrated exactly to give

$$w = \frac{-1}{\sqrt{1+\mathcal{E}^2/\gamma^2}} \quad . \tag{24c}$$

With a similar approximation relaxation can be accounted for, leading to absorption as explained elsewhere [16,17]. Introducing these results in (22) one finds, at low level ($\mathcal{E} \ll \gamma$), the phase velocity

$$\frac{c}{v_{ph}} = n + 4\pi\, N\mu^2/n\,\hbar\,\delta \quad , \tag{25a}$$

and the group velocity

$$\frac{c}{v_g} = n + 4\pi\, N\mu^2\omega/n\,\hbar\,\delta^2 \quad , \tag{25b}$$

where δ is the input value of γ. One notes that:

i) in moderately dense vapors n (the background index) is *very* close to 1 and c/v_{ph} is also close to 1 ;

ii) c/v_g however can be much larger than 1 because ω/δ can easily approach 10^5;

iii) within the approximation used in deriving (25b) the group velocity is related tc the phase velocity by the usual low-level dispersion result

$$\frac{c}{v_g} = \frac{\partial}{\partial\omega}\left(\omega\frac{c}{v_{ph}}\right) \cong n + \omega\frac{\partial}{\partial\omega}\left(\frac{c}{v_{ph}}\right) \quad . \tag{25c}$$

This has been verified by delay measurements of the kind shown in Fig.14 at various frequency offsets δ, and various densities N. The results for v_g, plotted vs. N/δ^2, fall on a branch of hyperbola as shown in Fig.15 [15].

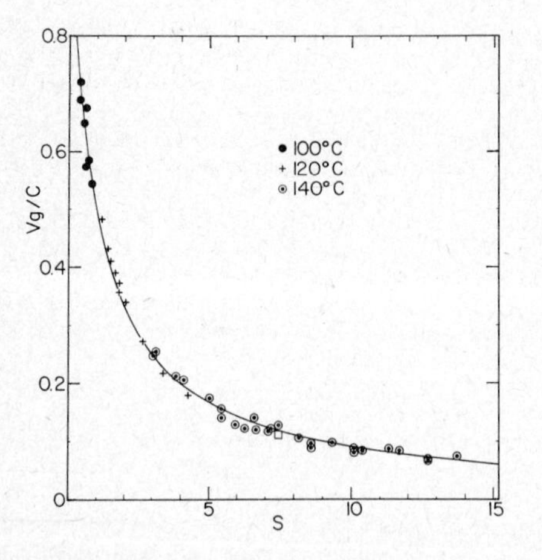

Fig.15. Measured group velocities at various cell temperatures plotted versus the quantity S defined in (26c). S can be calculated from the known vapor pressure and frequency offsets.(From Phys. Rev.[15].)

Introducing (24) in (22a-c), without restriction on $\mathcal{E}$, one obtains the desired propagation equations. They can be put into an elegant form with the definitions

$$\zeta \equiv \frac{n}{c} z \quad , \tag{26a}$$

$$\tau \equiv t - \frac{z}{v_g} \quad \text{(local time)} \quad , \tag{26b}$$

$$S \equiv \frac{4\pi\mu^2 \omega N}{n^2 \hbar \delta^2} \quad \text{(coupling strength)} \ . \tag{26c}$$

They are:

$$\frac{\partial \mathcal{E}}{\partial \zeta} = M \frac{\partial \mathcal{E}}{\partial \tau} + R\mathcal{E} \frac{\partial \gamma}{\partial \tau} \quad , \tag{27a}$$

$$\frac{\partial \gamma}{\partial \zeta} = M \frac{\partial \mathcal{E}}{\partial \tau} - R\mathcal{E} \frac{\partial \mathcal{E}}{\partial \tau} \quad , \tag{27b}$$

where:

$$M \equiv S \left\{ 1 - \frac{\delta^2}{\gamma^2} \frac{1}{(1 + \mathcal{E}^2/\gamma^2)^{3/2}} \right\} \quad , \tag{28a}$$

$$R \equiv S \frac{\delta^2}{\gamma^3} \frac{1}{(1 + \mathcal{E}^2/\gamma^2)^{3/2}} \quad . \tag{28b}$$

This is a system of two coupled first-order partial differential equations for $\mathcal{E}$ and γ . It shows a number of interesting features:

i) Initially ($\gamma \approx \delta$, $\mathcal{E} << \gamma$) M is zero, which simply means that the group velocity (25b) is the proper propagation velocity. This can change due to frequency modulation ($\gamma \neq \delta$) , or high excitation ($\mathcal{E}/\gamma$ no longer negligible).

ii) R always exists, also in the absence of strong non-linearity ($\mathcal{E} << \gamma$). In (27a) it plays the role of the usual dispersion term. In (27b) it causes self-phase modulation.

iii) The discriminant of the system $(b^2 - 4ac) = - 4 R^2\mathcal{E}^2$ is always negative, meaning that the system is elliptic, though the boundary conditions (initial conditions) are those appropriate to a hyperbolic problem. Such systems are commonly termed "improperly-posed" or "ill-posed" and always lead to singularities.

iv) The evolution is extremely sensitive to the second derivative of the envelope. This is easily seen by taking $M = 0$ in (27), integrating (27b) approximately, which gives

$$\gamma \simeq \delta - \frac{R}{2} (\zeta - \zeta_o) \frac{\partial \mathcal{E}^2}{\partial \tau} \quad , \tag{29a}$$

where ζ_o is the position of the cell input face. Using (29a) in (27a) one finally obtains

$$\frac{\partial \mathcal{E}^2}{\partial \zeta} \simeq - R^2\mathcal{E}^2 (\zeta - \zeta_o) \frac{\partial^2 \mathcal{E}^2}{\partial \tau^2} \quad . \tag{29b}$$

This shows that initially the envelope distorts proportionally to its second derivative, and linearly with distance. This behavior is confirmed by an integration of the full system (27), as shown in Fig.16. The input pulse is taken to be half a sinusoid, and the conditions are typical for propagation in Rb vapor. The output pulse after 1 meter cell length is shown in Fig.16b, with the associated phase modulation in Fig.16c. If one adds a minute gaussian bump to the input, as shown fifty times enlarged in Fig.16a, the output shows sizeable perturbations both on the phase and on the envelope. That on the phase goes like the first derivative of the gaussian, that on the envelope like the second derivative.

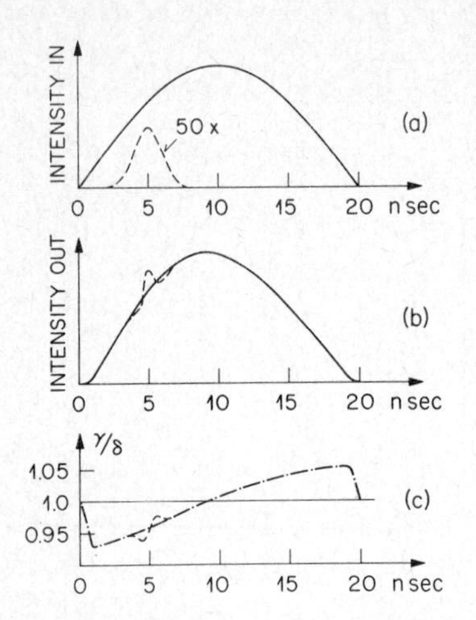

Fig.16. (a) Input pulse, without and with a small gaussian bump. The gaussian bump alone is shown 50 times enlarged (dashed line); (b) output pulse, without and with the bump, which is shown now with its real size riding on the otherwise smooth envelope; (c) output frequency normalized to the input value.

v) This extreme sensitivity to second derivatives requires great care in the integration of numerical data. An experimental input pulse cannot be entered directly point by point; it first requires fitting to a smooth function.

vi) The extreme sensitivity both to phase and to envelope shapes suggests that this may be a very useful technique for pulse shaping, in particular in the case of the very powerful pulses that are used for laser-produced fusion. For example, a minute amount of phase modulation, which is relatively simple to produce by a Pockels cell, can be translated after passage through a vapor into considerable amplitude modulation.

c. Experimental results

Figure 17 shows typical pulse reshaping for $\mathcal{E}/\delta \simeq 0.3$. More intense or more nearly-resonant pulses develop complicated envelopes with abrupt rises and multiple peaks. The pulse (d) is far off resonance and shows little reshaping or attenuation; it propagated with $v_g \simeq 0.8c$. The other pulses have a velocity $\sim c/4$ and an attenuation of $\sim 30\%$. One notices the great sensitivity to input amplitude and frequency: Pulse (b), which is slightly more resonant than (a), suffered more severe reshaping; pulse (c), which is slightly more powerful than (a), developed a complicated envelope. Figure 18 shows the result of numerical integration on the pulse of Fig.17a. The integrated equations are a modified form of the system (27). The modifications account for: i) the hyperfine splitting of the line, ii) the absorption caused by T_2, iii) the diffraction of the wave [17]. For comparison to the output, the input has been drawn as if it had gone through the cell at the low-level group velocity subject only to linear absorption and diffraction. With this normalization (and except for the intensity dependence of α) the energies of the input and output pulses are equal meaning that the output peak is higher than the input due to time compression. This agrees with the calibration deduced from Fig.17d. Note that there is no adjustable parameter in this fit, except for the exact power calibration.

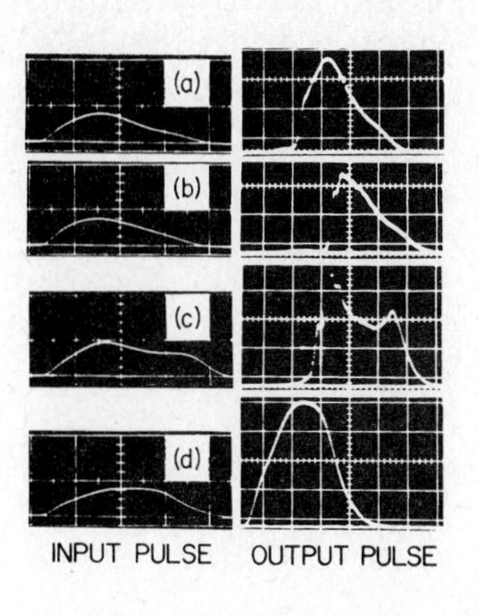

Fig.17. A typical experimental run showing input pulses (300 W/div and 5 nsec/div) and corresponding output pulses (5 nsec/div). The oscilloscope monitoring the input pulse also triggered that monitoring the output pulse, so that the delays observed are significant. The rise-time of the output in (b) is the instrumental resolution. (a) (a) δ = 0.24 cm^{-1} ; (b) δ = 0.20 cm^{-1}; (c) δ = 0.23 cm^{-1}; (d) no magnetic field applied δ = 0.78 cm^{-1} , which means that this last pulse can be used as reference for amplitude and delay. (From Phys. Rev. Letters [16].)

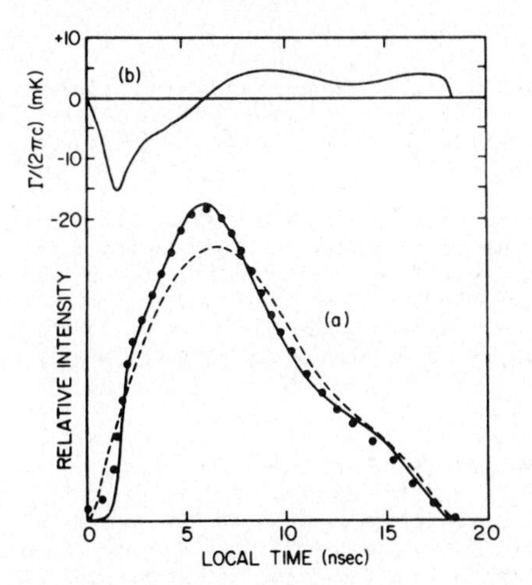

Fig.18.(a) The output pulse (dots) of Fig.17(a), compared with the calculated pulse (solid line) and the normalized input (dashed line). (b) Calculated self-phase modulation $\Gamma = \partial\phi/\partial t$ in $10^{-3}cm^{-1}$. (From Phys. Rev. Letters [16] .)

Numerical integration on pulse (a) indicates that the pulse was still steepening at the end of the cell. Similar integration on pulse (b) leads to a singularity at 70 cm in the cell. Beyond this point the integration of (27) cannot be continued. This suggests that an optical shock propagated over 30 cm in the actual experiment. By shock one means an extremely sharp front propagating with little further reshaping due to a dynamic balance between the steepening mechanisms of (27) and a smoothing mechanism, not taken into account by these equations, and operating only during extremely rapid variations of the field.

The great sensitivity to phase modulation has also been recently tested [20]. In the regime $\mathcal{E} \ll \gamma$ an analytic solution to pulse propagation was found. A slight sinusoidal frequency modulation (~ 200 MHz modulation depth) of the input, with an otherwise smooth pulse, leads to a spiky output envelope which is reminiscent of a mode-locked train. It is worthwhile noting that even with $\mathcal{E} << \gamma$ new frequencies are eventually generated in this process. The experiments confirm in all points the theory.

[1] There has been a large number of articles written on the subject. A fairly recent review was written by the author for the "Laser Handbook", F.T. Arecchi and E.O. Schulz-DuBois, eds., North-Holland Publ. Co., Amsterdam 1972, Chapter E5, pp. 1259-1322.

[2] For a definition of atomic correlation see, for example, F.T. Arecchi, E. Courtens, R. Gilmore and H. Thomas, Phys. Rev. A6 (1972) 2211.

[3] F. Bloch, Phys. Rev. 70 (1946) 460.

[4] F.T. Arecchi and E. Courtens, Phys. Rev. A2 (1970) 1730.

[5] S.L. McCall and E.L. Hahn, Phys. Rev. Lett. 18 (1967) 908.

[6] R.H. Dicke, Phys. Rev. 93 (1954) 99.

[7] M.M.T. Loy, Phys. Rev. Lett. 32 (1974) 814.

[8] A. Abragam, The Principles of Nuclear Magnetism, Oxford Clarendon Press 1961.

[9] R.G. Brewer and R.L. Shoemaker, Phys. Rev. Lett. 27 (1971) 631.

[10] R.L. Shoemaker and R.G. Brewer, Phys. Rev. Lett. 28 (1972) 1430.

[11] R.G. Brewer and R.L. Shoemaker, Phys. Rev. A6 (1972) 2001.

[12] R.G. Brewer and E.L. Hahn, Phys. Rev. A8 (1973) 464.

[13] J. Schmidt, Paul R. Berman and R.G. Brewer, Phys. Rev. Lett. 31 (1973) 1103.

[14] E.L. Hahn, Phys. Rev. 80 (1950) 580.

[15] D. Grischkowsky, Phys. Rev. A7 (1973) 2096.

[16] D. Grischkowsky, E. Courtens and J. Armstrong, Phys. Rev. Lett. 31 (1973) 422.

[17] D. Grischkowsky, E. Courtens and J. Armstrong in the Proceedings of the Laser Spectroscopy Conference, Vail, Colorado, June 1973, to be published by Plenum Press.

[18] D. Grichkowsky, Phys. Rev. Lett. 24 (1970) 866.

[19] D. Grichkowsky and J. Armstrong, Phys. Rev. A6 (1972) 1566.

[20] D. Grischkowsky, to be published.

Cooperative Phenomena, H. Haken, ed.

THE SINGLE-MODE MODEL OF SUPERRADIANCE

Roy J.Glauber

Lyman Laboratory of Physics, Cambridge, Massachusetts, USA

and

Fritz Haake

Universität Essen-Gesamthochschule, Fach Physik, 43 Essen, Germany

I. INTRODUCTION

Superradiance is a big word for the following simple thing. Consider a bunch of N atoms initially prepared in some excited state. In the course of returning to its ground state such a system will, in general, emit parts of its excitation energy in the form of electromagnetic radiation. The total radiated intensity I will depend on the number of atoms N. If $I \sim N^2$, the radiation pulse is called a superradiant pulse. If, in contrast, $I \sim N$ only, one speaks of normal radiation.

Radio waves and other man-produced radiations at still larger wavelengths are usually superradiant. Such radiation pulses can be understood, as is well known, in terms of the classical picture of N identical dipoles radiating in phase with oneanother. The electric field thus generated is the superposition of the contributions of all dipoles,

$$\vec{E} = \sum_{i=1}^{N} \vec{E}_i \sim N$$

and the resulting intensity

$$I \sim |\vec{E}|^2 \sim N^2$$

is quadratic in the number of atoms.

Pulses of visible light, however, are usually normal. The reason simply is that spontaneous emission is, under ordinary circumstances, the predominant radiation mechanism at wavelengths corresponding to visible light or shorter ones. Spontaneous emission acts in an N-atom system give rise to contributions $\vec{E}_i$ to the total electric field $\vec{E}$ without phase correlations between those. Because of destructive interference the total mean intensity then cannot pick up an N^2 term,

$$\bar{I} \sim \sum_{i,j} \overline{\vec{E}_i \vec{E}_j^*} = \sum_i |\vec{E}_i|^2 + \sum_{i \neq j} \vec{E}_i \vec{E}_j^*$$
$$= \sum_i |\vec{E}_i|^2 \sim N .$$

In 1954 Dicke [1] realized that a light pulse spontaneously generated by N identical free atoms is superradiant, if these atoms are (i) concentrated in a volume with linear dimensions small compared to the wavelength corresponding to their transition frequency and (ii) prepared to an initial state of excitation displaying an electric polarization proportional to N. While Dicke's considerations have opened up superradiance as a field of research, they have not by themselves shown the way towards observing the effect experimentally in the visible or even infrared spectrum. In a volume as small as required by condition (i) in the case of visible light no sizable number of atoms can be assembled in practice.

This paper reviews some of the more recent efforts to understand cooperative spontaneous radiation processes. The most important result of such efforts has been the realization that superradiance can take place under conditions much weaker than those originally given by Dicke. Infact, in accord with these new conditions infrared superradiant pulses have just lately been produced experimentally [2].

We will exclusively be concerned with the simplest nontrivial model of a superradiant system [3]. This model accounts for N identical[+] two-level atoms interacting with a single mode of the electromagnetic field within the active volume[++] and simulates the leakage of light from the active volume in terms of a damping of the field mode. Although not fully adequate for a quantitative discussion of the experiments mentioned above this model displays all essential characteristics of a superradiant system. Moreover, it allows for an exact quantum-mechanical treatment[+++] and thus gives insight into the dynamics and statistics of cooperative radiation processes.

[+] Inhomogeneous broadening of the atomic line is thus discarded.

[++] This amounts to neglecting propagation effects within the active volume.

[+++] To be precise, asymptotic expressions, valid for a large number of atoms, can be given for all observables.

In Sec. II we give a quick and nonrigorous derivation of our model from the basic equations of electrodynamics[+]. The main point of this derivation consists in ascertaining the conditions under which a system of many initially excited atoms emits superradiant pulses. In order to avoid complications irrelevant at this stage we use semiclassical theory, that is Maxwell's and Bloch's equations. The simplified versions of these equations pertinent to our model are then solved and demonstrated to imply superradiance [3,4].

Sec. III presents the quantum-mechanical version of our model in terms of a master equation for the density operator of the atoms and an adiabatic relation between atomic and field observables.

Sec. IV prepares for solving the superradiance master equation by introducing a diagonal representation of the atomic density operator in terms of directed angular momentum states. We also briefly review the most important properties of these states[++] which are particularly well suited for the description of any cooperative behavior of systems of many two-level atoms [5]

By applying the techniques developped in Sec. IV we construct, in Sec. V, asymptotic expressions, valid in the case of a large number of atoms N, for atomic and field observables. The following remarkable properties of superradiant pulses are thus elucidated [3,6] : (i) The radiated field is fully coherent, if the atoms initially show a nonvanishing electric polarization. (ii) If the atoms are very highly excited initially and therefore do not display any noticable polarization, the pulses are accompanied by very large quantum fluctuations.

II. SEMICLASSICAL TREATMENT

We consider a plane-wave radiation pulse moving along the x-direction through a medium of identical two-level atoms. The carrier frequency ω of the pulse is in resonance with the atomic transition frequency. The system is described by Maxwell's equations for the electromagnetic field and Bloch's equations for the atomic observables polarization density $\vec{P}$ and inversion density D. These equations take on a rather simple form in the following limit for the pulse duration τ :

$$\omega^{-1} \ll \tau \ll T_1 , T_2 \qquad (2.1)$$

The left hand condition allows a carrier frequency to be assigned to the pulse. The right hand condition requires the pulse to be shorter than the relaxation times T_1 for the atomic inversion and T_2 for the polarization.This latter condition is necessary since a superradiant pulse can be generated by in-phase

---

+ A more detailed justification is given by R.Bonifacio at this school.

++ These states, often called Bloch states or atomic coherent states, are discussed in more detail by T.Arecchi at this school.

cooperation of all atoms only and since phase correlations between the atoms cannot persist for times longer than the relaxation times T_1 and T_2. In this limit we can use slowly varying field variables

$$\vec{E}(x,t) = \vec{E}^*(x,t) = \vec{e}\, i\sqrt{2\pi\hbar\omega/V}\left\{ b(x,t)\, e^{-i(\omega t-kx)} - b^+(x,t)\, e^{+i(\omega t-kx)} \right\}$$

$$\vec{P}(x,t) = \vec{P}^*(x,t) = -\vec{e}\, i(\mu/V)\left\{ S_-(x,t)\, e^{-i(\omega t-kx)} - S_+(x,t)\, e^{+i(\omega t-kx)} \right\} \qquad (2.2)$$

$$D(x,t) = D^*(x,t) = (1/V)\; 2S_z(x,t)$$

with

$$\left\{ \begin{matrix} \partial/\partial t \\ \partial/\partial x \end{matrix} \right\} \left\{ b, b^+, S_-, S_+, S_z \right\} \ll \left\{ \begin{matrix} \omega \\ \omega/c \end{matrix} \right\} \left\{ b, b^+, S_-, S_+, S_z \right\}. \qquad (2.3)$$

Here we have assumed, for the sake of simplicity, linear polarization in the direction of the unit vector $\vec{e}$. μ is the component of the atomic dipole moment in the direction $\vec{e}$. V is the volume filled by the atoms. The normalization chosen for the slowly varying amplitudes b, b^+, S^α serves to make these variables dimensionless and is a matter of convention. Since there are N atoms in the volume V we have $|S^\alpha| \leq N/2$.

Maxwell's and Bloch's equations imply the following well-known equations of motion [7, 8, 9]

$$\left(\frac{\partial}{\partial t} + c\frac{\partial}{\partial x} + \kappa\right) b^+(x,t) = i g\, S_+(x,t)$$

$$\frac{\partial}{\partial t} S_+(x,t) = -i\, 2g\, b^+(x,t)\, S_z(x,t) \qquad (2.4)$$

$$\frac{\partial}{\partial t} S_z(x,t) = ig\left\{ b^+(x,t)\, S_-(x,t) - b(x,t)\, S_+(x,t) \right\}$$

with the coupling constant

$$g = \mu\sqrt{2\pi\omega/\hbar} \; . \qquad (2.5.)$$

Note the field damping term involving the damping constant κ in the first of eqs. (2.4). This term is included as a simulation of field losses by the leakage of light through the right end face of the active volume. Accordingly the constant κ is chosen as the inverse flight time of a photon through the length ℓ of the atomic system[+],

$$\kappa = c/\ell \,. \tag{2.6}$$

Eqs. (2.4) can be solved by introducing the so called Bloch angle $\theta(x,t)$

$$\begin{aligned} S_-(x,t) = S_+(x,t) &= \frac{N}{2} \sin \theta(x,t) \\ S_z(x,t) &= \frac{N}{2} \cos \theta(x,t) \\ b^+(x,t) = -b(x,t) &= (i/2g) \frac{\partial \theta(x,t)}{\partial t} \,. \end{aligned} \tag{2.7}$$

By inserting this ansatz in (2.4) we find

$$\left(\frac{\partial^2}{\partial t^2} + \frac{\partial^2}{\partial x \partial t} + \kappa \frac{\partial}{\partial t} \right) \theta(x,t) = N g^2 \sin \theta(x,t) \,. \tag{2.8}$$

The parameter $g\sqrt{N}$ occuring here is a frequency characteristic for the interaction of the atoms with the field. Roughly speaking it is the frequency at which energy would oscillate back and forth between the atoms and the field mode if no field damping were present. Its dependence on the number of atoms shows that the atom-field interaction is a cooperative effect involving all atoms coherently. It can be expressed in terms of measurable quantities as

$$g\sqrt{N} = c/\ell_c \,, \quad \ell_c = \sqrt{2\pi c/\rho\gamma\lambda^2} \,, \quad \rho = \frac{N}{V} \,, \quad \gamma = (8\pi^2/3)(\mu^2/\hbar\lambda^3) \,, \tag{2.9}$$

where γ is the natural line width of the atomic transition. ℓ_c is the cooperation length introduced by Arecchi and Courtens [10]. It gives, according to (2.8), the scale of length on which the pulse would change due to the atom-field coupling if the field damping were absent.

We now have to solve the equation of motion for the Bloch angle (2.8) with a suitable initial condition $\theta(x,0) = \theta_0(x)$, $\dot{\theta}(x,0) = \dot{\theta}_0(x)$. We consider the simplest initial situation

$$\left. \begin{aligned} \theta_0(x) &= \theta_0 = \text{const} \\ \dot{\theta}_0(x) &= 0 \end{aligned} \right\} \text{ in } 0 \le x \le \ell \,. \tag{2.10}$$

[+] A justification for this schematic way of dealing with leakage effects, which we have taken over from laser theory [7], is given by R.Bonifacio at this school.

This corresponds, according to (2.7), to a spatially homogeneous initial state of atomic excitation and vanishing initial field. Then the solution of eq. (2.8) will be spatially homogeneous within the active volume for $t \geqslant 0$ too and can thus be obtained from the ordinary differential equation+

$$\ddot{\Theta} + \kappa \dot{\Theta} = g^2 N \sin \Theta \quad . \tag{2.11}$$

This is the well-known equation of motion of a damped nonlinear pendulum. Damped oscillations of the Bloch angle imply a similar behavior of the field amplitude $b \sim \dot{\Theta}$ within the active volume. The radiated intensitity

$$I(t) = 2\kappa \hbar \omega \, b^+(t) \, b(t) \tag{2.12}$$

will therefore have the appearance of a sequence of pulses of decreasing magnitude. We can expect this intensity to be superradiant at least at some times since our limit (2.1) ensures in-phase cooperation of all atoms.

For the sake of simplicity we will concern ourselves with the special case where the motion of the Bloch angle corresponds to a heavily overdamped pendulum, i.e. where

$$\kappa \gg g\sqrt{N} \quad \overset{(2.6,9)}{\Longleftrightarrow} \quad \ell \ll \ell_c \quad . \tag{2.13}$$

This limit can be realized experimentally by adjusting the length ℓ of the active volume so as to be small compared to the material constant ℓ_c. Then the rate of escape κ of photons from the active volume is much larger than the rate of their being fed back into the atoms through the cooperative interaction, $g\sqrt{N}$. In view of eq. (2.4) the condition (2.13) also implies that the field amplitude $b^+(t)$ follows the motion of the atomic polarization $S^+(t)$ adiabatically,

$$b^+(t) = i\,(g/\kappa)\, S_+(t) \, . \tag{2.14}$$

Furthermore, the Bloch angle can now be determined from the much simpler equation

$$\dot{\Theta}(t) = \frac{1}{\tau} \sin \Theta(t) , \qquad \tau = \kappa / g^2 N \, . \tag{2.15}$$

This has the solution

$$\tan \frac{\Theta(t)}{2} = e^{t/\tau} \tan \frac{\Theta_0}{2} \tag{2.16}$$

which leads to the following expression for the radiated intensity++

++For full initial excitation, $\Theta_0 = 0$, this semiclassical result becomes meaningless, since formally $t_{max}\,(\Theta_0 = 0) = \infty$; the pendulum has an unstable equilibrium position at $\Theta_0 = 0$; fully excited atoms leave the excited state by spontaneous emission; this is a quantum effect not included in Bloch's and Maxwell's equations; the quantummechanical treatment in the next section will correct the semiclassical result for highly excited atomic initial states.

+ The pertinence of this equation to the more general situation is ephasized and will be discussed by R.Bonifacio at this school.

$$
\begin{aligned}
I(t) &= 2\kappa\hbar\omega\, b^{\dagger}(t) b(t) = \hbar\omega\, I_1\, S_+(t) S_-(t) \\
&= \hbar\omega\, I_1\, (N/2)^2 \sin^2\Theta(t) \\
&= \hbar\omega\, I_1\, (N/2)^2\, \mathrm{sech}^2[(t-t_{max})/\tau]
\end{aligned}
\tag{2.17}
$$

with $\quad I_1 = 2g^2/\kappa$

and

$$
\begin{aligned}
t_{max} &= -(\tau/2)\ln\tan(\Theta_0/2) \\
&= \tau \ln\left[\left(\tfrac{N}{2}+S_z(0)\right)/\left(\tfrac{N}{2}-S_z(0)\right)\right].
\end{aligned}
\tag{2.18}
$$

As expected for the limit (2.13) the initial atomic excitation energy $\hbar\omega(S_z(0)+N/2)$ is radiated into a single pulse of temporal width τ. Clearly, the radiated intensity is of order N^2, i.e. superradiant, if the initial atomic energy is of order N.

In order to find the physical meaning of the constants τ and I_1 we now specify the shape of the active volume as that of a long thin cylinder of length ℓ and diameter d such that

$$
\lambda = 2\pi c/\omega \ll d \ll \ell . \tag{2.19}
$$

On the other hand it is easily checked with the help of eqs. (2.6, 9) that the expressions (2.15) and (2.18) for τ and I_1 can be rewritten as

$$
\frac{1}{\tau} = N\gamma\,\Delta\Omega/4\pi \quad , \quad I_1 = \gamma\,\Delta\Omega/4\pi \tag{2.20}
$$

where $\quad \Delta\Omega = \lambda^2/d^2 \ll 1$

is the diffraction solid angle of the right end face of the active volume. This suggests the following intuitive interpretation. The radiated pulse has an angular aperture $\Delta\Omega$. A single excited atom produces spontaneously a photon at the rate γ. Only the fraction $\Delta\Omega/4\pi$ of the transition rate γ refers to emissions into the solid angle $\Delta\Omega$ around the x-axis. Only those photons with wavevectors in $\Delta\Omega$ around the x-direction can be part of the cooperatively generated superradiant pulse. Therefore the bandwidth $1/\tau$ of the pulse is the natural linewidth of the atomic transition reduced by the geometry factor $\Delta\Omega/4\pi$ and enhanced by the factor N. Correspondingly, I_1 is the fraction of the intensity of a single-atom emission going into the solid angle $\Delta\Omega$. The precise justification of this interpretation cannot be given within the framework of our present model but follows from a more general analysis to be presented by R.Bonifacio at this school[+].

We conclude this section by rewriting our conditions (2.1, 13) for superradiance in the simple sense of (2.17) to occur

$$
\begin{aligned}
&\lambda \ll d \ll \ell \\
&\ell_{abs} = 4\pi/T_2\,\rho\gamma\lambda^2 \ll \ell \ll \ell_c = \sqrt{2\pi c/\rho\gamma\lambda^2}\,.
\end{aligned}
\tag{2.21}
$$

[+] See also [11, 12].

The atomic medium must have an absorption length l_{abs} so much smaller than the cooperation length l_c that the length of the active cylinder can be comfortably fitted in between these two material constants.

III. QUANTUM-MECHANICAL VERSION OF THE MODEL

We consider the system characterized by the conditions (2.21). The amplitudes b and b^+ of the quasimonochromatic light field now become the annihilation and creation operators of photons obeying the Bose commutation rules

$$[b, b^+] = 1 \quad , \quad [b, b] = [b^+, b^+] = 0 \tag{3.1}$$

The polarization variables S_+, S_- and the inversion S_z are also operators and satisfy the angular momentum commutation rules[+]

$$[S_+, S_-] = 2 S_z \quad , \quad [S_z, S_\pm] = \pm S_\pm \ . \tag{3.2}$$

The state of the system is now characterized by the density operator W (t). The initial situation again to be considered is one with no photons present and therefore corresponds to the density operator

$$W(0) = |0\rangle\langle 0| \otimes \varrho(0) , \tag{3.3}$$

where $|0\rangle$ is the photon vacuum, $b|0\rangle = 0$, and $\varrho(0)$ the as yet unspecified initial value of the reduced density operator of the atoms alone,

$$\varrho(t) = tr_F \ W(t) \ . \tag{3.4}$$

The state of the coupled system atoms + field at later times is then determined by the equation of motion for $W(t)$[++]

$$\dot{W}(t) = -i L \, W(t) \quad , \quad L = L_{AF} + i\Lambda \ . \tag{3.5}$$

The two terms in the Liouvillian L refer to the atom-field interaction[+++] (L_{AF}) and the field damping (Λ). They are defined as usual by

$$L_{AF} W = \frac{1}{\hbar} [H_{AF}, W] \quad , \quad H_{AF} = \hbar g (b S_+ + b^+ S_-) \tag{3.6}$$

$$\Lambda W = \kappa \{ [b, W b^+] + [b W, b^+] \} . \tag{3.7}$$

[+] A single two-level atom is formally equivalent to a spin $\frac{1}{2}$ system; the vector operator $\vec{S} = \{S_x, S_y, S_z\} = \{\frac{1}{2}(S_+ + S_-), \frac{1}{2i}(S_+ - S_-), S_z\}$ is the sum of all N atomic spin operators and thus formally an angular momentum operator.

[++] In the interaction picture .

[+++] The constants g and κ have the same meaning as before.

H_{AF} is the usual Hamiltonian for a system of identical two-level atoms interacting with a single resonant field mode [7]. Λ formally is the damping Liouvillian for a damped harmonic oscillator moving under the influence of a zero temperature heat bath [4,7,13]. We shall need the following property of Λ

$$\mathrm{tr}\, b^{+\ell} b^{\ell'} \exp(\Lambda t)\, X = \exp[(\ell+\ell')\kappa t]\; \mathrm{tr}\, b^{+\ell} b^{\ell'} X; \tag{3.6}$$

X arbitrary; $\ell, \ell' = 0, 1, 2, \ldots$,

which is easily proved by differentiating with respect to the time t and using the definition (3.5) and the Bose rules (3.1).

Since we still stick to the limit (2.21) the field degrees of freedom will, as we have seen in the semiclassical treatment, adiabatically follow the atomic degrees of freedom in eq. (3.5). The problem stated by (3.3, 5) can therefore be considerably simplified by adiabatically eliminating the field degrees of freedom from (3.5). To achieve this goal we adopt a technique developped by Zwanzig [14] and Nakajima [15] and first decompose the atom-field density operator W (t) as

$$W(t) = \mathcal{P} W(t) + (1-\mathcal{P}) W(t), \qquad \mathcal{P} = |0\rangle\langle 0| \otimes \mathrm{tr}_F , \qquad |0\rangle = \text{photon vacuum.} \tag{3.7}$$

The projector $\mathcal{P}$ $(\mathcal{P}^2 = \mathcal{P})$ acts on the density operator so as factor it into the vacuum field density operator $|0\rangle\langle 0|$ and the atomic density operator $\rho(t)$,

$$\mathcal{P} W(t) = |0\rangle\langle 0| \otimes \rho(t). \tag{3.8}$$

The first term on the right hand side in (3.7) thus contains the full information about the atoms while the second one, $(1-\mathcal{P}) W(t)$, takes up the information concerning the field and atom-field correlations. Only the first one is of relevance to us for the moment being. For it we get an equation of motion in the following way. We insert the decomposition (3.7) into (3.5) and act on the resulting equation from the left with $\mathcal{P}$ and $(1-\mathcal{P})$, alternatively. The following two equations result

$$\mathcal{P}\dot{W}(t) = -i\mathcal{P}L\mathcal{P}W(t) - i\mathcal{P}L(1-\mathcal{P})W(t)$$

$$(1-\mathcal{P})\dot{W}(t) = -i(1-\mathcal{P})L\mathcal{P}W(t) - i(1-\mathcal{P})L(1-\mathcal{P})W(t).$$

A formal integral of the second of these,

$$(1-\mathcal{P})W(t) = \exp\{-i(1-\mathcal{P})Lt\}(1-\mathcal{P})W(0) - i\int_0^t dt' \exp\{-i(1-\mathcal{P})Lt'\}(1-\mathcal{P})L\mathcal{P}W(t-t'), \tag{3.9}$$

is then inserted in the first one

$$\begin{aligned}\mathcal{P}\dot{W}(t) = & -i\mathcal{P}L\mathcal{P}W(t) \\ & - \int_0^t dt' \mathcal{P}L \exp\{-i(1-\mathcal{P})Lt'\}(1-\mathcal{P})L\mathcal{P}W(t-t') \\ & - i\mathcal{P}L\exp\{-i(1-\mathcal{P})Lt\}(1-\mathcal{P})W(0).\end{aligned} \tag{3.10}$$

This is an inhomogeneous integrodifferential equation for the relevant part $\mathcal{P}W = |0\rangle\langle 0| \otimes \rho$ of the density operator. By simply taking the trace tr_F over the field we now obtain an equation of motion for the atomic density operator $\rho(t)$. Before writing down this equation let us note that the inhomogeneity in (3.10), that is the third term on the right hand side, vanishes because of our initial condition (3.3). Some further simplifications follow from the identities $\mathcal{P}\Lambda = 0$, $\Lambda\mathcal{P} = 0$, and $\mathcal{P}L_{AF}\mathcal{P} = 0$ which are easily proved by using the definitions of Λ, L_{AF}, and $\mathcal{P}$. We thus get

$$\dot{\rho}(t) = \int_0^t dt' \, K(t')\, \rho(t-t'), \tag{3.11}$$

$$K(t) = -tr_F \, L_{AF} \exp\{[\Lambda - i(1-\mathcal{P})L_{AF}]t\} \, L_{AF} \, |0\rangle\langle 0| .$$

The field variables are now formally eliminated. Further progress relies on the perturbation expansion of the integral kernel $K(t)$ in powers of the interaction Liouvillian L_{AF},

$$K(t) = -tr_F \, L_{AF} \Big\{ \sum_{n=0}^{\infty} \int_0^t dt_n \int_0^{t_n} dt_{n-1} \ldots \int_0^{t_2} dt_1 \, U(t-t_n)[-i(1-\mathcal{P})L_{AF}] \times$$
$$\times U(t_n - t_{n-1})[-i(1-\mathcal{P})L_{AF}] \ldots U(t_2-t_1)[-i(1-\mathcal{P})L_{AF}] \, U(t_1) \Big\} L_{AF} \, |0\rangle\langle 0|, \tag{3.12}$$

where $U(t) = \exp \Lambda t$.

It is now important to realize that this series goes in terms of the small dimensionless parameter $g\sqrt{N}/\kappa \ll 1$ and can therefore be replaced by the first $(n=0)$ term

$$K(t) = -tr_F \, L_{AF} \, e^{\Lambda t} \, L_{AF} \, |0\rangle\langle 0| . \tag{3.13}$$

To see this we recall from the semiclassical treatment that the observables of our system tend to change in time at a rate $g\sqrt{N}$ according to the atom-field interaction, while the field damping tends to impose the time rate of change κ. Formally[+],

$$O(L_{AF}) = g\sqrt{N} , \qquad O(\Lambda) = \kappa . \tag{3.14}$$

Therefore the n-th term in the series (3.12) is of order $(g\sqrt{N})^2 (g\sqrt{N}/\kappa)^n$; there are $(n+2)$ factors L_{AF} and each of the n time integrals throws a factor κ in the denominator. In the adiabatic limit we have $g\sqrt{N}/\kappa \ll 1$.

With the kernel $K(t)$ approximated by (3.13) the equation of motion (3.11) for the atomic density operator reads, after using (3.6) for $\exp \Lambda t$,

$$\dot{\rho}(t) = \int_0^t dt' \, \kappa \, e^{-\kappa t'} \, \Lambda_c \, \rho(t-t') \tag{3.15}$$

[+] We here appeal to our experience with the semiclassical theory. The order-of-magnitude estimates (3.14) can, however, also be corroborated by quantum-mechanical arguments. $O(\Lambda) = \kappa$ is immediately obvious from (3.6). As for $O(L_{AF}) = g\sqrt{N}$, an analysis of the eigenvalue spectrum of H_{AF} [16] yields the eigenvalues to be nearly equidistant (for $N \gg 1$) with a spacing of order $g\sqrt{N}$; therefore, if L_{AF} were the only term in L, observables would tend to display a quasiperiodicity with quasiperiod $1/g\sqrt{N}$.

with the collective decay Liouvillian

$$\Lambda_c \varrho = (1/N\tau)\{[S_-, \varrho S_+] + [S_-\varrho, S_+]\} \quad , \quad \frac{1}{\tau} = \frac{g^2 N}{\kappa} . \tag{3.16}$$

The order of magnitude of Λ_c can be found as

$$O(\Lambda_c) = O\left(\int_0^\infty dt\, L_{AF}\, e^{\Lambda t} L_{AF}\right) = g^2 N \int_0^\infty dt\, e^{-\kappa t} = \frac{g^2 N}{\kappa} = \frac{1}{\tau} \tag{3.17}$$

which is just the inverse pulse duration known from the semiclassical treatment. Since we have $\kappa \gg 1/\tau$ we may neglect retardation effects in (3.15) and thus arrive at the following "superradiance master equation" [3]

$$\dot{\varrho}(t) = \Lambda_c \varrho(t) . \tag{3.18}$$

We have now carried out the adiabatic elimination of the field variables. Eq. (3.18) is the quantum-mechanical analog of the pendulum equation (2.15) for the Bloch angle Θ. It describes the collective atomic behavior during the radiation process. The construction and discussion of its solution will be the subject of the next sections.

We still have to translate the adiabatic correspondence (2.14), $b^+ = (ig/\kappa) S_+$ into quantum-mechanics. To this end we use the decomposition (3.7) of the atom-field density operator in calculating the normally ordered intensity moments.

$$\begin{aligned} \langle b^{+\ell}(t)\, b^{\ell'}(t) \rangle &= tr_A\, tr_F\; b^{+\ell} b^{\ell'} \{ \mathcal{P} W(t) + (1-\mathcal{P}) W(t) \} \\ &= tr_A\, tr_F\; b^{+\ell} b^{\ell'} (1-\mathcal{P}) W(t) \quad , \quad \ell + \ell' \neq 0 . \end{aligned} \tag{3.19}$$

Clearly, this expression involves $(1-\mathcal{P}) W(t)$, since this part of the density operator $W(t)$ carries the information about the dynamics of the field. We now use eq. (3.9) which expresses $(1-\mathcal{P}) W$ in terms of the atomic part $\mathcal{P} W = |0\rangle\langle 0| \otimes \varrho$

$$\begin{aligned} \langle b^{+\ell}(t)\, b^{\ell'}(t) \rangle &= -i \int_0^t dt'\; tr_A\, tr_F\; b^{+\ell} b^{\ell'} \exp\{[\Lambda - i(1-\mathcal{P})L_{AF}]t\} L_{AF} |0\rangle\langle 0| \varrho(t-t') \\ &= \sum_{n=0}^{\infty} (-i)^{n+1} \int_0^t dt' \int_0^{t'} dt_n \int_0^{t_n} dt_{n-1} \cdots \int_0^{t_2} dt_1\; tr_A\, tr_F\; b^{+\ell} b^{\ell'} \times \\ &\quad \times U(t'-t_n)\, Q L_{AF}\, U(t_n - t_{n-1})\, Q L_{AF} \cdots \times \\ &\quad \times U(t_2 - t_1)\, Q L_{AF}\, U(t_1)\, L_{AF}\, |0\rangle\langle 0|\, \varrho(t-t') , \end{aligned} \tag{3.20}$$

$$Q = 1 - \mathcal{P} .$$

Here we have immediately written out a perturbation expansion similar to the above one for the kernel $K(t)$. Again this series goes in terms of the small parameter $g\sqrt{N}/\kappa$ and may thus be replaced with its first nontrivial term. A somewhat lengthy but elementary calculation[+] shows this term to arise for $n = \ell + \ell' - 1$ and to read

$$\langle b^{+\ell}(t)\, b^{\ell'}(t)\rangle = (-ig)^{\ell+\ell'}(-1)^{\ell}(\ell+\ell')\int_0^t dt'\, e^{-(\ell+\ell')\kappa t'}\left(e^{\kappa t'}-1\right)^{\ell+\ell'-1}\langle S_+^{\ell}(t-t')\, S_-^{\ell'}(t-t')\rangle .$$

Since the atomic expectation value $\langle S_+^{\ell}\, S_-^{\ell'}\rangle$ occuring here changes in time on a scale $\tau \gg 1/\kappa$ retardation effects can again be neglected whereupon we finally get the quantum-mechanical adiabatic relation between field and atomic observables

$$\langle b^{+\ell}(t)\, b^{\ell'}(t)\rangle = \mu^{*\ell}\mu^{\ell'}\langle S_+^{\ell}(t)\, S_-^{\ell'}(t)\rangle , \quad \mu = -ig/\kappa = -i\ell/\ell_c N . \qquad (3.21)$$

This correspondence law is a quite intuitive generalization of (2.14) and allows the determination of the statistical properties of the superradiant pulse once the superradiance master equation (3.18) is solved.

IV. DIAGONAL REPRESENTATION OF ρ IN TERMS OF DIRECTED ANGULAR MOMENTUM STATES

a) Review of directed angular momentum states[++]

A convenient representation of the atomic density operator ρ is provided by the angular momentum eigenstates $|j,m\rangle$. These states have the properties

$$S_z|j,m\rangle = m|jm\rangle \quad , \quad \vec{S}^2|j,m\rangle = j(j+1)|j,m\rangle ,$$

$$S_\pm|j,m\rangle = \{(j\mp m)(j\pm m+1)\}^{1/2}|j,m\pm 1\rangle , \qquad (4.1)$$

$$\vec{S}^2 = S_x^2 + S_y^2 + S_z^2 = S_+S_- + S_z^2 - S_z .$$

With respect to a state $|j,m\rangle$ the "transverse" components S_x, S_y of $\vec{S}$ have dispersions

$$\langle j,m|S_\alpha^2|jm\rangle - \langle j,m|S_\alpha|j,m\rangle^2 = \tfrac{1}{2}\{j(j+1)-m^2\};\ \alpha = x,y . \qquad (4.2)$$

We see that these dispersions are smallest, for a fixed value of j for $m = \pm j$. The states $|j,\pm j\rangle$ thus specify the direction of $\vec{S}$ with lesser uncertainty than all other states $|j,m\rangle$ with $m \neq \pm j$. Clearly, the directions assigned to $\vec{S}$ in this sense are the positive and negative z-directions for $m=+j$ and $m=-j$, respectively.

[+] One uses (3.6) for $U(t) = \exp \Lambda t$, the vacuum property $\langle 0|b^{+\ell}b^{\ell'}|0\rangle = 0$ and the definitions of and. See, if necessary, [3] or [4].

[++] An exhaustive discussion of these states is given in [5] and will be presented by T.Arecchi at this school. When used in the context of cooperatively radiating atoms the states $|\gamma\rangle$ have also been called atomic coherent states; this name alludes to certain formal analogies to the coherent states of the radiation field [17].

By a simple rotation we can transform the minimum uncertainty state $|j,j\rangle$ into another state $|j,\theta,\varphi\rangle$ which has the angular momentum vector oriented in the direction specified by the two angles of rotation θ and φ. (See fig. 1 at the end of this paper). The rotation is defined by the unitary operator

$$U(\theta,\varphi) = \exp\{i\theta(S_x \sin\varphi - S_y \cos\varphi)\} \tag{4.3}$$

$$= (\exp \gamma S_-)(\exp - S_z \ln[1+\gamma\gamma^*])(\exp - \gamma^* S_+)$$

$$\equiv U(\gamma,\gamma^*) \quad , \quad \gamma = e^{i\varphi} \tan\frac{\theta}{2} .$$

The states thus generated

$$|j,\theta,\varphi\rangle \equiv |j,\gamma\rangle = U(\gamma,\gamma^*)|j,j\rangle = (1+|\gamma|^2)^{-j} \sum_{\nu=0}^{2j} \gamma^\nu \sqrt{\binom{2j}{\nu}} |j,j-\nu\rangle \tag{4.4}$$

may be called directed angular momentum states since they assign to $\vec{S}$ the direction (θ,φ) with the same (maximum) precision with which the state $|j,j\rangle$ has $\vec{S}$ oriented in the positive z -direction. In the following we shall label the directed angular momentum states by the complex variable $\gamma = e^{i\varphi} \tan\frac{\theta}{2}$. A geometric interpretation of γ is given in fig. 2. While the angles θ and φ locate a point on the unit sphere, $Re\,\gamma$ and $Im\,\gamma$ are the cartesian coordinates of the stereographic projection of this point onto the plane tangential to the sphere in the point $\theta = 0$, the projection being taken from the point $\theta = \pi$ on the sphere. This geometric interpretation of γ is also suggested by the following expectation values of $\vec{S}$ with respect to a state $|j,\gamma\rangle$

$$\langle j,\gamma|S_z|j,\gamma\rangle = j\frac{1-|\gamma|^2}{1+|\gamma|^2} \quad , \quad \langle j,\gamma|S_\pm|j,\gamma\rangle = j\frac{2}{1+|\gamma|^2} \cdot \begin{Bmatrix}\gamma \\ \gamma^*\end{Bmatrix} . \tag{4.5}$$

In a subspace of fixed total angular momentum quantum number j, where $\vec{S}^2 = j(j+1)$, we may use the states $|j,\gamma\rangle \equiv |\gamma\rangle$ with γ variable in the complex plane as a basis.

Note that the states $|\gamma\rangle$ are normalized to unity but not orthogonal

$$\langle\alpha|\gamma\rangle = [(1+\alpha^*\gamma)^2/(1+|\alpha|^2)(1+|\gamma|^2)]^j . \tag{4.6}$$

Furthermore, the states $|\gamma\rangle$ form an overcomplete set. They give rise to the following resolution of unity

$$\int d^2\gamma \frac{2j+1}{\pi(1+|\gamma|^2)^2} |\gamma\rangle\langle\gamma| = \sum_{m=-j}^{+j} |j,m\rangle\langle j,m| = 1 , \tag{4.7}$$

where the integral covers the whole complex plane,

$$\int d^2\gamma = \iint_{-\infty}^{+\infty} d(Re\,\gamma)\, d(Im\,\gamma) \tag{4.8}$$

Any state $|\,\rangle = \sum_m c_m |j\,m\rangle$ can thus be expanded in terms of the directed angular momentum states. More importantly, any operator $\hat{O} = \sum_{m,m'} |j,m\rangle O_{mm'} \langle j,m|$ can be given a diagonal representation[+] as

$$\hat{O} = \int d^2\gamma \; O(\gamma,\gamma^*) \, |\gamma\rangle\langle\gamma| \, . \tag{4.9}$$

b) Diagonal representation of the atomic density operator

It is easily checked that the cooperative decay of atomic excitation according to our superradiance master equation (3.18) is such that the squared length of the angular momentum vector is conserved: $tr_A \vec{S}^2 \Lambda_c \rho = 0$. Therefore the atomic density operator $\rho(t)$ will remain a mixture of the kind (4.9),

$$\rho(t) = \int d^2\gamma \; P(\gamma,\gamma,t) \, |\gamma\rangle\langle\gamma| \, , \tag{4.10}$$

if it was one initially or if, especially, the atoms were initially prepared in a directed angular momentum state $|j,\alpha\rangle$,

$$\rho(0) = |j,\alpha\rangle\langle j,\alpha| \quad , \text{ i.e. } P(\gamma,\gamma^*;0) = \delta^{(2)}(\gamma-\alpha) \, . \tag{4.11}$$

Such initial states are physically significant since they can be realized experimentally. If a system of identical two-level atoms is first brought to the ground state and then illuminated by an intense pulse with fixed amplitude and phase, the atomic state produced is just one like (4.11) with $j = N/2$ and γ depending on the duration, the amplitude, and the phase of the exciting pulse 5 . If, on the other hand, the atoms are pumped incoherently, the state produced is well approximated by $\rho(0) = |j,\gamma=0\rangle\langle j,\gamma=0| = |j,m=j\rangle\langle j,m=j|$ with $j \approx \langle S_z\rangle \le N/2$, if $\langle S_z^2\rangle \gg N/2 \gg 1$ [1].

With the density operator represented as in (4.10) we have the following expression for expectation values of angular momentum operators

$$\begin{aligned} \langle S_+^{\ell}(t) S_-^{\ell'}(t)\rangle &= \int d^2\gamma \; P(\gamma,\gamma^*;t) \, \langle\gamma| S_+^{\ell} S_-^{\ell'} |\gamma\rangle \\ &= \int d^2\gamma \; P(\gamma,\gamma^*;t) \, (1+|\gamma|^2)^{-2j} \frac{\partial^{\ell+\ell'}}{\partial\gamma^{*\ell}\partial\gamma^{\ell'}} (1+|\gamma|^2)^{2j} \, . \end{aligned} \tag{4.12}$$

This identity is easily verified by using the definition (4.4). It can be further evaluated by carrying out the differentiations. Obviously , the expectation value $\langle S_+^{\ell} S_-^{\ell'}\rangle$ is a polynomial in j of order $(\ell+\ell')$.

Let us note that the weight function P is real since the density operator is hermitian. It need not, however, be positive. It can be looked upon as a quasiprobability distribution over the orientation of the angular momentum vector of the system, the orientation being specified by the two angles θ and φ or, equivalently, by the stereographic projection variable $\gamma = e^{i\varphi} \tan\frac{\theta}{2}$.

[+] The weight function $O(\gamma,\gamma^*)$ is not uniquely defined by eq. (4.9). It is shown in [5], however, that $O(\theta,\varphi)$, defined by $O(\gamma,\gamma^*)d^2\gamma = O(\theta,\varphi) \sin\theta d\theta d\varphi$, is made unique by the additional requirement that its expansion in terms of spherical harmonics $Y_{\ell m}(\theta,\varphi)$ contain no terms with $\ell > 2j$.

c) Equation of motion for the weight function $P(\gamma,\gamma^*;t)$ [+]

The superradiance master equation (3.18) implies an equation of motion for the weight function $P(\gamma,\gamma^*;t)$. To construct this equation it is convenient to extract a normalization factor from the directed angular momentum state

$$|\gamma\rangle = (1+|\gamma|^2)^{-j}\,\|\gamma\rangle \quad , \quad \|\gamma\rangle = (\exp \gamma S_-)\,|j,j\rangle \tag{4.13}$$

and to write the diagonal representation (4.10) of ϱ as

$$\varrho(t) = \int d^2\gamma\; P(\gamma,\gamma^*;t)\,(1+|\gamma|^2)^{-2j}\;\|\gamma\rangle\langle\gamma\| \,. \tag{4.14}$$

The non-normalized states $\|\gamma\rangle$ are, according to (4.13), analytic functions of γ throughout the whole complex plane (except for $\gamma=\infty$). We further have from (4.13) [++]

$$S_-\|\gamma\rangle = \frac{\partial}{\partial\gamma}\|\gamma\rangle \quad , \quad S_+\|\gamma\rangle = \gamma\,(2j-\gamma\frac{\partial}{\partial\gamma})\|\gamma\rangle \,. \tag{4.15}$$

Equipped with eqs. (4.13-15) we use the superradiance master equation (3.18) and find $(N=2j)$:

$$\begin{aligned}
\dot\varrho(t) &= \int d^2\gamma\; \dot P(\gamma,\gamma^*;t)\,|\gamma\rangle\langle\gamma| \\
&= \frac{1}{N\tau}\int d^2\gamma\; P(\gamma,\gamma^*;t)\,(1+|\gamma|^2)^{-N} \times \\
&\quad \times\left\{ 2\,S_-\|\gamma\rangle\langle\gamma\|S_+ - S_+S_-\|\gamma\rangle\langle\gamma\| - \|\gamma\rangle\langle\gamma\|S_+S_-\right\} \\
&= \frac{1}{N\tau}\int d^2\gamma\; P(\gamma,\gamma^*;t)\,(1+|\gamma|^2)^{-N} \times \\
&\quad \times\left\{ \frac{\partial^2}{\partial\gamma\partial\gamma^*} + \frac{\partial^2}{\partial\gamma^2}\gamma^2 - (N+2)\frac{\partial}{\partial\gamma}\gamma + c.c.\right\}\|\gamma\rangle\langle\gamma\| \,.
\end{aligned}$$

After partial integration on the right hand side we obtain

$$\begin{aligned}
\int d^2\gamma\; \dot P(\gamma,\gamma,t)\,|\gamma\rangle\langle\gamma| &= \int d^2\gamma\,|\gamma\rangle\langle\gamma|\,\frac{1}{N\tau}\left\{-(N+2)\frac{\partial}{\partial\gamma}\gamma\right. \\
&\quad \left. + \frac{\partial^2}{\partial\gamma^2}\gamma^2 + \frac{\partial^2}{\partial\gamma\partial\gamma^*} + c.c.\right\} P(\gamma,\gamma^*;t) .
\end{aligned}$$

[+] An equivalent equation for $P(\theta,\varphi,t)$ will also be discussed by L.Narducci at this school [18].

[++] These identities are obtained by using $\frac{\partial}{\partial\gamma}e^{\gamma S_-} = \gamma e^{\gamma S_-}$ and $[S_+, e^{\gamma S_-}] = -\gamma^2 S_- + 2\gamma S_z$.

It is sufficient for this equation to hold that the weight function P obey the partial differential equation

$$\dot{P}(\gamma,\gamma^*;t) = \frac{1}{N\tau}\left\{-(N+2)\frac{\partial}{\partial\gamma}\gamma + \frac{\partial^2}{\partial\gamma^2}\gamma^2 + \frac{\partial^2}{\partial\gamma\partial\gamma^*} + c.c.\right\}P(\gamma,\gamma^*;t). \tag{4.16}$$

The statistical properties of superradiant pulses can now be obtained by first solving the above equation of motion for P for the initial condition (4.11) and then evaluating the expectation values $\langle S_+^{\ell} S_-^{\ell'}\rangle \sim \langle b^{+\ell} b^{\ell'}\rangle$ according to eq. (4.12). Although this programm can be carried out rigorously it is much more economic and interesting to first use the fact that we are concerned with a large number of atoms N. In the case $N \gg 1$ both the equation of motion (4.16) and the expression (4.11) for the expectation values of polarization operators can be considerably simplified.

V. QUANTUM STATISTICS OF SUPERRADIANT PULSES

a) Asymptotic expressions for the polarization expectation values

Let us investigate the expectation value $\langle\gamma|S_+^{\ell}S_-^{\ell'}|\gamma\rangle$ with respect to a directed angular momentum state for $j = N/2 \gg 1$. From (4.12) we have

$$\langle\gamma| S_+^{\ell} S_-^{\ell'} |\gamma\rangle = (1+|\gamma|^2)^{-N} \frac{\partial^{\ell+\ell'}}{\partial\gamma^{*\ell}\partial\gamma^{\ell'}}(1+|\gamma|^2)^N. \tag{5.1}$$

The right hand side in this equation is a polynomial of order $(\ell+\ell')$ in N with coefficients depending on the amplitude γ. Asymptotically, for $N\to\infty$, this polynomial can be replaced by its monomial of order $\ell+\ell'$, provided that γ is finite, i.e. $\gamma \neq 0,\infty$. This asymptotic form of (5.1) is easily obtained by noting that each of the $\ell+\ell'$ differentiations has to generate an explicit factor N if a term $N^{(\ell+\ell')}$ is to be produced. We find, using

$$N!/(N-\ell-\ell')! \longrightarrow N^{\ell+\ell'} \quad \text{for } N\to\infty,$$

$$\begin{aligned}\langle\gamma|S_+^{\ell}S_-^{\ell'}|\gamma\rangle &= \left\{\frac{N\gamma}{1+|\gamma|^2}\right\}^{\ell}\left\{\frac{N\gamma^*}{1+|\gamma|^2}\right\}^{\ell'} \\ &= \left\{\frac{N}{2}e^{i\varphi}\sin\theta\right\}^{\ell}\left\{\frac{N}{2}e^{-i\varphi}\sin\theta\right\}^{\ell'} \\ &= \left\{\langle\gamma|S_+|\gamma\rangle\right\}^{\ell}\left\{\langle\gamma|S_-|\gamma\rangle\right\}^{\ell'}\end{aligned} \tag{5.2}$$

Of course, this is exact for $\ell+\ell' = 0,1$. For $\ell+\ell' > 1$ the asymptotic validity requires that the modulus of γ, if it depends on N at all, approach neither zero nor infinity as $N\to\infty$. For states $|\gamma\rangle$ with γ asymptotically close to 0 or ∞, that is with θ asymptotically close to 0 or π, i.e. to the fully excited or the ground state of the system the right hand side in (5.2) vanishes. This does not imply, of course, that $\langle\gamma|S_+^{\ell}S_-^{\ell'}|\gamma\rangle$ vanishes but rather that this expectation value is of lower than $(\ell+\ell')$-th order in N.

If the density operator ϱ is a mixture of the kind (4.10), (5.2) generalizes to

$$< S_+^{\ell} S_-^{\ell'} > / N^{\ell+\ell'} = \int d^2\gamma \, P(\gamma,\gamma^*) \left\{ \frac{\gamma}{1+|\gamma|^2} \right\}^{\ell} \left\{ \frac{\gamma^*}{1+|\gamma|^2} \right\}^{\ell'} . \tag{5.3}$$

Eq. (5.3) is correct if the integral on the right hand side is finite for $N \to \infty$. That is the quasiprobability P must not accumulate all its weight on $\gamma \approx 0$ or $\gamma \approx \infty$.

b) The asymptotic weight function $P(\gamma,\gamma^*;t)$

The first-order-derivate terms in the equation of motion (4.16) for $P(\gamma,\gamma^*,t)$ describe a drift of the quasiprobability in the complex γ-plane directed radially away from the origin. The drift coefficient can obviously be simplified as $(1+2/N) \to 1$ as N grows large. Although the second-order-derivatives all carry an explicit factor $1/N$ it would be wrong to say that they are asymptotically small for $N \to \infty$. We will, however, show that the effect of

$$L_1 \equiv \frac{\lambda}{N} \left(\frac{\partial^2}{\partial\gamma^2} \gamma^2 + c.c. \right) \quad , \quad \lambda = 1 \tag{5.4}$$

on finite order moments of the quasiprobability P is indeed negligable for N large. To this end we simply extract from eq. (4.16) equations of motion for the moments

$$\overline{\gamma^n \gamma^{*m}}(t) \equiv \int d^2\gamma \, \gamma^n \gamma^{*m} \, P(\gamma,\gamma^*;t) \quad ; \quad n,m = 0,1,2,\dots . \tag{5.5.}$$

We obtain

$$\tau \frac{d}{dt} \overline{\gamma^n \gamma^{*m}}(t) = \left\{ n+m + \frac{\lambda}{N} [n(n-1) + m(m-1)] \right\} \overline{\gamma^n \gamma^{*m}}(t) + \{ 2nm/N \} \overline{\gamma^{n-1} \gamma^{*m-1}}(t) , \quad \lambda = 1. \tag{5.6}$$

Even without solving these equations it is immediately obvious that L_1 has no effect on the moments as long as

$$(n^2+m^2)/N \longrightarrow 0 \quad \text{for} \quad N \to \infty . \tag{5.7}$$

The equations (5.6) can, however, be solved explicitly. Starting from $n = m = 0$, for which case we have $\overline{1} = 1$, we can recursively construct all moments (5.5). For the sake of illustration let us write down the result for the special case $m = 0$

$$\overline{\gamma^n}(t) = \overline{\gamma^n}(0) \exp\left\{ \frac{t}{\tau} [n + \lambda n(n-1)/N] \right\} , \quad \lambda = 1. \tag{5.8}$$

The asymptotic smallness of L_1 stated above can again be inferred.

Inasmuch as the physically interesting properties of the superradiant system are determined by finite order moments of P in the sense (5.7) we can drop the term L_1 from the equation of motion (4.16) which thus becomes

$$\tau \dot{P}(\gamma,\gamma^*;t) = \left\{ - \left(\frac{\partial}{\partial\gamma} \gamma + \frac{\partial}{\partial\gamma^*} \gamma^* \right) + (2/N) \frac{\partial^2}{\partial\gamma \partial\gamma^*} \right\} P(\gamma,\gamma^*;t) . \tag{5.9}$$

A remarkable simple picture now emerges for the statistical behavior of the atomic polarization. According to the asymptotically correct equation (5.9) the variable γ which specifies the direction of the vector $\vec{S}$ behaves like the complex amplitude of a twodimensional harmonic oscillator subject to Gaussian white noise and

---

+ The unimportance of L_1 with respect to the expectation values $< S_+^{\ell} S_-^{\ell} >$ can also be proved a posteriory by a perturbation expansion in terms of L_1.

linear amplitude amplification+. The solution of the Fokker Planck equation (5.9) is well-known [28]

$$P(\gamma,\gamma^*,t|\alpha,\alpha^*) = \{\pi(e^{2t/\tau}-1)/N\}^{-1} \exp\left[-|\gamma-\alpha e^{t/\tau}|^2\{(e^{2t/\tau}-1)/N\}^{-1}\right]$$

with

$$P(\gamma,\gamma^*,0|\alpha,\alpha^*) = \delta^{(2)}(\gamma-\alpha) \tag{5.10}$$

We see that the quasiprobability P drifts, starting at its initial location $\gamma = \alpha$, towards infinity in the complex γ-plane; being sharply peaked initially according to the initial condition (4.11), the quasiprobability spreads out until it covers, at $t = \infty$, the whole complex γ-plane with uniform and therefore vanishing density; the normalization to unity is, as has to be the case, retained at all times.

c) Photon statistics

As we have indicated in the introduction superradiant pulses are characterized by

$$\langle b^{+\ell} b^{\ell} \rangle \sim \langle S_+^{\ell} S_-^{\ell} \rangle \sim N^{2\ell} \tag{5.11}$$

Let us note that the normally ordered intensity moments $\langle b^{+\ell} b^{\ell}\rangle$ can, in principle, be measured in photon counting experiments and therefore determine the statistical behavior of the superradiant pulse. We are now equipped to evaluate these moments for the case $N \gg 1$ as

$$\begin{aligned} M_\ell(\alpha,t) &= \langle S_+^{\ell}(t) S_-^{\ell}(t)\rangle / (N/2)^{2\ell} \\ &= \int d^2\gamma \left\{ \frac{2|\gamma|}{1+|\gamma|^2} \right\}^{2\ell} \{\pi(e^{2t/\tau}-1)/N\}^{-1} \times \\ &\quad \times \exp\left[-|\gamma-\alpha e^{t/\tau}|^2 \{(e^{2t/\tau}-1)/N\}^{-1}\right] \end{aligned} \tag{5.12}$$

Here we have used the asymptotic from (5.10) of the quasiprobability P and the asymptotic expression (5.3) for the angular momentum expectation values. Let us emphasize that (5.12) is correct for initial amplitudes γ and times such that the integral on the right hand side is finite for $N \to \infty$. For such initial amplitudes and times only superradiance is possible.

We first identify the range of initial amplitudes γ for which superradiance does occur while the width of the quasiprobability is still asymtotically small, i.e. for which $(\exp 2t/\tau - 1)/N \to 0$ as $N \to \infty$, or, roughly,

$$0 \leq t/\tau < \tfrac{1}{2}\epsilon \ln N \quad \text{with} \quad \epsilon \lesssim 1. \tag{5.13}$$

In this time interval we have

$$P(\gamma,\gamma^*;t|\alpha,\alpha^*) = \delta^{(2)}(\gamma - \alpha e^{t/\tau}) \tag{5.14}$$

and

$$M_\ell(\alpha,t) = \{M_1(\alpha,t)\}^{\ell} = \left\{ \operatorname{sech} \frac{t-t_{max}}{\tau} \right\}^{2\ell} \tag{5.15}$$

+ Eq. (5.9) differs from the well-known Fokker Planck equation for a damped harmonic oscillator [28] by the sign of the drift term; this change of sign turns the damping into an amplification.

$$t_{max} = -\tau \ln|\alpha| = \frac{\tau}{2} \ln \frac{\frac{N}{2} + \langle S_z(0)\rangle}{\frac{N}{2} - \langle S_z(0)\rangle} \quad . \tag{5.15}$$

This expression holds and describes superradiance if $M_1(\alpha,t) = O(1)$ for times within the interval (5.13). This is the case under two different circumstances. Either t_{max} is nonnegative, i.e. $|\alpha| \leq 1$ and $\langle S_z(0)\rangle \geq 0$. Then the time of maximum intensity t_{max} must lie in the interval (5.13); equivalently, the initial amplitude $|\alpha|$ and the initial inversion $\langle S_z(0)\rangle$ have to obey

$$|\alpha| > N^{-\epsilon/2} \ , \quad \langle S_z(0)\rangle < \frac{N}{2}(1-2N^{-\epsilon/2}) \ , \quad \epsilon \lesssim 1. \tag{5.16}$$

Or we have $t_{max} < 0$, i.e. $|\alpha| > 1$ and $\langle S_z(0)\rangle < 1$. Then the radiated intensity consists of part of the decaying wing of $M_1(\alpha,t)$ only and, obviously, the initial amplitude has to be such that

$$M_1(\alpha,0) = \mathrm{sech}^2 t_{max} = \left(\frac{2|\alpha|}{1+|\alpha|^2}\right)^2 = \frac{|\langle S_+(0)\rangle|^2}{N^2/4} =$$
$$= 1 - \frac{\langle S_z(0)\rangle^2}{N^2/4} = O(1) \ , \tag{5.17}$$

$$\text{or} \quad \frac{\langle S_z(0)\rangle}{N} \to 0 \quad \text{for} \quad N \to \infty \ .$$

We conclude that the atomic system will radiate a superradiant pulse if its initial excitation energy above the ground state, $\hbar\omega(\frac{N}{2} + \langle S_z(0)\rangle)$ is of order N. If, moreover, the atomic initial state is not asymptotically close to the fully excited state in the sense of (5.16) the superradiant pulse is described by (5.15) and thus behaves fully classically.

We now investigate the behavior of our system for very highly excited atomic states, that is for

$$\frac{N}{2} - N^{(1-\epsilon/2)} \lesssim \langle S_z(0)\rangle \leq \frac{N}{2} \ , \quad \epsilon \lesssim 1$$
$$\text{or} \quad 0 \leq |\alpha| \lesssim N^{-\epsilon/2} \ . \tag{5.18}$$

After inserting

$$\alpha = e^{i\varphi}\sqrt{\frac{m}{N}} \quad \text{with } 0 \leq \frac{m}{N} \lesssim N^{-\epsilon} \tag{5.19}$$

in the general expression (5.12) we see that superradiance sets in after long delay times of order $\ln N$ since only for such times does $P(\gamma,\gamma^*,t|\alpha,\alpha^*)$ grow to finite weight over finite amplitudes γ. For such times, however, the quasiprobability P has aquired a finite width $(\exp 2t/\tau - 1)$. It is clear, therefore, that the superradiant pulses produced from the very highly excited atomic system will be accompanied by large quantum fluctuations. To evaluate the integral (5.12) in this case we use the representation (5.19) for α and the rescaled time variable

$$z = N \exp(-2t/\tau) \tag{5.20}$$

We obtain, for finite z,

$$M_\ell\left(\alpha = e^{i\varphi}\sqrt{\frac{m}{N}},\ t = \frac{\tau}{2}\ln\frac{N}{z}\right) = \int d^2\gamma \left(\frac{2|\gamma|}{1+|\gamma|^2}\right)^{2\ell} \left(\frac{z}{\pi}\right) \exp\left\{-z\left|\gamma - e^{i\varphi}\sqrt{\frac{m}{z}}\right|^2\right\}.$$

The integral over the phase of γ can be carried out and yields a Bessel function of imaginary argument and order zero. We transform the remaining integral over the modulus $|\gamma|$ by introducing the integration variable $y = z(1+|\gamma|^2)$ and thus get

$$M_\ell\left(\alpha = e^{i\nu}\sqrt{\frac{m}{N}},\ t = \frac{\tau}{2}\ln\frac{N}{z}\right) = \tag{5.21}$$
$$= z^\ell e^z e^{-m} \int_z^\infty dy\, \frac{(y-z)^\ell}{y^{2\ell}}\, e^{-y}\, I_0\left(2\sqrt{m(y-z)}\right).$$

For the fully excited atomic initial state, $m = 0$ or $\alpha = 0$ this reduces to a result first obtained by Degiorgio [20]. For very highly excited atomic initial states, defined in (5.18), the expression (5.21) is equivalent to a result of ours [6] obtained for angular momentum states $|\frac{N}{2},\frac{N}{2}-\nu\rangle$, $\nu = 0,1,2,\ldots$ as initial states. The equivalence is proved in the appendix.

The transition of the statistical properties of superradiant pulses from the large-fluctuation behavior for very highly excited atomic initial states to the classical behavior for initial states with amplitudes α not asymptotically close to zero in the sense (5.16) has been studied by numerically integrating some of the integrals (5.21). The results are shown in figs. 3 - 5. The transition in the behavior of the intensity is displayed in figs. 3 and 4 where the "time" $z_{max}(m)$ of maximum intensity and the deviation of the maximum intensity from its classical value 1 are given. Finally, fig. 5 shows, as a direct measure of quantum fluctuations in the pulse, the dispersion of the intensity at the time $z_{max}(m)$ of maximum intensity,

$$\sigma(m) = \left\{M_2(m, z_{max}(m)) - M_1(m, z_{max}(m))^2\right\} / M_1(m, z_{max}(m))^2 \tag{5.22}$$

These numerical results substantiate earlier conclusions by Bonifacio, Schwendimann, and Haake [3] drawn from a direct numerical integration of the superradiance master equation (3.18) for $N = 10^3, 10^4$.

Let us note that the large fluctuation pertinent to pulses emitted by a very highly excited atomic system are easily understood qualitatively. For such initial states the pulses are triggered by spontaneously emitted quanta. Since the sequence of a few such spontaneous quanta can be looked upon as initial noise the pulses finally produced appear as amplified noise. On the other hand, atomic initial states with amplitudes not asymptotically small display an electric polarization not asymptotically small. Such an initial polarization acts as a classical source of radiation und thus entails pulses behaving classically.

APPENDIX

For angular momentum states $|\frac{N}{2},\frac{N}{2}-\nu\rangle$, $\nu = 0,1,2,\ldots$ as atomic initial states we had previously obtained [6] the following expression for the normalized moments

$$\langle S_+^\ell(t) S_-^\ell(t)\rangle / \frac{N^2}{4} \equiv M_\ell(\nu,t) = z^\ell e^z \int_z^\infty dy\, \frac{(y-z)^\ell}{y^{2\ell}}\, e^{-y}\left[\frac{(y-z)^\nu}{\nu!}\right] \tag{A.1}$$

with z as in (5.20). By using the decomposition (4.4) of the directed angular momentum states $|\alpha\rangle$ in terms of the states $|\frac{N}{2},\frac{N}{2}-\nu\rangle$ we can construct the intensity moments $M_\ell(\alpha,t)$ as

$$M_\ell(\alpha,t) = (1+|\alpha|^2)^{-N} \sum_{\nu=0}^{N} \binom{N}{\nu} |\alpha|^{2\nu} M_\ell(\nu,t). \tag{A.2}$$

We here insert (A.1) with $\alpha = e^{i\psi}\sqrt{m/N}$ and obtain

$$M_\ell(\alpha,t) = z^\ell e^z \int_z^\infty dy \, \frac{(y-z)^\ell}{y^{2\ell}} e^{-y} \times \tag{A.3}$$

$$\times \left[\left(1+\frac{m}{N}\right)^{-N} \sum_{\nu=0}^{N} \frac{N!}{\nu!\nu!(N-\nu)!} \left(\frac{m(y-z)}{N}\right)^\nu \right].$$

For $N \to \infty$, $(1+m/N)^{-N} \to \exp(-m)$ and the sum becomes the power series expansion of the Bessel function $I_0(2\sqrt{m(y-z)})$ whereupon (5.21) is recovered.

We acknowledge discussions with Rodolfo Bonifacio and Paolo Schwendimann. The material of sections I-III of this first emerged from a collaboration with them. We are indepted to Wolfgang Hendörfer for carrying out the numerical integration of eq. (5.21).

REFERENCES

1. R.M.Dicke, Phys.Rev. 93, 493 (1954)
2. N.Skribanovitz et al., Phys.Rev.Letters 30, 309 (1973)
3. R.Bonifacio, P.Schwendimann and F.Haake, Phys.Rev. A4, 302(1971) and ibid. A4, 854 (1971)
4. F.Haake, Springer Tracts in Modern Physics, vol. 66 (1973)
5. F.T.Arecchi, E.Courtens, R.Gilmore, and H.Thomas, Phys.Rev. A6, 2211 (1972)
6. F.Haake and R.Glauber, Phys.Rev. A5, 1457 (1972)
7. H.Haken, Handbuch der Physik, Bd. XXV/2c, Berlin, Springer(1970)
8. P.Kryukov and V.S.Letokhov, Sov.Phys. 12, 641 (1970)
9. G.L.Lamb, Jr., Rev.Mod.Phys. 43, 99 (1971)
10. F.T.Arecchi and E.Courtens, Phys.Rev. A2, 336 (1970)
11. V.Ernst and P.Stehle, Phys.Rev. 176, 1456 (1968)
12. N.W.Rehler and J.H.Eberly, Phys.Rev. A3, 1735 (1971)
13. W.Weidlich and F.Haake, Z.Physik 186, 203 (1965)
14. R.Zwanzig, J.Chem.Phys. 33, 1338 (1960)
15. S.Nakajima, Progr.Theor.Phys. 20, 948 (1958)
16. G.Scharf, Helv.Phys.Acta 43, 806 (1970)
17. R.Glauber, Phys.Rev. 130, 2529 (1963) and ibid. 131, 2766 (1963)
18. L.M.Narducci et al., Phys.Rev. A9, 829 (1974)
19. N.Ch.Wang and G.E.Uhlenbeck, Rev.Mod.Phys. 17, 323 (1945)
20. V.Degiorgio, Opt.Comm. 2, 326 (1971)

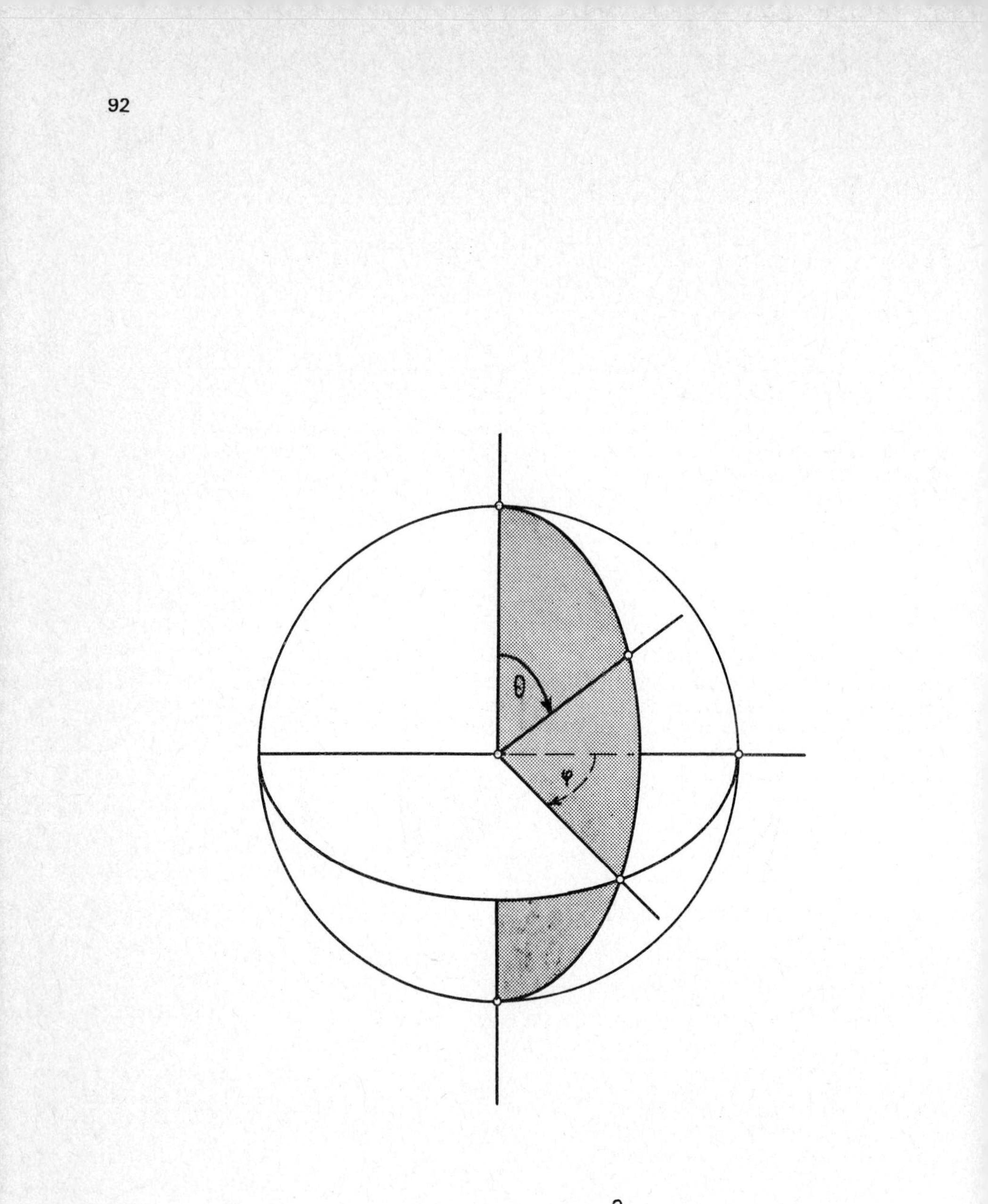

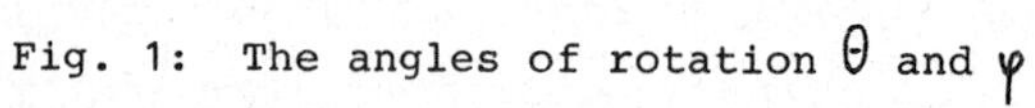
Fig. 1: The angles of rotation θ and φ

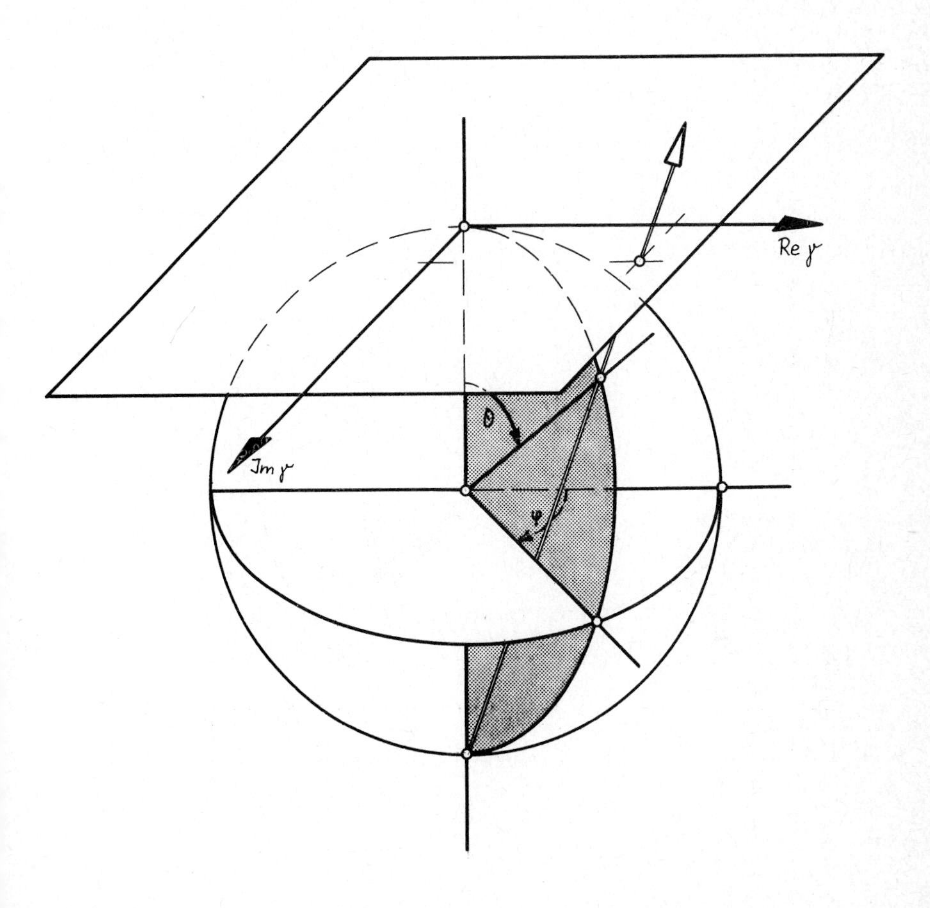

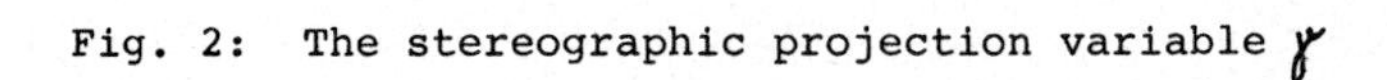
Fig. 2: The stereographic projection variable γ

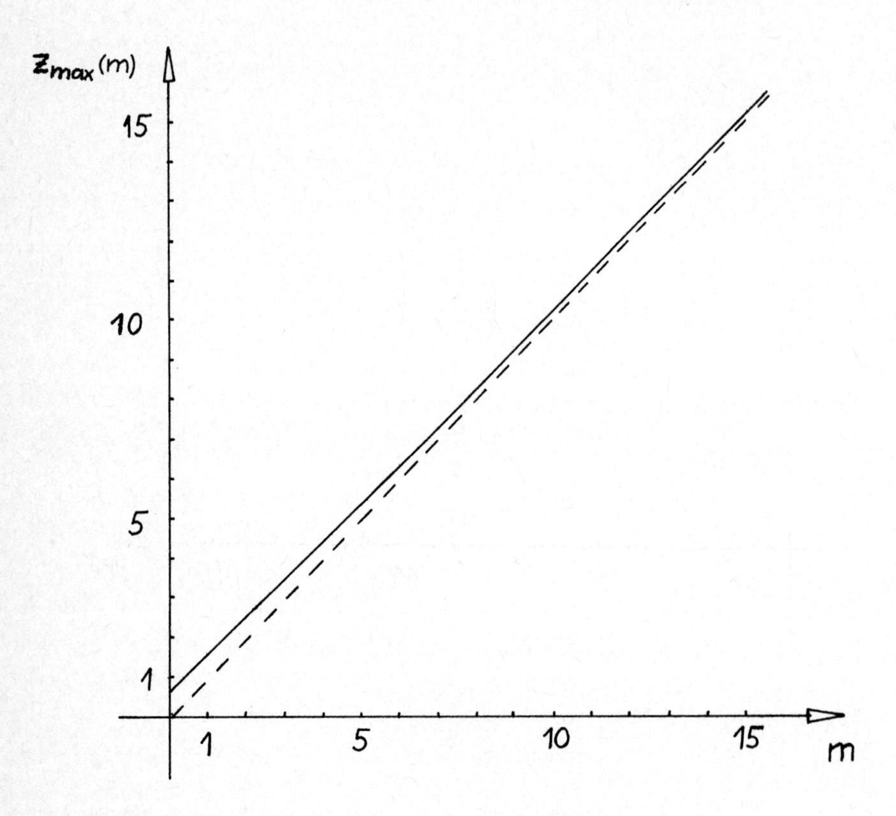

Fig. 3: Time $z_{max}(m)$ of maximum intensity as a function of the deviation m of the initial inversion $\langle S_z(0)\rangle$ from total inversion N/2; the dashed line shows the classical result $z_{max,class}(m) = m$ (from eqs. (2.18), (4.5), (5.19)).

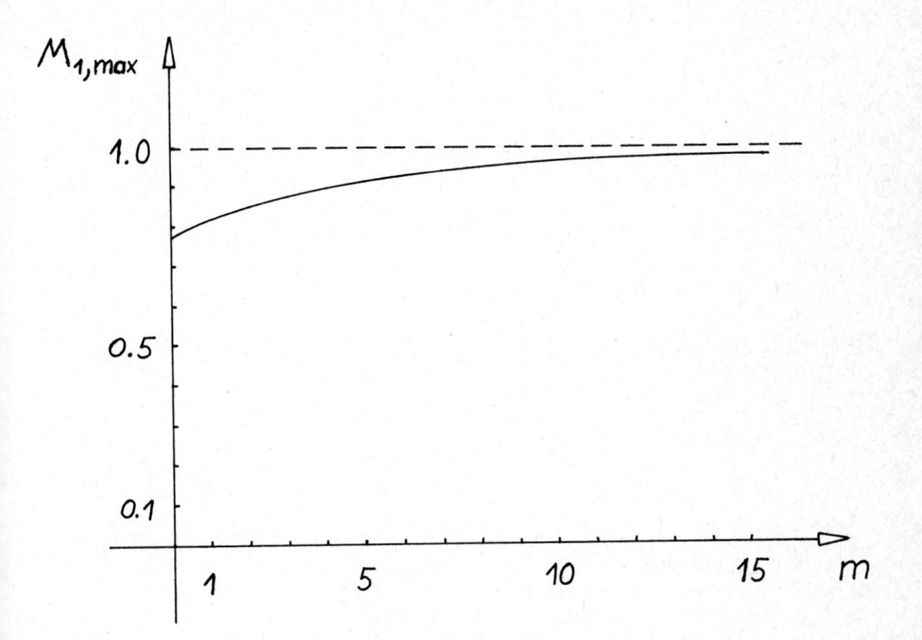

Fig. 4: Maximum intensity $M_1(m, z_{max}(m)) \equiv M_{1,max}$; the dashed line gives the classical result $M_{1,max} = 1$.

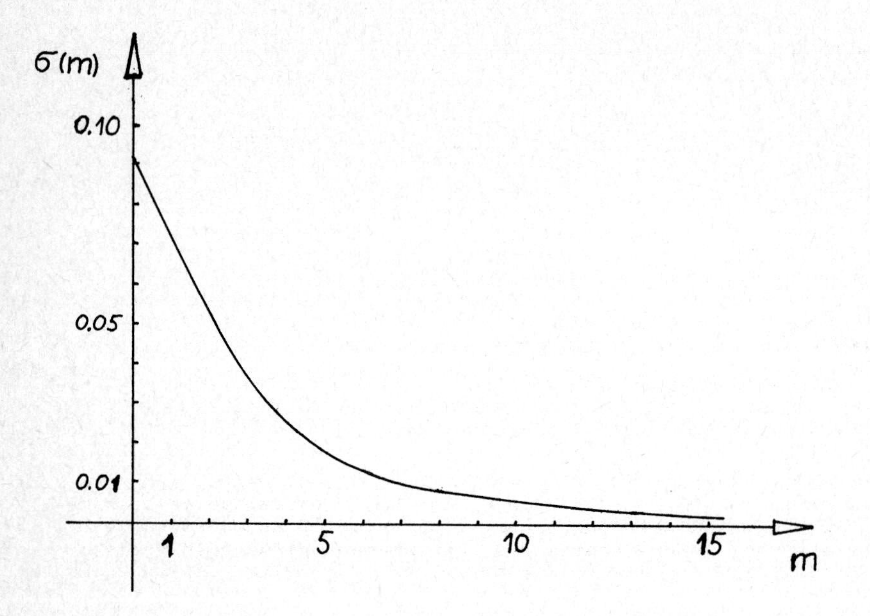

Fig. 5: Dispersion of the intensity σ(m) at the time z_{max} of maximum intensity

Cooperative Phenomena, H. Haken, ed.

NON-MARKOVIAN THEORY OF SUPERRADIANCE *

R. Bonifacio
Istituto di Fisica Università di Milano - Italy

Introduction

The aim of these lectures is to give a general theory of cooperative effects in spontaneous emission by a system of N excited two-level atoms (spins). Our main purpose is to remove the limitations of a previous theory [1] in order to describe the main features of the Skribanowitz [2] et al. experiment on Superradiance. Such limitations are essencially the following:

1. The Master Equation which is the starting point of the analysis of ref. [1] is not derived from first principles. Especially the assumption about the existence of a preferred "damped mode" in the mirrorless volume, which contains the active atoms, needs to be clarified.
2. In ref. [1] it is considered only the case in which the rate K at which photons escape from the active volume, is much larger than the rate $g\sqrt{N}$ at which photons and the atomic system exchange energy. This amounts to assume that the atoms emit practically in the vacuum of photons, so that no appreciable reaction of the field on the atoms, i.e. stimulated emission and absorption, occurs. On the other hand, in the recent experiments [2] on Superradiance $g\sqrt{N}$ is of the same order of magnitude of K.
3. In ref. [1] it is assumed that all the atoms have the same transition frequency (homogeneous line) whereas in the experiments [2] the line is inhomogeneous.
4. In ref. [1] the atom-field interaction has been described neglecting antiresonant terms (Rotating Wave Approximation); this approximation has to be removed to describe correctly frequency shifts.

Furthermore our theory does not contain the perturbative assumptions of reff. [3] and [4], and our results are not limited to a small number of atoms or to a small volume as it is in reff. [5] and [6].

The central ideas underlying our calculations are the following:

i. We quantize the e.m. field in a volume V much larger than the volume v which contains the active atoms. Hence we find that the field inside v can be described by mean of Bose Quasi-Mode (Q.M.) operators which represent wave packets localized in v. In the limit $V \to \infty$ the Quasi-Modes are damped as a consequence of the escape of photons toward the external volume V.

ii. We describe the N-atoms system by mean of collective dipole operators (C.D.) which are defined as a linear superposition of single atom dipole's operators.

We find that the time evolution of the density operator for the coupled atom-field system in the limit $V \to \infty$ is given by an irreversible M.E. provided the system is described in terms of Q.M. and C.D. operators.

This general M.E. can be used to rediscuss many-modes Laser theory with inhomogeneous broadening from a more fundamental view-point. In particular, under special conditions, this M.E. reduces to the M.E. of ref. [1].

The results we find are the following: under condition $g\sqrt{N} \ll K$ and for a pencil-shaped volume we rediscover the results of ref. [1] adding to them a complete description of the radiation pattern and of frequency shifts. More precisely: below some threshold of excitation, which coincides with the threshold for Laser action, we find the Wigner-Weisskopf's exponencial decay of the atomic excitation and of the radiated intensity. The radiation pattern is typically isotropic (normal fluorescence).

Above threshold, cooperative decay takes place: the system radiates

the well-known superradiant pulse whose maximum intensity is proportional to N^2. The radiation pattern is no more isotropic but it is condensed into the diffraction lobes of the axial modes. This is also a cooperative effect. Furthermore the frequency does not coincide with the atomic transition frequency but there is a time-dependent frequency shift (frequency chirping), whose maximum amplitude is proportional to the total number of atoms. This cooperative time-dependent frequency shift is superimposed to the normal constant Lamb shift and cannot be eliminated by renormalization. This cooperative frequency shift is the macroscopic consequence of an effective dipole-dipole interaction which takes place between different atoms via the exchange of virtual photons of the Superradiant pulse [6]. This effect is experimentally unexplored and has been previously described semiclassically only for a single atom [7] or for a volume smaller than a wavelenght [8].
Removing the condition $g\sqrt{N} \ll K$ and allowing $g\sqrt{N} \gtrsim K$ stimulated effects take place. As a consequence non-Markovian processes become relevant during the cooperative decay: the atomic excitation decay becomes oscillatory as well as the radiated intensity (ringing). Furthermore the time-delay for the formation of Superradiant pulses becomes much larger than the one given in reff. [1] and [9].
The appearing of ringing and of very large time-delays are the most peculiar feature of the experimental observation of Superradiance described in ref. [2].

Hamiltonian model

We consider N two level atoms, with position x_i, and resonant frequency ω_i coupled with a quantized e.m. field by an electric dipole transition. We quantize the e.m. field in a volume V much larger than the volume v containing the atoms. At the end of the calculations we will let $V \to \infty$ obtaining irreversibility. In the rotating wave approximation (R.W.A.) the Hamiltonian reads [10]:

$$H = \sum_i \hbar\omega_i r_{3i} + \sum_{\underline{k}} \hbar ck\, a^+_{\underline{k}} a_{\underline{k}} + \frac{i\hbar}{\sqrt{V}} \sum_{\underline{k},i} g_{\underline{k}} \left(a^+_{\underline{k}} r^-_i e^{-i\underline{k}\cdot x_i} - h.c. \right) \tag{1}$$

Here $a_{\underline{k}}, a^+_{\underline{k}}$ are the field creation and annihilation operators and $r^{\pm}_i$, r_{3i} are spin operators which describe the two level i-atom.

$$[a_{\underline{k}}, a^+_{\underline{k}'}] = \delta_{\underline{k},\underline{k}'} \tag{2}$$

$$[r^+_i, r^-_j] = 2r_{3i}\,\delta_{i,j};\ [r_{3i}, r^{\pm}_j] = \delta_{i,j}\, r^{\pm}_j \tag{3}$$

$2r_{3i}$ represents the population difference which can be +1 or -1. $r^{\pm}_i$ represents the dipole of the i-atom. In the so-called Neo-Classical Theory (N.C.T.) of Janes [7] et al. $r^{\pm}_i$ are c-numbers because the atoms are not described in second quantization.

Rotating frame picture

We find convenient to describe the dynamic of the coupled system in terms of the explicitely time-dependent rotated operators:

$$\tilde{a}_{\underline{k}}(t) = a_{\underline{k}}\, e^{i\omega_0 t} \qquad \tilde{r}^+_j = r^+_j\, e^{-i\omega_j t}$$

or in general

$$\tilde{B}(t) = e^{iAt}\, B\, e^{-iAt} \qquad \text{with}$$

$$A = \sum_{i=1}^{N} \omega_i r_{3i} + \omega_0 \sum_{\underline{k}} a^+_{\underline{k}} a_{\underline{k}}$$

Here ω_0 is a reference frequency.
The equation of motion of $\tilde{B}$ in the Heisenberg picture is:

$$\frac{d\tilde{B}}{dt} = \frac{1}{i\hbar}\,[\tilde{B},H] + \frac{\partial\tilde{B}}{\partial t} = \frac{1}{i\hbar}\,[\tilde{B},\tilde{H}]$$

Here $\tilde{H}$ is the Hamiltonian in the rotated frame:

$$\tilde{H} = \sum_k \hbar\,(ck - \omega_0)\,\tilde{a}^+_k\,\tilde{a}_k + \frac{\hbar}{\sqrt{V}} \sum_{\underline{k},i} g_{\underline{k}} \left(\tilde{a}^+_{\underline{k}}\,\tilde{r}^-_i\, e^{-i\underline{k}\cdot\underline{x}_i}\, e^{-i\delta_i t} - hc\right) \tag{4}$$

$$\delta_i = \omega_i - \omega_0\ ,\ R_3 = \sum_{i=1}^{N} r_{3i}$$

From now on we will refer to $\sim$ operators omitting $\sim$.
A very parallel argument holds in the Schrödinger picture.

Collective Dipole Operators and Lattice Model

We find convenient to assume that the atoms are arranged on a regular cubic lattice with interatomic dinstances d_J and $N = N_x \cdot N_y \cdot N_z$ $v = L_x \cdot L_y \cdot L_z$ with $N_J = \frac{L_J}{d_J}$ $(j = x, y, z)$.

The continuous limit eventually can be performed at the end of the calculations. We define the cavity modes $\underline{\alpha}$ as:

$$\alpha_J = \frac{2\pi}{L_J}\, n_J \qquad n_J = 0, 1 \ldots\ldots N_J - 1 \qquad J = x, y, z \tag{5}$$

It is easy to verify that:

$$\frac{1}{N} \sum_{\underline{\alpha}} e^{i\underline{\alpha}\cdot(\underline{x}_i - \underline{x}_J)} = \delta_{i,J} \tag{6}$$

We define collective dipole operators as:

$$R^{\pm}(\underline{\alpha}) = \sum_{J=1}^{N} r^{\pm}_J\, e^{\pm i\underline{\alpha}\cdot\underline{x}_J} \tag{7}$$

From (3) one gets:

$$\begin{aligned} &[R^+(\underline{\alpha})\ ,\ R^-(\underline{\alpha})] = 2R_3\,(\underline{\alpha} - \underline{\alpha}') \\ &[R^{\pm}(\underline{\alpha})\ ,\ R_3] = \pm R^{\pm}(\underline{\alpha});\ \text{with}\ R_3(\underline{\alpha}) = \sum_i r_{3i}\, e^{i\underline{\alpha}\cdot\underline{x}_i} \end{aligned} \tag{8}$$

Hence $R^{\pm}(\underline{\alpha})$ are angular momentum operators for each $\underline{\alpha}$. In particular R_3, whose eigenvalues go from $\frac{N}{2}$ to $-\frac{N}{2}$, represents the half population difference of the atomic system.
Relation (7) can be inverted using (6):

$$r^{\pm}_i = \frac{1}{N} \sum_{\underline{\alpha}} R^{\pm}(\underline{\alpha})\, e^{\mp i\underline{\alpha}\cdot\underline{x}_i} \tag{9}$$

Substituting into (4):

$$H = \sum_k \hbar\,(ck - \omega_0)\, a^+_{\underline{k}}\, a_{\underline{k}} + \frac{i\hbar}{\sqrt{V}} \sum_{\underline{k},\underline{\alpha}} g_{\underline{k}} \left[a^+_{\underline{k}}\, R^-(\underline{\alpha})\, f(\underline{k} - \underline{\alpha}, t) - hc\right]$$

where:

$$f(\underline{\eta}, t) = \frac{1}{N} \sum_{J=1}^{N} e^{i\underline{\eta}\cdot x_J}\, e^{i\delta_i t} \equiv \langle e^{i\underline{\eta}\cdot\underline{x}}\, e^{i\delta t} \rangle$$

Here the average has to be taken over all the positions and frequencies. We now assume that:
<u>i.</u> The atoms are uniformely and continuously distributed $(d_J \ll \lambda)$
<u>ii.</u> The atomic frequencies are continuously distributed and $G(\omega)$ is the normalized frequency distribution (Gaussian or Lorentzian usually)
<u>iii.</u> $G(\omega)$ has center simmetry that we choose to coincide with the arbitrary reference frequency .
Hence $f(\eta, t)$ can be written as:

$$f(\underline{\eta}, t) = g(t)\, F(\underline{\eta}) \quad \text{where}$$

$$\begin{cases} g(t) = \int_{-\infty}^{+\infty} d\delta\, G(\delta)\, e^{i\delta t} \\ F(\underline{\eta}) = \frac{1}{V}\int_V d^3\underline{x}\, e^{i\underline{\eta}\cdot\underline{x}} \equiv \prod_J \operatorname{sinc} \eta_J \frac{L_J}{2} \qquad J = x, y, z \end{cases} \tag{10}$$

with $\operatorname{sinc} x = \frac{\sin x}{x}$

Hence H can be written as:

$$\begin{cases} H = \underbrace{\sum_{\underline{k}} \hbar (ck - \omega_0) a^+_{\underline{k}} a_{\underline{k}}}_{H_0} + \frac{i\hbar}{\sqrt{V}} \sum_{\underline{k},\underline{\alpha}} g_{\underline{k}}(t) \left[a^+_{\underline{k}} R^-(\underline{\alpha}) F(\underline{k}-\underline{\alpha}) - hc \right] \\ g_{\underline{k}}(t) = g_{\underline{k}}\, g(t) \end{cases} \tag{11}$$

The limiting case of homogeneous broadening is obtained from (10) putting $g(t) = 1$.

Here inhomogeneous broadening is taken into account simply by a time-dependent coupling constant.

The coupling between $\underline{k}$-e.m. modes and $\underline{\alpha}$-atomic modes takes place through the "diffraction function" $F(\underline{k} - \underline{\alpha})$ which depends on the geometry of the system. If one does not perform the continuous limit mantaining the lattice model

$$F(\underline{\eta}) = \frac{1}{N}\sum_{i=1}^{N} e^{i\underline{\eta}\cdot\underline{x}_i} = \prod_J \frac{\sin \eta_J \frac{L_J}{2}}{N_J \sin \frac{\eta_J L_J}{2 N_J}} \quad (J = x, y, z) \tag{12}$$

This is a periodic function over a Brillouin zone $\left(\frac{2\pi N_J}{L_J} = \frac{2\pi}{d_J}\right)$

However assuming that the interatomic dinstances $d_J = \frac{L_J}{N_J}$ are very small to respect to the wavelenght into play or, mathematically, taking the limit $N_J \to \infty$ in the denominator of $F(\underline{\eta})$ one gets again expression (10) which is no more periodic.

Periodicity in a Brillouin zone must be taken into account dealing with X-rays transitions.

In the continuous limit H does not depend on the lattice model. Hence we mantain all our definitions and equations in the continuous limit substituting sums with integrals as follows:

$$\sum_i \to \frac{N}{V}\int_V d^3\underline{x}\ ; \quad \sum_{\underline{\alpha}} \to \frac{V}{(2\pi)^3}\int d^3\underline{\alpha}\ ; \quad \sum_{\underline{k}} \to \frac{V}{(2\pi)^3}\int d^3\underline{k}$$

In particular the following orthogonality relation is important:

$$\frac{V}{(2\pi)^3}\int d^3\underline{k}\ F(\underline{k}-\underline{\alpha})\, F(\underline{k}-\underline{\alpha}') = \delta_{\underline{\alpha},\underline{\alpha}'} \tag{13}$$

Finally note that the e.m. $\underline{k}$-modes which are coupled with an atomic $\underline{\alpha}$-mode are only those included in a diffraction angle around $\underline{\alpha}$. This is because the coupling in Hamiltonian (11) takes place via the diffraction function $F(\underline{k} - \underline{\alpha})$

Diffraction-Limited Quasi-Modes (D.Q.) and Field Damping

The form of the interaction Hamiltonian in H suggests to define the D.Q. as follows:

$$A(\underline{\alpha}) = \sqrt{\frac{v}{V}} \sum_{\underline{k}} a_{\underline{k}}\, F(\underline{k}-\underline{\alpha});\quad A^+(\underline{\alpha}) = \sqrt{\frac{v}{V}} \sum_{\underline{k}} a^+_{\underline{k}}\, F(\underline{k}-\underline{\alpha}) \tag{14}$$

By relations (2) and (13) we obtain:

$$\left[A(\underline{\alpha}),\ A^+(\underline{\alpha}')\right] = \delta_{\underline{\alpha},\underline{\alpha}'} \tag{15}$$

Hence D.Q. are Bose creation and annihilation operators. Only $\underline{k}$-modes contained in a diffraction angle around $\underline{\alpha}$ contribute to the quasi-mode $A(\underline{\alpha})$

If one assumes that g_k is a slowly varying function of $\underline{k}$ to respect to F $(\underline{k} - \underline{\alpha})$ one can replace $g_{\underline{k}} \rightarrow g_{\underline{\alpha}}$ in eq. (11). Hence the interaction Hamiltonian can be written as:

$$H = \frac{i\hbar}{\sqrt{v}} \sum_{\underline{\alpha}} g_{\underline{\alpha}}(t) \left[A^+(\underline{\alpha}) R^-(\underline{\alpha}) - hc\right] \tag{16}$$

Hence atomic collective polarization $R^-(\underline{\alpha})$ is coupled only with the diffraction quasi-mode $A^+(\underline{\alpha})$.
Note that going from (11) to (16) the quantization volume V has been replaced by the "cavity" volume v i.e. by the interaction volume containing the active atoms.

Field Damping

Relation (14) cannot be inverted so that it is not possible to write H_0 in terms of $A(\underline{\alpha})$ in (11). This has a fundamental reason: as we shall see in this section $A(\underline{\alpha})$-modes are damped i.e. they have an intrinsic linewidth. Hence their time evolution cannot be completely unitary i.e. cannot be represented simply by an Hamiltonian. In this sense they are quasi-modes and not real modes as the k-modes. The intuitive reason of the damping is the following: let $\mathcal{E}(\underline{x})$ be

$$\mathcal{E}(\underline{x}) = \sum_{\underline{\alpha}} A(\underline{\alpha})\, e^{i\underline{\alpha}\cdot\underline{x}}$$

By eq. (14), using the convolution theorem, we get:

$$\mathcal{E}(\underline{x}) \simeq E(\underline{x}) \cdot F(\underline{x}) \quad \text{with} \quad E(\underline{x}) \simeq \sum_{\underline{k}} a_{\underline{k}}\, e^{i\underline{k}\cdot\underline{x}}$$

Here $E(\underline{x})$ is the Field operator which obeys to Maxwell equations and $F(\underline{x})$ is a function which is 1 for $\underline{x}$ inside v and zero outside. Hence $A(\underline{\alpha})$ are the modes of the "internal field" $\mathcal{E}(\underline{x})$ which coincides with the real field $E(\underline{x})$ inside v and is zero outside. Furthermore the free field evolution which is simply the undeformed propagation for $E(\underline{x})$ will appear authomatically as a damping for the internal field as far as $E(\underline{x})$ gets out of the volume v. Clearly the order of magnitude of this damping time will be the transit time $\frac{L}{c}$. This argument can be made precise [11] showing that the continuous of $a_{\underline{k}}$-modes of the external field acts as a zero temperature heathbath on the $A(\underline{\alpha})$-modes. As a consequence the density operator in the Hilbert space of $A(\underline{\alpha})$-modes obeys to the following Master Equation:

$$\frac{dW}{dt} = \frac{1}{i\hbar}\left[H_0, W\right] + \frac{1}{i\hbar}\left[H', W\right] + \Lambda_F W \tag{17}$$

where

$$H_0 = \hbar \sum_{\underline{\alpha}} \left(c|\underline{\alpha}| - \omega_0\right) A^+(\underline{\alpha})\, A(\underline{\alpha}) \tag{18}$$

$$H' = \frac{i\hbar}{\sqrt{v}} \sum_{\underline{\alpha}} g_{\underline{\alpha}}(t) \left[A^+(\underline{\alpha}) R^-(\underline{\alpha}) - hc\right] \tag{19}$$

$$\Lambda_F W = \sum_{\underline{\alpha}} K(\hat{\alpha}) \left\{\left[A(\underline{\alpha}), W A^+(\underline{\alpha})\right] + hc\right\} \tag{20}$$

Where the damping constant $K(\hat{\alpha})$ is given by

$$K(\hat{\alpha}) = \frac{c}{2}\left(\frac{\hat{\alpha}_x}{L_x} + \frac{\hat{\alpha}_y}{L_y} + \frac{\hat{\alpha}_z}{L_z}\right) \qquad \left(\hat{\alpha} = \frac{\underline{\alpha}}{|\alpha|}\right) \tag{21}$$

We have now succeeded in the program of describing the full time evolution in terms of $A(\underline{\alpha})$ operators instead of $a_{\underline{k}}$. The price we did pay in changing the Hilbert space is the appearance of the non-unitary damping terms Λ_F which is well known in the theory of the damped harmonic oscillator and in Laser theory [10]. The meaning of the various terms of eq. (17) can be easily seen from the equation of motion of $<A(\underline{\alpha})>$:

$$\frac{d<A(\underline{\alpha})>}{dt} = \mathrm{Tr}\, A(\underline{\alpha})\dot{W} = -i\left(c|\alpha|-\omega_0\right)<A(\underline{\alpha})> + \frac{g_\alpha}{\sqrt{v}} <R^-(\underline{\alpha})> - K(\hat{\alpha})<A(\underline{\alpha})> \qquad (22)$$

In particular we see that Λ_F gives rise to an exponencial damping for $<A(\underline{\alpha})>$ in a time $[K(\hat{\alpha})]^{-1}$ which corresponds to the transit time of a photon traveling in the $\hat{\alpha}$ direction. We emphasize that irreversibility here has been obtained from first principles as a consequence of the "contraction" from the large (infinite) volume V to the finite open volume v whose characteristic operators are the "quasi-mode" operators $A(\underline{\alpha})$. Eq. (22) is derived from first principles in App. A.

Generalization of the M.E.

All our calculations can be very easily generalized to remove the R.W.A. In fact if we do not perform this approximation we have to add to Hamiltonian (1) the "antiresonant terms":

$$\frac{i\hbar}{\sqrt{V}} \sum_{\underline{k},i} g_{\underline{k}} \left(a^+_{\underline{k}}\, r^+_i\, e^{-i\underline{k}\cdot x_i} - hc\right) \qquad (23)$$

In such a case, mantaining all previous definitions of $R^{\pm}(\underline{\alpha})$ and $A(\underline{\alpha})$ operators, one would obtain the M.E. (17) generalized to include the rapidly oscillating terms:

$$\frac{i\hbar}{\sqrt{v}} \sum_{\underline{\alpha}} g_{\underline{\alpha}}(t) \left[A^+(\underline{\alpha})\, R^+(\underline{\alpha})\, e^{2i\omega_0 t} - hc\right] \qquad (24)$$

This term has to be added to the expression of H' as given by eq. (19)

Non Markovian Cooperative Emission

We now study eq. (17) referring to a specific initial condition and for a special geometry.

i. Initial condition

We assume that at $t = 0$ all the atoms are uniformely excited with some positive population difference. For sake of simplicity we assume that at $t = 0$ all the atoms are prepared in the excited state. This is a tipically uncorrelated initial state in which the atomic system does not have a macroscopic polarization. This is characteristic of an "incoherent" preparation like in the experiment of ref. [2]. The radiation process from such initial state requires a quantum-mechanical description because the initial radiation is just ordinary spontaneous emission.

ii. Pencil-shaped geometry

The active volume v has a pencil-shape in the sense that we assume

$$L_x = L_y = D; \quad L_z = L \quad \text{with } L \gg D \gg \lambda$$

With this geometry $K(\hat{\alpha})$ given by eq. (21) takes the form

$$K(\hat{\alpha}) \simeq \frac{c}{2}\left[\frac{|\sin\theta|\,(|\cos\varphi| + |\sin\varphi|)}{D} + \frac{|\cos\theta|}{L}\right] \sim \frac{c}{2L}\left[|\cos\theta| + \frac{L}{D}|\sin\theta|\right]$$

where we have defined θ and φ as the polar and azimuthal angle of $\hat{\alpha}$ and we made the unessential semplification $|\cos\varphi| + |\sin\varphi| \sim 1$.
It is very important for the rest of the discussion to see how rapidly $K(\hat{\alpha})$ changes as a function of θ.
For the axial modes $\underline{\alpha}_0(\theta = 0, \pi)$ we find:

$$K_0 = \frac{c}{2L} \qquad (25)$$

From the definition (5) of α-modes we see that the first axial modes are tilted of an angle $\theta \simeq \lambda/D$ on the $\hat{z}$ direction. Hence, if $\lambda \ll D$ they have a damping K_1 given by

$$K_1 \simeq K_0 \left[1 + \frac{1}{\mathcal{N}}\right] ; \quad \mathcal{N} = \frac{D^2}{L\lambda} \tag{25'}$$

$\mathcal{N}$ is the well-known Fresnel number which is the ratio between the "geometrical angle" D/L and the "diffraction angle" $\frac{\lambda}{D}$.
From (25') we see that if

$$\mathcal{N} \leq 1 \Rightarrow K_1 \geq 2K_0 \tag{26}$$

That is if $\mathcal{N} \leq 1$ the damping of non-axial modes is, at least, twice the one of the axial modes. This argument gives a quantitative support to the Dicke's [3] statement that off-axial modes are irrelevant to describe cooperative radiation processes. Clearly off-axial modes and axial modes are equally relevant to describe the normal spontaneous radiation process which is isotropic. However assuming that the cooperative and the normal processes take place on two different time-scales, respectively τ_R and τ_0 and assuming that

$$\tau_R \ll \tau_0 \tag{27}$$

we can neglect off-axial modes. Hence under conditions (26) and (27) we consider the M.E. (17) restricting $\sum_\alpha$ only to axial modes. Furthermore let us assume that the length L is such that two α-modes exist, say $\pm\alpha_0$, which satisfy the resonance condition $|\alpha_0| = \omega_0/c$. A simple inspection of eqs. (18) and (22) shows that only these resonant modes give rise to secular terms in a perturbation expansion Hence we consider only the two resonant axial modes $\pm\alpha_0$ i.e. the so-called end-fire modes [3].
In conclusion in eq. (17) we consider only the two modes which are resonant with the atomic transition and have the slowest damping. We still have a two-modes M.E. which is different from the one-mode M.E. which was the starting point of ref. [1]. However if one imagines to have a mirror on one end-face of the volume, clearly the radiation process will take place only through the α-mode leaving from the other face. Hence, specializing the M.E. (17) to a single axial mode, one obtains the M.E. of ref. [1]:

$$\begin{cases} \dfrac{dW}{dt} = \dfrac{1}{i\hbar}\left[H, W\right] + \Lambda_F W \\ \text{where} \\ H' = \dfrac{i\hbar}{\sqrt{V}}\, g_0\, g(t) \left[A^+ R^- - h.c\right] \\ \Lambda_F W = K_0 \left\{\left[A\, W, A_-^+\right] + h.c\right. \end{cases} \qquad g_0 = \left(\frac{\omega_0}{2\hbar}\,\mu^2\right)^{\frac{1}{2}} \qquad K_0 = \frac{c}{2L} \tag{28}$$

Here A, A^+ and $R^\pm$ are operators relative to the α_0 mode under conside ration and μ is the dipole moment of the atomic transition. We empha size that eq. (28) with $g(t) = 1$ is the fundamental equation of the one-mode Laser theory [10]. In such theory the presence of two mirrors with equal reflection coefficient R is taken into account phenomenologically defining the losses as $K_0 = \frac{c(1-R)}{L}$. Here the one mirror's assumption has been done only for sake of simplicity.
In fact, without any mirror it can be shown [11] that, with our initial condition, the radiation processes into the two axial modes are statistically independent and each of them is described by a M.E. which is formally identical to (28). However the approximations we made for deriving eq. (28) will appear more justified by the Markovian many-modes theory that we shall consider later. Furthermore in ref. [11] the many-modes M.E. (17) is discussed in the semiclas-

sical approximation deriving a system of generalized Maxwell-Bloch equations for the field and polarization envelopes which includes both direction of propagation and inhomogeneous broadening. With our initial condition these equations lead without approximations to the elimination of non-resonant modes. The solutions of these semiclassical equations are in a complete agreement with the mean-values solutions of M.E. (28) that we are going to discuss in the next section. [The generalized M.-B.eqs.are devided and discussed in Appendix B.]

Mean Values Equations

Equation (28) has been discussed in ref. [1] in the limit $g_o(t)$=const, $g_o\sqrt{N} \ll K_o$ (See the lectures of Prof. Haake). Let us repeat that in this limit the atoms radiate spontaneously in the vacuum of photons, i.e. both stimulated emission and absorption are forbidden. We now show that removing this limit one describes simultaneously spontaneous and stimulated processes. As a consequence one gets typical non-Markovian oscillations (ringing) in the atomic decay and in the radiated intensity. The analysis will be here given only in terms of mean values. A fully quantum statistical treatment in terms of a suitable Master Equation has been done by Lugiato and myself [11] and will be published elsewhere.
Let us note that the total angular momentum associated to $R^{\pm}$ operators is a constant of motion to respect to eq. (28):

$$R^+ R^- + R_3^2 - R_3 = \text{const} = \frac{N}{2}\left(\frac{N}{2}+1\right) \tag{29}$$

Here the fully excited initial condition has been taken into account. Mean values equations for an operator B of the system can be easily derived in the following way:

$$\frac{d\langle B\rangle}{dt} = \frac{d}{dt}\,\mathrm{Tr}\, W(t)\, B = \mathrm{Tr}\, \dot{W}(t)\, B$$

where the right-hand side of M.E. (28) has to be inserted in place of $\dot{W}$. In this way, using the angular momentum commutation relations, we obtain

$$\frac{d}{dt}\left(\langle A^+A\rangle + \langle R_3\rangle\right) = -2K_o\langle A^+A\rangle \tag{30}$$

This equation appears as a simple energy balance between the variation of the total internal energy ($\langle A^+A\rangle + \langle R_3\rangle$) and the energy output $2K_o\langle A^+A\rangle$ both measured in $\hbar\omega_o$ units. Hence we define the total number of photons radiated per unit time I(t) as:

$$I(t) = 2K_o\langle A^+A\rangle \tag{31}$$

We now derive an equation for $\langle R_3\rangle$ which, remember, is the half population difference $\frac{N^+ - N^-}{2}$ between the excited and the fundamental state.

$$\frac{d\langle R_3\rangle}{dt} = -\frac{g_o(t)}{\sqrt{V}}\left(\langle A^+R^-\rangle + \langle AR^+\rangle\right)$$

Differenciating both sides and using M.E. (28) for calculating $\frac{d\langle A^+R^-\rangle}{dt}$ we find after some calculations:

$$\langle\ddot{R}_3\rangle + \left(K_o + \frac{1}{2T_2^*}\right)\langle\dot{R}_3\rangle + \frac{2g_o^2(t)}{V}\left[\langle R^+R^-\rangle + 2\langle A^+AR_3\rangle\right] = 0 \tag{32}$$

Here we have assumed that $g_o(t) = g_o\, e^{-\frac{t}{2T_2^*}}$ where the "dephasing time" T_2^* is the reciprocal of a Lorentzian inhomogeneous linewidth. The equation for $\langle R_3\rangle$ of the theory of ref. [1] can be obtained from eq. (32) neglecting $\langle\ddot{R}_3\rangle$ (Markov approximation) and $\langle A^+A\, R_3\rangle$ (vacuum state for the field) and assuming $T_2^* = \infty$ (homogeneous line). The term $\langle A^+A\, R_3\rangle$ represents stimulated processes whereas the term

$\langle R^+R^-\rangle$ is the well-known cooperative spontaneous emission term described in ref. [1]. As we shall see both Markov approximation and the "vacuum state" approximation have the same limit of validity. In other words non-Markovian effects are strictly connected to stimulated processes. We now show that under suitable approximations eqs. (29), (30) and (32) become a closed system of equations for $\langle R^+R^-\rangle$, $\langle R_3\rangle$ and $\langle A^+A\rangle$.

i. Neo-Classical Approximation (N.C.A.)

We assume that

$$\langle R_3^2\rangle = \langle R_3\rangle^2 \tag{33}$$

This approximation is typical of the so-called neo-classical theory of Janes[7] et al.

ii. Self-Consistent Field Approximation (S.C.F.A.)

We assume that

$$\langle A^+AR_3\rangle = \langle A^+A\rangle\left(\langle R_3\rangle - \frac{1}{2}\right) \tag{34}$$

This approximation is characteristic of all semi-classical theories of atom-field interaction [10].

We assume that these approximations hold for $N \gg 1$.

Under these approximations eqs. (29), (30) and (32) can be simultaneously solved in the following way. Define the "modified Bloch angle" φ as:

$$\langle R_3\rangle - \frac{1}{2} = \left(\frac{N}{2} + \frac{1}{2}\right)\cos\varphi(t) \tag{35}$$

Using (33) and (35), eq. (29) gives:

$$\langle R^+R^-\rangle = \left(\frac{N}{2} + \frac{1}{2}\right)^2 \sin^2\varphi(t) \tag{35'}$$

Substituting eqs. (34), (35) and (35') into eqs. (30) and (32) we find that they are exactly equivalent to

$$\langle A^+A\rangle = \frac{v}{4g_o^2}(\dot{\varphi}^2)\, e^{\frac{t}{T_2^*}} \tag{36}$$

$$\ddot{\varphi} + \left(K_o + \frac{1}{2T_2^*}\right)\dot{\varphi} - \frac{g_o^2}{v}\, e^{\frac{t}{T_2^*}}(N+1)\sin\varphi = 0 \tag{36'}$$

Equation (36) represents a pendulum with a friction $K_o + (T_2^*)^{-1}$ and with a frequency which decreases exponencially in time. Here $\varphi = 0$ corresponds to the unstable equilibrium situation. If one makes a completely classical theory the Bloch angle comes out to be defined by eq. (35) but without the $\frac{1}{2}$ terms. As a consequence the initial situation correspondent to $\langle R_3\rangle = \frac{N}{2}$ would be the unstable equilibrium point $\varphi_o = 0$. Here we find

$$\varphi_o = \arccos\frac{\frac{N}{2} - 1}{\frac{N}{2} + 1} \sim \frac{2}{\sqrt{N}} \tag{37}$$

This finite value of φ_o is essential for moving the pendulum far away from the unstable equilibrium point. Once we solved eq. (36') with the initial condition (37) we can easily calculate $\langle R_3\rangle$, $\langle A^+A\rangle$ and the radiated intensity I using respectively eqs. (35), (36) and (31). Now we discuss eq. (36') showing that for particular values of the parameters one obtains all previous results on cooperative decay.

No Damping - Homogeneous line ($K_o = 0$, $T_2^* = \infty$)

In this case eq. (36') gives exactly the results obtained in ref. [12]. I(t) is given by a periodic elliptic function which, for $N \gg 1$, can be represented by a train of pulses whose time duration τ_c and repetition time T_M are given by

$$\tau_c = \left(g_0 \sqrt{\frac{N}{V}}\right)^{-1} ; \quad T_M = \tau_c \lg\sqrt{N} \tag{37}$$

Each pulse within a good approximation has an hyperbolic sechant shape with temporal width τ_c.
In this case the atom and the field are continuously exchanging energy in a time τ_c. Since the peak of each pulse is proportional to N and not to N^2 one cannot speak of Superradiance. Note that τ_c is inversely proportional to the atomic density and coincides with the Arecchi [13] Courtens cooperation time. The pendulum equation can be written in terms of τ_c as follows:

$$\ddot{\varphi} + \left(K_0 + \frac{1}{2T_2^*}\right)\dot{\varphi} - \frac{1}{\tau_c^2}\, e^{-\frac{t}{T_2^*}} \sin\varphi = 0 \tag{38}$$

A detailed analytical and numerical study of eq. (38) shows that Superradiance is possible only if the following inequalities are satisfied:

$$K_0^{-1} \lesssim \tau_c \ll T_2^* ; \quad T_2^* \gg K_0^{-1} \tag{39}$$

This condition can be heuristically understood on the basis of the following time-scale argument: the characteristic time for cooperative atom-field interaction is clearly τ_c. Hence τ_c has to be much smaller than T_2^* if we want the decreasing exponencial of eq. (38) to be irrelevant. This explains the right-hand side condition (39). Furthermore if we are in a condition such that $K_0 + (T_2^*)^{-1} \ll \tau_c^{-1}$, the damping is completely unimportant. Hence we would obtain a train of pulses proportional to N that we have described before. Hence we need $\tau_c^{-1} \leqslant K_0 + (T_2^*)^{-1}$ which together with $\tau_c \ll T_2^*$ leads to (39). Note that the necessary condition for Superradiance $T_2^* \gg K_0^{-1} = L/2c$ is never satisfied in an usual Laser also because the presence of mirrors has the effect of increasing the effective length of the sample $L/(1 - R)$. This condition, on the contrary, is well satisfied in the experiment [2]. Let us now distinguish the two particularly important cases: $K_0\tau_c \gg 1$ and $K_0\tau_c \sim 1$.

Pure Superradiance: $K_0\tau_c \gg 1$.

This is the case considered in ref. [1]. In this case the pendulum is overdamped, in the sense that it does not oscillate around the equilibrium position $\varphi = \pi (\langle R_3 \rangle = -\frac{N}{2})$ but it stops there at $t = +\infty$. This non-oscillating motion can be described neglecting $\ddot{\varphi}$ in eq. (38) so that one obtains:

$$\dot{\varphi} = \frac{1}{\tau_R}\, e^{-\frac{t}{T_2^*}} \sin\varphi ; \quad \tau_R = K_0 \tau_c^2 = \frac{V}{g_0^2 N K_0} \tag{40}$$

The time τ_R coincides with the time duration of Superradiant pulses described in reff. [1] and [4] and with the "characteristic time" of the experiment of ref. [2]. Note that (39) implies $\tau_R \gtrsim \tau_c$. Hence the real condition for dephasing to be ineffective is $\tau_R \ll T_2^*$ which is more restrictive than $\tau_c \ll T_2^*$ whereas $K_0\tau_c \geqslant 1$ is equivalent to $K_0\tau_R \gtrsim 1$ because $K_0\tau_R = (K_0\tau_c)^2$. In this way inequality (39) becomes:

$$K_0^{-1} \lesssim \tau_R \ll T_2^* ; \quad T_2^* \gg K_0^{-1} \tag{41}$$

Let us remark that eq. (40) can be obtained from eq. (32) neglecting $\langle \ddot{R}_3 \rangle$ and $\langle A^+ A\, R_3 \rangle$ and viceversa. Hence, as we have already stated, the limit of validity of these two approximations is the same, i.e. $K_0\tau_R \gg 1$. We now discuss the particular situation in which $K_0\tau_R = (K_0\tau_c)^2 \gg 1$. Eq. (40), if $T_2^* = \infty$, gives the well-known hyperbolic sechant Superradiant pulses whose peak intensity is proportional to N^2. In general, for finite T_2^*, eq. (40) can be exactly solved as follows.

Define the reduced time τ as

$$\tau = T_2^* \left(1 - e^{-\frac{t}{T_2^*}}\right) \tag{42}$$

τ is a reduced time in the sense that τ goes from zero to T_2^* as t goes from zero to infinite. Eq. (40) can be written as:

$$\frac{d\varphi}{d\tau} = \frac{1}{\tau_R} \sin\varphi ; \; \varphi(0) \simeq \frac{2}{\sqrt{N}} \tag{42'}$$

Eq. (42) has the solution

$$\sin\varphi(\tau) = \mathrm{sech}\, \frac{1}{\tau_R}\left(\tau - \tau_M\right) ; \; \tau_M = \tau_R \lg \sqrt{N} \tag{43}$$

Here τ_M coincides with the delay time of ref. [1]. The radiated intensity can be obtained using eqs. (31) and (36):

$$I(t) = \frac{1}{\tau_R^2} e^{-\frac{t}{T_2^*}} \mathrm{sech}^2 \frac{1}{\tau_R}\left(\tau - \tau_M\right) \tag{44}$$

where τ is the function of t given by (42). Let us discuss this solution. Keeping in mind that the physically possible values of τ go from zero to T_2^* , it is easy to see the following general features of the solution as a function of t:

i. $\tau_M << T_2^*$

The system behaves as if $T_2^* = \infty$. In particular $\tau \sim t$ and the picture in the reduced time coincides with the one in the real time. This is the case of ref. [1].

ii. $\tau_M < T_2^*$

The pulse gets broad in time. In particular the hyperbolic sechant reaches a peak at a time t_M given by

$$t_M = T_2^* \lg \frac{T_2^*}{T_2^* - \tau_M} > \tau_M \tag{45}$$

iii. $\tau_M > T_2^*$

In this case the peak occurs at a value of τ which lies outside the physical region. Hence the pulse does not reach a maximum but it decays in a time T_2^* as it can be seen by eq. (44).

Summarizing we can say that pure Superradiance occurs only if

$$K_0^{-1} << \tau_R << \frac{T_2^*}{\lg \sqrt{N}}$$

The situation described by eq. (40) is the one considered in ref. [9]. In the approximation $K_0\tau_R \gg 1$ one can qualitatively explain the increasing of the decay time t_M due to the inhomogeneous broadening (T_2^* finite), as it is observed in the experiment [2]. However one does not describe quantitatively the experimental t_M which is still larger than the one given by (45) and furthermore one does not describe the oscillation in the radiated intensity observed experimentally. Both these features can be taken into account dropping the condition $K_0\tau_R >> 1$.

Superradiance with Oscillations: $K_0\tau_R \sim 1$

This is the case of experiment [2]. The motion of the pendulum is no more overdamped and we have to take into account the second derivative $\ddot{\varphi}$. Obviously in this way one gets oscillation in the radiated intensity, which is proportional to $(\dot{\varphi})^2$. Furthermore a detailed numerical analysis [14] show that τ_M (time at which the first maximum is reached) increases strongly as far as the condition $K_0\tau_R \gg 1$ is relaxed. This increasing is added to the one due to the finite T_2^*. In conclusion, conditions (41) are necessary and sufficient for observing Superradiance. If $K_0\tau_R \gg 1$ one obtains pure Superradiance. If $K_0\tau_R \sim 1$ one observes non-Markovian oscillations because stimulated effects become relevant. This is the experimental situation of

ref.[2]. The detail of the numerical analysis and its comparison with the experiment [2] will be given elsewhere [14].
Let us remark that our theory is in very good agreement with the experimental results but not with the theoretical interpretation given in ref.[2]. There the authors claim that the right-hand side of inequality (41) is the only condition for Superradiance. In our opinion the left-hand side condition (41) is fundamental to distinguish a stimulated emission amplifier from a Superradiant device. In fact even if $\tau_R << T_2^*$ but $\tau_R << K_0^{-1}$ the system radiates by stimulated emission proportionally to N, like when $K_0 = 0$ and $T_2^* = \infty$.
Finally let us remark that, if other relaxations are present, T_2^* has to be replaced by τ_0 in (41) where τ_0^{-1} is the total incoherent relaxation rate. In this way all relaxation processes can be neglected.
Furthermore on can give an intuitive explanation of the increasing of τ_M: the overdamped pendulum ($\ddot{\varphi} = 0$) is, strictly speaking, a zero mass particle. In our case ($\ddot{\varphi} \neq 0$) the pendulum has a finite mass, i.e. a finite inertia, so that it takes more time to move from the initial position which is very near to the unstable equilibrium point.

Many Modes Theory of Cooperative Spontaneous Emission

We have studied the radiation process by a N-atoms system concentrating on the cooperative emission into the end-fire axial modes. Now, in the pure Superradiance limit $K_0\tau_R << 1$ we want to give a complete description of the radiation process which includes: cooperative and non cooperative decay, radiation pattern, frequency shifts.
A many modes theory has been given by Lehmberg [5] and Agarwal [6]. However these authors do not introduce collective variables so that they are able to give explicit solutions only for a value smaller than a wavelength or for N very small (say 1, 2, 3). Assuming that $\tau_R << \tau_0$ we will neglect all atomic relaxations including T .

The Many Modes Markovian M.E.

Since we want to describe frequency shifts we remove the Rotating Wave Approximation (R.W.A.) generalizing M.E. (17) to include the antiresonant terms (24) in H. In the pure Superradiance limit $K_0\tau_R << 1$ we eliminate the field variables with the same procedure and the same Born and Markov approximations as in ref.[1]. The only difference is that here we have the sum of $\underline{\alpha}$-modes instead of having a single mode and we have antiresonant terms. However these differences do not give any complication in the derivation of the M.E. for the density operator W_A of the atomic system alone. This M.E. reads:

$$\frac{dW_A}{dt} = \sum_{\underline{\alpha}} \left\{ \frac{\Gamma^-(\underline{\alpha})}{2} \left[R^-(\underline{\alpha}) W_A, R^+(\underline{\alpha}) \right] + h.c \right\} + \sum_{\underline{\alpha}} \left\{ \frac{\Gamma^+(\underline{\alpha})}{2} \left[R^+(\underline{\alpha}) W_A, R^-(\underline{\alpha}) \right] + h.c \right\} \quad (46)$$

where the transport coefficients $\Gamma^\pm(\underline{\alpha})$ are given by

$$\Gamma^\pm(\underline{\alpha}) = \frac{1}{V} \frac{2g^2(\underline{\alpha})}{K(\underline{\alpha}) + i\,(c|\underline{\alpha}| \pm \omega_0)} \quad (47)$$

The one-mode Superradiance M.E. of ref.[1] is obtained from eq. (46) if one takes a single mode out of the sum on $\underline{\alpha}$ and if one puts $\Gamma^+(\underline{\alpha}) = 0$.
Here the Γ^+ terms come from the antiresonant terms (24). The M.E. (46) has been firstly derived by Banfi [15] and myself starting directly from the Hamiltonians (1) and (23) and using the method of ref.[1].
However in such a case one finds different transport coefficients. Precisely $\Gamma^\pm(\underline{\alpha})$ turns out to be:

$$\Gamma^\pm(\underline{\alpha}) = \frac{1}{(2\pi)^3} \int_0^{+\infty} d\tau \int d^3\underline{k}\; 2g_{\underline{k}}^2 F^2(\underline{k} - \underline{\alpha})\, e^{-i(ck \pm \omega_0)\tau} \quad (48)$$

However, approximating k with $\underline{k}\cdot\hat{\alpha}$ in the exponent and $F^2(\underline{k} - \underline{\alpha})$ with a Lorentzian expression (48) reduces to (47). [See Appendix A]

In order to study eq. (46) it is convenient to introduce the real and the imaginary part of $\Gamma^{\pm}(\underline{\alpha})$

$$\Gamma^{\pm}(\underline{\alpha}) = \gamma^{\pm}(\underline{\alpha}) - i\,\Omega^{\pm}(\underline{\alpha})$$

From the exact expression (48) we see that $\gamma^{+}(\underline{\alpha}) = 0$. The expression of the other coefficients can be obtained from eq. (47) or (48).

$$\gamma^{-}(\underline{\alpha}) = \gamma(\underline{\alpha}) = \frac{1}{8\pi^2}\int d^3\underline{k}\; 2g_{\underline{k}}^2\,\delta(ck-\omega_0)\,F^2(\underline{k}-\underline{\alpha}) \simeq \frac{2}{v}\,g^2(\underline{\alpha})\,\frac{K(\hat{\alpha})}{K^2(\hat{\alpha})+(c|\alpha|-\omega_0)^2} \quad (49)$$

$$\Omega^{\pm}(\underline{\alpha}) = \frac{PP}{(2\pi)^3}\int d^3\underline{k}\; 2g_{\underline{k}}^2\,\frac{F^2(\underline{k}-\underline{\alpha})}{(ck\pm\omega_0)} \simeq \frac{2g^2(\underline{\alpha})}{v}\,\frac{(c|\alpha|\pm\omega_0)}{K^2(\hat{\alpha})+(c|\alpha|\pm\omega_0)^2} \quad (50)$$

Writing $R^{\pm}(\underline{\alpha})$ operators in terms of single atom spin operator as given by eq. (7), eq. (46) becomes identical to the Lehmberg-Agarwal's M.E. M.E. (46) can be written as:

$$\frac{dW}{dt} = \sum_{\underline{\alpha}} \gamma(\underline{\alpha})\left[R^{-}(\underline{\alpha})W, R^{+}(\underline{\alpha})\right] + \frac{1}{i\hbar}\left[\mathcal{H}, W\right] \quad (51)$$

where the operator $\mathcal{H}$ is defined as

$$\mathcal{H} = \hbar\,\Omega_0 R_3 + \sum_{\underline{\alpha}} \Omega_{\underline{\alpha}}\, P_c(\underline{\alpha}) \quad (52)$$

with

$$\Omega(\underline{\alpha}) = \Omega^{+}(\underline{\alpha}) + \Omega^{-}(\underline{\alpha})\,;\quad \Omega_0 = \sum_{\underline{\alpha}} \Omega(\underline{\alpha}) \quad (53)$$

and

$$P_c(\underline{\alpha}) = R^{+}(\underline{\alpha})\,R^{-}(\underline{\alpha}) - \left(\frac{N}{2} + R_3\right) = \sum_{i\neq j} r_i^{+}\, r_j^{-}\, e^{i\underline{\alpha}(\underline{x}_i - \underline{x}_j)} \quad (54)$$

The time evolution of W is due to two kinds of terms. The γ-s terms give rise to a non-unitary time evolution and describe cooperative and non cooperative damping. The commutator with the pseudo-Hamiltonian $\mathcal{H}$ gives rise to unitary time evolution and describes both cooperative and non cooperative frequency shifts. Let us comment on $\mathcal{H}$. The first term describes a constant frequency shift of the single atom transition frequency ω_0 and it can be eliminated renormalizing the naked frequency ω_0 in Hamiltonian (1). Keeping in mind that $\sum_{\underline{\alpha}} F^2(\underline{k}-\underline{\alpha}) = 1$, one can easily see that Ω_0 coincides with the Bethe's part of the Lamb shift [5], [6].

The other term in $\mathcal{H}$ disappears for a single atom because from (54) we see that $P_c(\underline{\alpha}) = 0$ for one atom; it represents a cooperative effect due to atom-atom interaction. In fact, substituting (54) into (52) we obtain

$$\mathcal{H}' = \sum_{\underline{\alpha}} \Omega(\underline{\alpha})\, P_c(\underline{\alpha}) = \sum_{i\neq j} V_{ij}\, r_i^{+}\, r_j^{-}$$

Hence $\mathcal{H}'$ has the form of a dipole-dipole interaction with an "interaction potential" V_{ij} given by:

$$V_{ij} = \sum_{\underline{\alpha}} \Omega(\underline{\alpha})\, e^{i\underline{\alpha}(\underline{x}_i - \underline{x}_j)}$$

In ref. [5] and [6] V_{ij} is explicitely calculated showing that it is a long-range potential whose dominant terms for large y_{ij} are of the form:

$$V_{ij} \simeq \frac{\cos y_{ij}}{y_{ij}} \quad \text{where } y_{ij} = \frac{\omega_0}{c}\left|\underline{x}_i - \underline{x}_j\right|$$

Hence V_{ij} is not due to electrostatic interaction but it is due to an effective dipole-dipole interaction which takes place through the electromagnetic field which has been eliminated from the equation of motion of the atomic system.

Cooperative Time-Dependent Frequency Shift (Chirping)

It can be shown that the effect of $\mathcal{H}$ on $<R^{\pm}(\underline{\alpha},t)>$ in the S.C.F.A. is the following:

$$<R^{\pm}(\underline{\alpha},t)> = <R_0^{\pm}(\underline{\alpha},t)> \, e^{\pm i \Phi_{\underline{\alpha}}(t)} \tag{55}$$

where the time evolution of $<R_0^{\pm}(\underline{\alpha},t)>$ can be obtained from (51) neglecting $\mathcal{H}$. The effect of $\mathcal{H}$ is only in $\Phi_{\underline{\alpha}}(t)$ which is given by

$$\Phi_{\underline{\alpha}}(t) = \Omega_0(t) + \Omega(\underline{\alpha}) \int_0^t <R_3(t')> dt' \tag{56}$$

Hence the "instantaneous frequency shift" is given by

$$\omega_{\underline{\alpha}}(t) = \dot{\Phi}_{\underline{\alpha}}(t) = \Omega_0 + \Omega(\underline{\alpha}) <R_3(t)> \tag{56'}$$

Since R_3, as we shall see, goes from $N/2$ to $-N/2$ the instantaneous frequency has total variation $\Delta\omega = \Omega(\underline{\alpha})N$. This is a cooperative frequency chirping effect because the frequency variation is proportional to the total number of atoms. Janes [7] et al. have described this effect in N.C.T. for a single atom. However, as we have seen, the Q.E.D. approach shows that chirping is a cooperative effect which disappears for a single atom. We now show the procedure to obtain eqs. (55) and (56). Using M.E. (51) and performing the S.C.F.A. (N>>1) one gets

$$<\dot{R}^{\pm}(\underline{\alpha})> = \left[\gamma(\underline{\alpha}) \pm i\Omega(\underline{\alpha})\right] <R_3> <R^{\pm}(\underline{\alpha})> \pm i\Omega_0 <R^{\pm}(\underline{\alpha})>$$

One can immediately verify that $<R_0^{\pm}(\underline{\alpha})>$ defined by (55) satisfies the equation with real coefficients:

$$<\dot{R}_0^{\pm}(\underline{\alpha})> = \gamma(\underline{\alpha}) <R_3> <R_0^{\pm}(\underline{\alpha})>$$

provided $\Phi_{\underline{\alpha}}$ is given by eq. (56)

Cooperative and non Cooperative Damping

We now briefly describe the time evolution of the system in terms of mean values of R_3 and $P_c(\underline{\alpha})$. The detail of the calculation will be published elsewhere [15]. From M.E. (51) one sees that R_3 and $P_c(\underline{\alpha})$ commute with $\mathcal{H}'$, hence we limit our considerations to the non-unitary part. If N >>1, keeping only dominant terms in the equation for $<P_c(\underline{\alpha})> = \mathrm{Tr}\, W\, P_c(\underline{\alpha})$ one obtains:

$$<\dot{P}_c(\underline{\alpha})> = 2\gamma(\underline{\alpha}) <P_c(\underline{\alpha})> <R_3> - \gamma <P_c(\underline{\alpha})> \qquad \text{where } \gamma = \sum_{\underline{\alpha}} \gamma(\underline{\alpha}) \tag{57}$$

By eq. (49), keeping in mind that $\sum_{\underline{\alpha}} F^2(\underline{k}-\underline{\alpha}) = 1$ one sees that γ coincides with the single atom decay rate given by the Wigner-Weisskopf's theory. From now on we will assume only for simplicity of notations that $g_{\underline{k}}^2$ depends only on $|\underline{k}|$ so that the single atom decay rate is isotropic.
The equation for $<R_3>$ can be obtained without any approximation for M.E. (51):

$$<\dot{R}_3> = -\sum_{\underline{\alpha}} \gamma(\underline{\alpha}) P_c(\underline{\alpha}) - \gamma\left(\frac{N}{2} + <R_3>\right) \tag{58}$$

Hence $<R_3>$ is a decreasing function of time.
Let $\gamma_0 = 2g^2/v = \gamma(\underline{\alpha}_0)$ be the value of $\gamma(\underline{\alpha})$ for the resonant end-fire modes $\underline{\alpha}_0$ $(|\underline{\alpha}_0| = \omega_0/c,\ \hat{\alpha}_0 \equiv \pm\hat{z})$

From eq. (49) we see that $\gamma(\underline{\alpha}) < \gamma_0$ for all $\underline{\alpha}$-s. Keeping in mind that at t = 0 is $\langle R_3\rangle = N/2$ we can distinguish the following situations:

i. below threshold: $\gamma_0 N < \gamma$.

In this case $\langle \dot{P}_c(\alpha)\rangle < 0$ for all $\underline{\alpha}$ at all times. Hence the system does not generate a macroscopic polarization.
Eq. (56) reduces to:

$$\langle \dot{R}_3\rangle = -\gamma\left(\frac{N}{2} + \langle R_3\rangle\right)$$

which says that $\langle R_3\rangle$ decays exponencially to the value $\langle R_3\rangle = -\frac{N}{2}$ in a time γ^{-1}.

ii. above threshold: $\gamma_0 N > \gamma$.

In this case $P_c(\underline{\alpha}_0)$ and eventually all $P_c(\underline{\alpha})$ for which $\gamma(\underline{\alpha})N > \gamma$ starts increasing exponencially and the cooperative terms in eq. (58) become relevant. Suppose that $\mathcal{N} = D^2/L\lambda \leq 1$. In this case all non-resonant and non-axial modes have a gain coefficient $\gamma(\underline{\alpha}) \leq \gamma_0/2$. This can be easily seen from eqs. (49) and (26). Hence also if many modes satisfy the threshold condition, the axial modes will be dominant on the others. Hence one can neglect all $P_c(\underline{\alpha})$ excepting for $\underline{\alpha} = \pm\underline{\alpha}_0$ in eq. (58)

$$\langle \dot{R}_3\rangle = -2\gamma_0 \langle P_c(\underline{\alpha}_0)\rangle - \gamma\left(\frac{N}{2} + \langle R_3\rangle\right) \tag{59}$$

Here we took into account that $\gamma(\underline{\alpha}_0), \gamma(-\underline{\alpha}_0)$ and by eq. (57) $P_c(\underline{\alpha}_0) = P_c(-\underline{\alpha}_0)$ if they are initially equal. Using M.E. (51) one can show that if $\mathcal{N} \leq 1$ and $N \gg 1$

$$\langle R_3^2\rangle + 2\langle P_c(\underline{\alpha}_0)\rangle = \frac{N^2}{4}. \tag{60}$$

Using (59), eq. (60) can be written as

$$\langle \dot{R}_3\rangle = -\gamma_0\left[\frac{N^2}{4} - \langle R_3^2\rangle\right] - \gamma\left(\frac{N}{2} + \langle R_3\rangle\right) \tag{61}$$

which is the equation discussed in reff. [1] and [4].
In conclusion, if $\mathcal{N} \leq 1$ the equation for $\langle R_3\rangle$ is the same as the one derived in the one-mode model. Let us remark that $\gamma_0 N$ coincides with τ_R^{-1}. Hence, in general, the threshold condition has to be written

$$(\gamma_0 N)^{-1} = \tau_R \ll \tau_0$$

Radiation Pattern

We have been able [15] to obtain the following general expression for the intensity radiated in the $\hat{k}$ direction (number of photons per unit time radiated per unit solid angle around $\hat{k}$):

$$I(\hat{k},t) = \frac{\gamma}{4\pi}\sum_{\underline{\alpha}} F^2\left(\hat{k}\frac{\omega_0}{c} - \underline{\alpha}\right)\langle P_c(\underline{\alpha},t)\rangle + \frac{\gamma}{4\pi}\left(\frac{N}{2} + \langle R_3\rangle\right) \tag{62}$$

If one defines the total intensity $I_T = \int I(\hat{k})\, d\Omega\, (k)$ and one looks at the expression of $\gamma(\underline{\alpha})$ and γ it is easy to see by comparison with (58) that

$$I_T = \sum_{\underline{\alpha}} \gamma(\underline{\alpha})\langle P_c(\underline{\alpha})\rangle + \gamma\left(\frac{N}{2} + \langle R_3\rangle\right) = -\langle \dot{R}_3\rangle \tag{63}$$

This is the simple energy balance which is postulated in reff. [3] and [4], whereas here is deduced in the limit $K_0\tau_R \ll 1$. A more general energy balance is given by eqs. (30) and (31).
Let us remark that $I(\hat{k})$ contains two different contributions:
i. the isotropic term $\gamma/4\pi\,(N/2 + \langle R_3\rangle)$ which is the usual fluorescence radiation proportional to the population of the upper level;
ii. the cooperative term which is made up by a superposition of diffraction patterns $F^2(\hat{k}\frac{\omega_0}{c} - \underline{\alpha})$ centered on $\underline{\alpha}$-modes, each with a

"pound" $<P_c(\underline{\alpha})>$.
Below threshold $(\gamma_0 N < \gamma)$ all $P_c(\underline{\alpha})$ are zero so that the radiation pattern is isotropic. Above threshold and with $\mathcal{N} \leq 1$ the $\sum_{\underline{\alpha}}$ can be reduced to the two dominant axial modes $\underline{\alpha} = \pm \underline{\alpha}_0$ and using eq. (60) one can write

$$I(\hat{k},t) = \frac{1}{2}\frac{\gamma}{4\pi}\left(\frac{N^2}{4} - <R_3>\right)\sum_{\underline{\alpha}=\pm\underline{\alpha}_0} F^2\left(\hat{k}\frac{\omega_0}{c} - \underline{\alpha}\right) + \frac{\gamma}{4\pi}\left(\frac{N}{2} + <R_3>\right)$$

From this expression we see that the angular distribution varies on time in the following way: the radiation pattern is initially isotropic because it comes from the usual non-cooperative spontaneous emission term γ. As the time increases $<R_3>$ decreases from $N/2$ and the cooperative term N^2 becomes relevant. When $<R_3> = 0$ the cooperative term N^2 is dominant and all the Superradiant radiation is condensed into two opposite diffraction lobes centered on the axial direction. The theory can be extended [15] to the case in which $\mathcal{N} = \frac{D}{L} / \frac{\lambda}{D} > 1$. Also in this case eqs. (61), (62) and (63) turn out to be correct. The difference is that in this case the cooperative radiation processes take place with equal intensity in the $\mathcal{N}^2$ diffraction modes $\underline{\alpha}$-contained in the geometrical angle
In conclusion, for $K_0^{-1} \ll \tau_R \ll \tau_0$, the behaviour of R_3 and I_T, as described with the one-mode model of ref.[1], turns out to be correct. Here we have given a first principle justification of this model adding a complete description of radiation pattern and frequency shifts.

APPENDIX A.

Eq. (22) can be derived directly from Hamiltonian (11). In fact the Heisenberg equation of motion for $a_{\underline{k}}$ is:

$$\dot{a}_{\underline{k}} = -i\delta_k a_{\underline{k}} + \frac{\hbar}{\sqrt{V}}\sum_{\underline{\alpha}} g_\alpha(t)\, R^-(\underline{\alpha})\, F(\underline{k}-\underline{\alpha})$$

with $\delta_k = (ck - \omega_0)$ and $g_{\underline{\alpha}} \simeq g_k$
Integrating one gets:

$$a_{\underline{k}}(t) = a_{\underline{k}}(0)\, e^{-i\delta_k t} + \frac{\hbar}{\sqrt{V}}\int_0^t d\tau\, e^{-i\delta_k(t-\tau)}\, g_{\underline{\alpha}}(\tau)\, R^-(\underline{\alpha},\tau)\, F(\underline{k}-\underline{\alpha})$$

Using eq. (14) we obtain

$$A(\underline{\alpha},t) = E(\underline{\alpha},t) + \sum_{\underline{\alpha}'}\int_0^t \Gamma(\underline{\alpha},\underline{\alpha}',t-\tau)\, g_{\underline{\alpha}}(\tau)\, R^-(\underline{\alpha}',\tau)\, d\tau \qquad (A/1)$$

where

$$E(\underline{\alpha},t) = \sqrt{\frac{v}{V}}\sum_{\underline{k}} a_{\underline{k}}(0)\, e^{-i\delta_k t}\, F(\underline{k}-\underline{\alpha})$$

and

$$\Gamma(\underline{\alpha},\underline{\alpha}',\tau) = \frac{v}{(2\pi)^3}\int d^3\underline{k}\, F(\underline{\alpha}-\underline{k})\, F(\underline{k}-\underline{\alpha}')\, e^{-i\delta_k\tau}$$

Let us neglect terms with $\underline{\alpha} \neq \underline{\alpha}'$ which are always much smaller than diagonal terms $\underline{\alpha} = \underline{\alpha}'$. Note that by eq. (13) follows $\Gamma(\underline{\alpha},\underline{\alpha}',0) = 0$ for $\underline{\alpha} \neq \underline{\alpha}'$. Hence eq. (A/1) becomes

$$A(\underline{\alpha},t) = E(\underline{\alpha},t) + \int_0^t \Gamma(\underline{\alpha},t-\tau)\, g_{\underline{\alpha}}(\tau)\, R^-(\underline{\alpha},\tau)\, d\tau \qquad (A/2)$$

where

$$\Gamma(\underline{\alpha},\tau) = \frac{v}{(2\pi)^3}\int d^3\underline{k}\, F^2(\underline{k}-\underline{\alpha})\, e^{-i\delta_k\tau} \qquad (A/3)$$

In order to give an explicit expression of Γ we make the following approximation: since $F^2(\underline{k} - \underline{\alpha})$ is sharply peaked on $\underline{k} = \underline{\alpha}$, we approximate $|\underline{k}|$ in the exponent with the component of $\underline{k}$ along $\hat{\alpha}$

direction, that is

$$|k| \rightarrow \underline{k}\cdot\hat{\alpha} \qquad \left(\hat{\alpha} = \frac{\underline{\alpha}}{|\alpha|}\right) \tag{A/4}$$

Furthermore let us approximate each sinc^2 in F^2 with a Lorentzian. In this way we obtain finally:

$$\Gamma(\underline{\alpha},\tau) = e^{-i(c|\alpha|-\omega_0)\tau}\, e^{-K(\hat{\alpha})\tau} \tag{A/5}$$

where $K(\hat{\alpha})$) is given by (21).
We now assume that the initial state of the field is the vacuum state and we refer to mean values of $A(\underline{\alpha})$ and $a_{\underline{k}}$ operators. In this way eq. (A/1) reduces to:

$$< A(\underline{\alpha}, t)> \int_0^t d\tau\,\Gamma(\underline{\alpha}, t-\tau)\; g_{\underline{\alpha}}(\tau)\; < R^-(\alpha,\tau)>$$

Using the expression (A/5) for and differenciating in time we obtain equation (22).
Looking at Eq. (A/5) we see that the damping coefficient $K(\hat{\alpha})$ can be generally defined in terms of Γ without using the approximation (A/4) and the Lorentzian approximation for the sinc function:

$$[K(\hat{\alpha})]^{-1} = \int_0^{+\infty} d\tau\,\Gamma\left(\hat{\alpha}\,\frac{\omega_0}{c},\tau\right)d\tau$$

where the expression (A/3) for Γ has to be used.

APPENDIX B

Generalized Maxwell-Bloch equations.

In this appendix we shall discuss eq.(17) in the semiclassical approximation. In the following, for reasons of simplicity, we shall indicate the expectation value of any observable by the same symbol which denotes the quantum-mechanical observable.
From Eq. (17), using Eq. (8) we obtain the following equations for mean values:

$$\dot{A}(\underline{\alpha}) = -i(c|\alpha|-\omega_0)A(\underline{\alpha}) - K_0 A(\underline{\alpha}) + \frac{g_{|\alpha|}(t)}{\sqrt{V}}R^-(\underline{\alpha})\,; \quad K_0 = \frac{c}{2L} \tag{B/1}$$

$$\dot{R}^-(\underline{\alpha}) = \frac{2}{\sqrt{V}}\sum_{\underline{\alpha}'} g_{|\alpha'|}(t)\, A(\underline{\alpha}')\, R_3(\underline{\alpha}'-\underline{\alpha}) \tag{B/2}$$

$$\dot{R}_3(\underline{\alpha}) = -\frac{1}{\sqrt{V}}\sum_{\underline{\alpha}'} g_{|\alpha'|}(t)\, A^+(\underline{\alpha}')\, R^-(\underline{\alpha}'-\underline{\alpha}) + \mathrm{c.c} \tag{B/3}$$

Assuming that the duration of the pulse is much larger than the inverse of the carrier frequency ω_0, it is reasonable to write the internal field $\mathcal{E}(\underline{x},t)$ in the following way:

$$\mathcal{E}(\underline{x},t) = e^{ik_0 x}A_R(\underline{x},t) + e^{-ik_0 x}A_L(\underline{x},t) \qquad k_0 = \omega_0/c \tag{B/4}$$

with

$$A_R(\underline{x},t) = \frac{1}{\sqrt{V}}\sum_{\underline{\alpha}>0} e^{i(\underline{\alpha}-\underline{k}_0)\underline{x}}A(\underline{\alpha})$$
$$A_L(\underline{x},t) = \frac{1}{\sqrt{V}}\sum_{\underline{\alpha}<0} e^{i(\underline{\alpha}+\underline{k}_0)\underline{x}}A(\underline{\alpha}) \tag{B/5}$$

where the labels R and L mean "right" and "left" respectively, and $A_{R,L}(x,t)$ are slowly varying in time and space, i.e.

$$\begin{Bmatrix}\partial/\partial_t\\ \partial/\partial_{\underline{x}}\end{Bmatrix}\{A_R, A_L\} \ll \begin{Bmatrix}\omega_0\\ \underline{k}_0\end{Bmatrix}\{A_R, A_L\} \tag{B/6}$$

(B/6) is the usual slowly varying envelope approximation (S.V.E.A.). Similarly, defining the macroscopic polarization field

$$\mathcal{R}^-(\underline{x},t) = \frac{1}{V}\sum_{\underline{\alpha}} e^{i\underline{\alpha}\underline{x}}\, R^-(\underline{\alpha}) \tag{B/7}$$

we can write, in analogy to (B/4),

$$\mathcal{R}^-(\underline{x},t) = e^{i\underline{k}_0\underline{x}}\,\mathcal{R}^-_R(\underline{x},t) + e^{-i\underline{k}_0\underline{x}}\,\mathcal{R}^-_L(\underline{x},t) \tag{B/8}$$

where the envelopes $\mathcal{R}^-_{R,L}(\underline{x},t)$

$$\begin{aligned}\mathcal{R}^-_R(\underline{x},t) &= \frac{1}{V}\sum_{\underline{\alpha}>0} e^{i(\underline{\alpha}-\underline{k}_0)\underline{x}}\, R^-(\underline{\alpha})\\ \mathcal{R}^-_L(\underline{x},t) &= \frac{1}{V}\sum_{\underline{\alpha}<0} e^{i(\underline{\alpha}+\underline{k}_0)\underline{x}}\, R^-(\underline{\alpha})\end{aligned} \tag{B/9}$$

are slowly varying in space and time. Finally let us define a field for inversion of population

$$\mathcal{R}_3(\underline{x},t) = \frac{1}{V}\sum_{\underline{\alpha}} e^{-i\underline{\alpha}\underline{x}}\, R_3(\underline{\alpha}) \tag{B/10}$$

$\mathcal{R}_3(\underline{x},t)$ is slowly varying.
Let us assume that the length L is such that two $\underline{\alpha}$-modes exist satisfying the resonance condition $|\alpha| = \underline{k}_0$. Since $g_{|\alpha|}(t)$ varies slowly with $|\alpha|$, we replace it by the constant value $g(t) = g_{|\alpha|=k_0}(t)$.
Then we derive by easy calculations the following equations:

$$\left(\frac{\partial}{\partial t} + c\,\frac{\partial}{\partial \underline{x}} + K_0\right) A_R(\underline{x},t) = g(t)\,\mathcal{R}^-_R(\underline{x},t) \tag{B/11}$$

$$\left(\frac{\partial}{\partial t} - c\,\frac{\partial}{\partial \underline{x}} + K_0\right) A_L(\underline{x},t) = g(t)\,\mathcal{R}^-_L(\underline{x},t) \tag{B/12}$$

$$\frac{\partial \mathcal{R}^-}{\partial t}(\underline{x},t) = 2g(t)\,\mathcal{E}(\underline{x},t)\,\mathcal{R}_3(\underline{x},t) \tag{B/13}$$

$$\frac{\partial \mathcal{R}_3}{\partial t}(\underline{x},t) = -g(t)\,\mathcal{E}(\underline{x},t)\,\mathcal{R}^+(\underline{x},t) + \text{c.c} \tag{B/14}$$

From (B/11) and (B/12) we see that A_R and A_L propagate in the right and in the left direction respectively.
Using eqs. (B/4), (B/8), (B/13) and (B/14) and neglecting, consistently with the S.V.E.A., terms rapidly varying on a space scale K , one obtains

$$\frac{\partial \mathcal{R}^-_R}{\partial t}(\underline{x},t) = 2g(t)\,A_R(\underline{x},t)\,\mathcal{R}_3(\underline{x},t) \tag{B/15}$$

$$\frac{\partial \mathcal{R}^-_L}{\partial t}(\underline{x},t) = 2g(t)\,A_L(\underline{x},t)\,\mathcal{R}_3(\underline{x},t) \tag{B/16}$$

$$\frac{\partial \mathcal{R}_3}{\partial t}(\underline{x},t) = -g(t)\left\{A_R(\underline{x},t)\,\mathcal{R}^+_R(\underline{x},t) + A_L(\underline{x},t)\,\mathcal{R}^+_L(\underline{x},t) + \text{c.c}\right\} \tag{B/17}$$

We call the set of equations (B/11), (B/12), (B/15), (B/16) and (B/17) the "generalized Maxwell-Bloch equations" (G.M.B.E.) because they take into account both directions of propagation as well as inhomogeneous broadening. Our simple representation of inhomogeneous broadening via a time-dependent coupling constant g(t) clearly contains the assumption that all the atoms start interacting with the field at the same t = 0 regardless for their position. Even if we do not see quite generally which are the limits of validity of this representation, we will not find any inconsistency in using it in the case of a homogeneous preparation of the system and for times shorter or comparable to T_2^*.
The field outside the active volume is the solution of the Maxwell equation in the vacuum which continuously matches the internal

field on the boundary of the active volume. Hence we stress that even when A_R and A_L depend only on time, the external field will be a pulse varying in space and time.
With our initial condition the field and polarization envelopes appearing in the G.M.B.E., as well as $\mathcal{R}_3(\underline{x})$, are initially homogeneous. Due to the structure of these equations, the envelopes remain homogeneous during the time evolution. In such a situation we have from (B/5), (B/8) and (B/10)

$$A_R(\underline{x},t) = \frac{1}{V}\int_V d\underline{x}\; A_R(\underline{x},t) = \frac{1}{\sqrt{V}} A(\underline{k}_0) \tag{B/18}$$

$$A_L(\underline{x},t) = \frac{1}{\sqrt{V}} A(-\underline{k}_0), \quad \mathcal{R}_3(\underline{x},t) = \frac{1}{V} R_3$$

$$\mathcal{R}_R^-(\underline{x},t) = \frac{1}{V} R^-(\underline{k}_0), \quad \mathcal{R}_L^-(\underline{x},t) = \frac{1}{V} R^-(-\underline{k}_0)$$

Then the G.M.B.E. reduce to the following equations for the resonant modes $A(\pm\underline{k}_0)$, $R(\pm\underline{k}_0)$:

$$\dot{A}(\pm\underline{k}_0) = -K_0 A(\pm\underline{k}_0) + \frac{g(t)}{\sqrt{V}} R^-(\pm\underline{k}_0) \tag{B/19}$$

$$\dot{R}^-(\pm\underline{k}_0) = 2\frac{g(t)}{\sqrt{V}} A(\pm\underline{k}_0) R_3$$

$$\dot{R}_3 = -\frac{g(t)}{\sqrt{V}} \left\{ A^+(\underline{k}_0) R^-(\underline{k}_0) + A^+(-\underline{k}_0) R^-(-\underline{k}_0) + c.c \right\}$$

Let us assume for simplicity that $g(t)$ is a Lorentzian:

$$g_1(t) = \exp.\left\{-\frac{|t|}{2T_2^*}\right\} \tag{B/20}$$

$$g(t) = g_0 \exp.\left\{-\frac{t}{2T_2^*}\right\}, \quad g_0 = g_{\underline{k}_0} \tag{B/21}$$

Let us define the total quantities:

$$R_T = \left\{ R^+(\underline{k}_0) R^-(\underline{k}_0) + R^+(-\underline{k}_0) R^-(-\underline{k}_0) \right\}^{1/2} \tag{B/22}$$

$$A_T = \left\{ A^+(\underline{k}_0) A(\underline{k}_0) + A^+(-\underline{k}_0) A(-\underline{k}_0) \right\}^{1/2}$$

From eqs. (B/19) we derive the following closed set of equations in R_T, A_T, R_3:

$$\frac{d}{dt}\left\{ R_T^2 + R_3^2 \right\} = 0 \tag{B/23}$$

$$\frac{d}{dt}\left\{ A_T^2 + R_3 \right\} = -2K_0 A_T^2 \tag{B/24}$$

$$\ddot{R}_3 + \left(K + \frac{1}{2T_2^*}\right)\dot{R}_3 = -\frac{g_0^2}{V} e^{-\frac{t}{T_2^*}} \cdot \left(2R_T^2\right) + 4A_T^2 R_3 \tag{B/25}$$

The interest in the system (B/23, 24, 25) is that all the quantities are phase-independent; the phases can be even assumed at random. With our initial condition, the constant of motion (B/23) has the value

$$R_T^2 + R_3^2 = \frac{N^2}{4} \tag{B/26}$$

This conservation law is characteristic of cooperative radiation processes and in particular of Superradiance. In fact eq. (B/26) shows that if R_3 decays from $N/2$ to $-N/2$, the system develops a macroscopic polarization R_T proportional to N when $R_3 \approx 0$, even if initially $R_T \approx 0$.
By (B/26) we can introduce the Bloch angle $\varphi(t)$ as follows:

$$R_3(t) = \frac{N}{2}\cos\varphi(t); \quad R_T(t) = \frac{N}{2}\sin\varphi(t) \tag{B/27}$$

Substituting eqs. (B/27) into eqs. (B/24) and (B/25) we see that such equations are equivalent to:

$$A_T(t) = \frac{\sqrt{V}}{2g_o}\ \dot{\varphi}(t)\ \exp\left\{\frac{t}{2T_2^*}\right\} \tag{B/28}$$

$$\ddot{\varphi}(t) + \left(K + \frac{1}{2T_2^*}\right)\dot{\varphi}(t) - \frac{g_o^2 N}{V}\ e^{-\frac{t}{T_2^*}} \sin\varphi(t) = 0 \tag{B/29}$$

which are practically the same as eqs. (36) obtained eliminating all non-resonant modes for the M.E. (17).

A few comments are in order.

1. Due to the symmetry in the exchange of $A(k_o)$, $R^-(k_o)$ with $A(-k_o)$, $R^-(-k_o)$ in eqs. (B/19) and in the initial condition, one has

$$A^+(k_o)\,A(\underline{k}_o) = A^+(-\underline{k}_o)\,A(-\underline{k}_o) = \frac{1}{2}A_T^2 \tag{B/30}$$

$$R^+(\underline{k}_o)\,R^-(\underline{k}_o) = R^+(-\underline{k}_o)\,R^-(-\underline{k}_o) = \frac{1}{2}R_T^2$$

Then the left and right diffraction lobes of the radiation output have equal intensity.

2. As we said, eqs. (B/23), (B/24) and (B/25) hold also when the phases of $A(k_o)$, $A(-k_o)$, $R^-(k_o)$, $R^-(-k_o)$ are completely at random. In such a case the radiation field has a vanishing mean value. As we shall see in the fully quantum mechanical treatment of ref.[11], our initial condition prescribes precisely that these phases are at random, giving therefore a vanishing field at all times. In fact, the density operator for the radiation field turns out to be diagonal in the photon number representation.
3. Our initial condition corresponds to $\varphi(0) = \dot{\varphi}(0) = 0$; i.e. to the unstable equilibrium point of the pendulum. This would imply that the system does not radiate. This drawback is due to the fact that our semiclassical treatment does not take into account the quantum noise which initially starts the pendulum. Due to the quantum noise, the correct initial condition is

$$\varphi(0) = \frac{2}{\sqrt{N}}\quad,\quad \dot{\varphi}(0) = 0 \tag{B/31}$$

The quantum treatment modifies the present classical analysis associating to the initial fully excited state a quantum noise polarization $R_T = \sqrt{N}$, which leads to $\varphi(0) = (2/\sqrt{N})$ (see eq. (37) in the text). Therefore the Bloch vector is no more pointing exactly to the north pole, but is very near to it; this slight displacement from the north pole makes the pendulum to move in a finite time.

In conclusion the motion of the Bloch vector during the superradiant decay is the following. It moves on a sphere of radius N/2 and its polar angle φ obeys a pendulum equation. On the other hand, since we do not have any information on the phase of the initial noise polarization, the azimuthal angle is completely random giving $\langle R^{\pm}(\pm\underline{k}_o)\rangle = 0$ (as well as one has $\langle A^{\pm}(\pm\underline{k}_o)\rangle = 0$). This randomness of the phases is one of the most characteristic features of Superradiance by a fully excited state.

ACKNOWLEDGEMENTS

Most of the material of these lectures has been worked out while I was visiting the University of Stuttgart in the Academic Years 1972 and 1973 as a guest of Professors Haken and Weidlich. I am deeply indebited to them and to Doctors Haake and Schwendimann for very helpful and stimulating discussions.

REFERENCES

*. Work supported by the Deutsche Forshungsgemeinshaft (Richard Merton Funds)

1. R. Bonifacio, P. Schwendimann and F. Haake: Phys. Rev. A 4 (1971) 302 and Phys. Rev. A 4 (1971) 854. See also Haake's lectures in this volume.
2. N. Skribanowitz et al. Phys. Rev. Lett. 30 (1973) 309 and Proc. of the Vail Conference on Laser Spectroscopy, 1973 - Plenum Publishing Co.
3. R.H. Dicke, in Proc. of the Third International Conference on Quantum Electronics, Paris, 1963 - Columbia University Press, N.Y. 1964, p. 35 - R.H. Dicke, Phys. Rev. 93, (1954) 99.
4. N. Rehler, J.H. Eberly, Phys. Rev. A 3 (1971) 1735.
5. R.H. Lehmberg, Phys. Rev. A 2 (1970) 883.
6. G.S. Agarwal in Proc. Third Rochester Conference on Coherence and Quantum Optics, Edited by L. Mandel and E. Wolf - Plenum Publishing Co.(1973)
7. E.T. Jaynes, in Proc. Third Rochester Conference on Coherence and Quantum Optics, Edited by L. Mandel and E. Wolf - Plenum Publishing Co. (1973)
8. C.R. Stroud Jr. and J.H. Eberly; W.L. Lama and L. Mandel; Phys. Rev. A 5 (1972), 1094.
9. E. Ressayre and A. Tallet, Phys. Rev. Lett. 30 (1973) 1239.
10. H. Haken, in Flügge, S. (Ed.) Laser Theory, Vol. XXV/2C - New York - Springer 1970 - Hadbuch der Physik.
11. R. Bonifacio and L.A. Lugiato, in preparation.
12. R. Bonifacio and G. Preparata, Phys. Rev. A 2 (1970) 336.
13. F.T. Arecchi and E. Courtens, Phys. Rev. A 2 (1970) 1730.
14. A. Airoldi Crescentini and R. Bonifacio: in preparation.
15. G. Banfi and R. Bonifacio, in preparation.

Cooperative Phenomena, H. Haken, ed.

SOME APPLICATIONS OF THE ATOMIC COHERENT STATE REPRESENTATION TO SUPERRADIANCE AND STOCHASTIC ATOMIC PROCESSES

Lorenzo M. Narducci

Physics Department, Worcester Polytechnic Institute,
Worcester, Massachusetts, USA

INTRODUCTION

Our objective is to discuss a number of problems of current interest in Quantum Optics in the framework of the continuous atomic coherent state representation (ACGT representation). The main results of our calculations include:

a) The construction of a Fokker-Planck equation for the quasi-probability function associated with the atomic density operator of a supperradiant system,

b) the mapping of atomic correlation functions into integral forms over the phase-space of the atomic variables.

This work was originally motivated by the observation that a close formal similarity exists between the Glauber coherent state representation of the field (Glauber [1]) and the ACGT representation for atomic systems (Arecchi [2]). The field coherent states have been used quite successfully in connection with a large number of problems involving electromagnetic field operators. Precise rules of correspondence have been developed to map operator equations into c-number differential equations (Lax [3]) and field operator correlation functions into integral form bearing a striking resemblance to the classical correlation functions of Markoff systems (Bonifacio [4]).

It is natural to inquire whether or not analogous rules of correspondence can be found in the context of the ACGT representation, when dealing with observables obeying the angular momentum algebra [5]. In this work we give evidence that, indeed, the close similarity between the Glauber and the ACGT representation goes beyond the formal algebraic structure of the formalism.

In order to proceed in a systematic way, we first develop a set of rules of correspondence to map operators and operator equations into corresponding c-number forms (for want of a better name we refer to these rules of mapping as the $\mathcal{D}$ -operator calculus). Next we apply our formalism to the master-equation of Agarwal [6,7], and Bonifacio et al [8,9] and derive a Fokker-Planck equation evolving in the phase-space of the atomic variables (Narducci [10]). Finally, we analyze a class of correlation functions for angular momentum operators and construct the corresponding classical integral representations in terms of phase-space density functions.

Examples taken from the well known theory of the irreversible relaxation of a two-level system interacting with a thermal reservoir will be given as illustrations.

1. THE $\mathcal{D}$-OPERATOR CALCULUS

Following the notation of Ref. [2] we define the atomic coherent states $|\Omega\rangle$ as follows

$$|\Omega\rangle = \sum_{m=-J}^{J} |J,m\rangle \binom{2J}{m+J}^{1/2} (\sin\theta/2)^{J+m} (\cos\theta/2)^{J-m} e^{-i(J+m)\phi} , \quad (1.1)$$

where θ and ϕ are the angular parameters in a spherical coordinate system (the Bloch sphere), and where $\theta = 0$ corresponds to the south pole of the sphere.

The atomic coherent states defined by Eq. (1.1) form an overcomplete set in the (2J+1)-dimensional subspace of the angular momentum with respect to the element of measure

$$d\mu(\Omega) = \frac{2J+1}{4\pi} d\Omega = \frac{2J+1}{4\pi} \sin\theta \, d\theta \, d\phi \quad , \qquad (1.2)$$

in the sense that the identity operator in this subspace can be resolved as follows

$$\mathbb{1} = \frac{2J+1}{4\pi}\int d\Omega \; |\Omega\rangle\langle\Omega| = \sum_{m=-J}^{J} |J,m\rangle\langle J,m| \quad . \qquad (1.3)$$

In Eqs. (1.1) and (1.3) the states $|J,m\rangle$, ($|m|\leq J$) are eigenstates of the z component of the angular momentum operator. They can also represent collective states of two-level atoms in the Dicke representation [11]. No distinction will be made in the following between the eigenstates of the angular momentum and the Dicke states since they obey the same formal algebra.

As already mentioned in Professor Arecchi's lectures, the atomic coherent states are normalized to unity but not orthogonal; the inner product of two such states is given by

$$|\langle\Omega|\Omega'\rangle|^2 = (\cos \tfrac{1}{2}\Theta)^{4J} \quad , \qquad (1.4)$$

where $\cos\Theta = \cos\theta \cos\theta' + \sin\theta \sin\theta' \cos(\phi - \phi')$.

For our purposes, the most significant feature of the ACGT representation is the existence of the diagonal representation

$$G = \int d\Omega\, g(\Omega)\; |\Omega\rangle\langle\Omega| \quad , \tag{1.5}$$

where, as shown in Ref. [2], the weighting function $g(\Omega)$ can always be expressed as a linear superposition of at most $(2J+1)^2$ spherical harmonic functions.

It has been pointed out that the specification of the weighting function $g(\Omega)$ is not unique for a given operator G. This turns out to be an advantage for certain applications. For example, if G happens to be the density operator of a given pure state $|\Omega'\rangle$, it may be convenient to represent $g(\Omega)$ as a singular function $\delta(\Omega-\Omega')$ on the surface of the Bloch sphere rather than as a well behaved linear superposition of spherical harmonics. We shall take advantage of this flexibility in connection with the mapping of multi-time correlation functions into integral form.

For the applications to be discussed in the present work, it will be necessary to deal with operators of the form

$$B_n \ldots B_1\; |\Omega\rangle\langle\Omega|\; A_1 \ldots A_m \tag{1.6}$$

where A_i and B_j are operators in the 2J+1-dimensional subspace of the angular momentum. Our immediate goal is to establish the existence of a differential operator $\mathcal{D}(\Omega)$ acting on the angular variables θ and ϕ, such that the following identity holds

$$B_n \cdots B_1 |\Omega\rangle\langle\Omega| A_1 \ldots A_m = \mathcal{D}(\Omega)\; |\Omega\rangle\langle\Omega| \tag{1.7}$$

For this purpose, consider the following explicit representation of the atomic state projector $\Lambda(\Omega)$

$$\Lambda(\Omega) = \sum_{p,q=0}^{2J} \binom{2J}{p}^{1/2} \binom{2J}{q}^{1/2} e^{-i(p-q)\phi}\, (\sin\theta/2)^{p+q}\, (\cos\theta/2)^{4J-(p+q)} \times$$

$$\times\, |J,p\rangle\, \langle J,q| \equiv \sum_{p,q=0}^{2J} \Gamma_{p,q}(\Omega) \quad , \tag{1.8}$$

and of the first order derivatives

$$\frac{\partial\Lambda}{\partial\theta} = -2J \text{ tg } \theta/2 \, \Lambda \, (\Omega) + \frac{1}{\sin\theta} \sum_{p,q} (p+q) \, \Gamma_{p,q}(\Omega) \tag{1.9}$$

$$\frac{\partial\Lambda}{\partial\phi} = -i \sum_{p,q} (p-q) \, \Gamma_{p,q}(\Omega) \tag{1.10}$$

The operators

$$\Sigma_p(\Omega) \equiv \sum_{p,q} p \, \Gamma_{p,q}(\Omega)$$

$$\Sigma_q(\Omega) \equiv \sum_{p,q} q \, \Gamma_{p,q}(\Omega) = \Sigma_p^{+}(\Omega) \tag{1.11}$$

can be expressed in terms of $\Lambda(\Omega)$ and its first order derivatives using Eqs. (1.9) and (1.10). One finds at once

$$\Sigma_p(\Omega) = \frac{1}{2} \sin\theta \frac{\partial\Lambda}{\partial\theta} + J(1-\cos\theta) + \frac{i}{2} \frac{\partial\Lambda}{\partial\phi} \tag{1.12}$$

$$\Sigma_q(\Omega) = \frac{1}{2} \sin\theta \frac{\partial\Lambda}{\partial\theta} + J(1-\cos\theta) - \frac{i}{2} \frac{\partial\Lambda}{\partial\phi} \quad . \tag{1.13}$$

Consider now the special case of Eq. (1.7) with $B_1 = J^+$, and $B_2 = \ldots = B_n = A_1 = \ldots = A_m = 1$. From the algebraic properties of the step operator J^+, we find

$$J^+ |\Omega\rangle\langle\Omega| = \text{cotg } \theta/2 \; e^{i\phi} \, \Sigma_p(\Omega) \quad , \tag{1.14}$$

whence it follows that

$$\begin{aligned} J^+ |\Omega\rangle\langle\Omega| &= e^{i\phi}\left(J \sin\theta + \cos^2 \theta/2 \, \frac{\partial}{\partial\theta} + \frac{i}{2} \text{ cotg } \theta/2 \, \frac{\partial}{\partial\phi}\right)\Lambda(\Omega) \\ &\equiv \mathcal{D}_{J^+}(\Omega) \, \Lambda(\Omega) \quad . \end{aligned} \tag{1.15}$$

A similar calculation leads to the explicit representation of the differential operators $\mathcal{D}_{J^-}$ and $\mathcal{D}_{J_3}$, namely

$$J^- \, |\Omega\rangle\langle\Omega| = 2J \, \mathrm{tg}\, \theta/2 \, e^{-i\phi} \, \Lambda(\Omega) - \mathrm{tg}\,\theta/2 \, \sum_p (\Omega) \, e^{-i\phi}$$

$$= e^{-i\phi} \left(J \sin\theta - \sin^2 \theta/2 \frac{\partial}{\partial\theta} - \frac{i}{2} \,\mathrm{tg}\,\theta/2 \frac{\partial}{\partial\phi}\right) \Lambda(\Omega) \tag{1.16}$$

$$= \mathcal{D}_{J^-}(\Omega) \, \Lambda(\Omega) \quad ,$$

and

$$J_3 \, |\Omega\rangle\langle\Omega| = -J \, \Lambda(\Omega) + \sum_p (\Omega)$$

$$= \left(-J \cos\theta + \frac{1}{2} \sin\theta \frac{\partial}{\partial\theta} + \frac{i}{2} \frac{\partial}{\partial\phi}\right) \Lambda(\Omega) \tag{1.17}$$

$$\equiv \mathcal{D}_{J_3}(\Omega) \, \Lambda(\Omega) \quad .$$

We shall need also the differential operators corresponding to $\Lambda(\Omega)J^\pm$ and $\Lambda(\Omega)J_3$. These can be calculated from Eqs. (1.8) and (1.11) using the algebraic properties of the angular momentum operators. A more direct approach is based on the identities

$$\Lambda(\Omega)J^\pm = \left[J^\pm \, \Lambda(\Omega)\right]^+ = \mathcal{D}^*_{J^\mp}(\Omega) \, \Lambda(\Omega) \tag{1.18}$$

$$\Lambda(\Omega)J_3 = \left[J_3 \, \Lambda(\Omega)\right]^+ = \mathcal{D}^*_{J_3}(\Omega) \, \Lambda(\Omega) \tag{1.19}$$

which follow from the Hermitian character of the projector $\Lambda(\Omega)$. The operators $\mathcal{D}^*_{J^\pm}$ and $\mathcal{D}^*_{J_3}$ are the complex conjugate of the $\mathcal{D}$ operators defined in Eqs. (1.15), (1.16), and (1.17).

It should be clear by now how to deal with more complicated expressions such as $B_2 B_1 \Lambda$ and $\Lambda A_1 A_2$. By successive transformations of the A and B operators into the corresponding differential forms we can establish the identities

$$B_2 B_1 \Lambda(\Omega) = \mathcal{D}_{B_1} \mathcal{D}_{B_2} \Lambda(\Omega)$$

$$\Lambda(\Omega) A_1 A_2 = \mathcal{D}^*_{A_1^+} \mathcal{D}^*_{A_2} \Lambda(\Omega) \quad . \tag{1.20}$$

It remains to consider operators of the form $B\Lambda(\Omega)A$. At first sight there seems to be some ambiguity in the form of the equivalent differential operator; we can operate with B first, and arrive at the result

$$B\Lambda(\Omega)A = \mathcal{D}_B \mathcal{D}^*_{A^+} \Lambda(\Omega) \quad .$$

On the other hand, if we operate with A first, we find

$$B\Lambda(\Omega)A = \mathcal{D}^*_{A^+}\mathcal{D}_B \Lambda(\Omega) \quad .$$

It is, however, a simple matter to verify that the above expressions are identical to one another.

The explicit calculation of the $\mathcal{D}$ operators for the applications to follow involves a substantial amount of algebraic manipulations. For convenience and ready reference, we list below the few differential forms of direct interest in this work:

$$J^+J^-\Lambda(\Omega) = \mathcal{D}_{J^-} \mathcal{D}_{J^+}\Lambda(\Omega)$$

$$= \left[J^2\sin^2\theta + \frac{J}{2}(1-\cos\theta)^2\right]\Lambda(\Omega) + \left[J\sin\theta\cos\theta + \frac{1}{4}\sin\theta(2-\cos\theta)\right]\frac{\partial\Lambda}{\partial\theta}$$

$$+ i\left(J\cos\theta + \frac{1}{2}\right)\frac{\partial\Lambda}{\partial\phi} - \frac{i}{2}\sin\theta\frac{\partial^2\Lambda}{\partial\theta\partial\phi} + \frac{1}{4}\frac{\partial^2\Lambda}{\partial\phi^2} - \frac{1}{4}\sin^2\theta\frac{\partial^2\Lambda}{\partial\theta^2} \quad , \qquad (1.21)$$

$$J^-J^+\Lambda(\Omega) = \mathcal{D}_{J^+} \mathcal{D}_{J^-}\Lambda(\Omega)$$

$$= \left[J^2\sin^2\theta + \frac{J}{2}(1+\cos\theta)^2\right]\Lambda(\Omega) + \left[J\sin\theta\cos\theta - \frac{1}{4}\sin\theta(2+\cos\theta)\right]\times$$

$$\frac{\partial\Lambda}{\partial\theta} + i\left(J\cos\theta - \frac{1}{2}\right)\frac{\partial\Lambda}{\partial\phi} - \frac{i}{2}\sin\theta\frac{\partial^2\Lambda}{\partial\theta\partial\phi} + \frac{1}{4}\frac{\partial^2\Lambda}{\partial\phi^2} - \frac{1}{4}\sin^2\theta\frac{\partial^2\Lambda}{\partial\theta^2} \quad , \qquad (1.22)$$

$$\Lambda(\Omega)J^+J^- = \mathcal{D}^*_{J^-}\mathcal{D}^*_{J^+}\Lambda(\Omega) \quad , \qquad (1.23)$$

$$\Lambda(\Omega)J^-J^+ = \mathcal{D}^*_{J^+}\mathcal{D}^*_{J^-}\Lambda(\Omega) \quad , \qquad (1.24)$$

$$J^- \Lambda(\Omega) J^+ = \mathcal{D}_{J^-} \mathcal{D}^*_{J^-} \Lambda(\Omega) = \mathcal{D}^*_{J^-} \mathcal{D}_{J^-} \Lambda(\Omega)$$

$$= \left[J^2 \sin^2\theta + \frac{J}{2} (1-\cos\theta)^2 \right] \Lambda(\Omega)$$

$$+ \left[-J\sin\theta \; (1-\cos\theta) + \frac{1}{4} \operatorname{cotg}\theta \; (1-\cos\theta \right] \frac{\partial\Lambda}{\partial\theta}$$

$$+ \frac{1}{4} (1-\cos\theta)^2 \frac{\partial^2\Lambda}{\partial\theta^2} + \frac{1}{4} \operatorname{tg}^2 \theta/2 \frac{\partial^2\Lambda}{\partial\phi^2} \quad , \qquad (1.25)$$

$$J^+ \Lambda(\Omega) J^- = \mathcal{D}_{J^+} \mathcal{D}^*_{J^+} \Lambda(\Omega) = \mathcal{D}^*_{J^+} \mathcal{D}_{J^+} \Lambda(\Omega)$$

$$= \left[J^2 \sin^2\theta + \frac{J}{2} (1+\cos\theta)^2 \right] \Lambda(\Omega)$$

$$+ \left[J\sin\theta \; (1+\cos\theta) + \frac{1}{4} \operatorname{cotg}\theta \; (1+\cos\theta)^2 \right] \frac{\partial\Lambda}{\partial\theta}$$

$$+ \frac{1}{4} (1+\cos\theta)^2 \frac{\partial^2\Lambda}{\partial\theta^2} + \frac{1}{4} \operatorname{cotg}^2 \theta/2 \frac{\partial^2\Lambda}{\partial\phi^2} \quad . \qquad (1.26)$$

2. FOKKER-PLANCK EQUATION FOR A MODEL OF SUPERRADIANCE

The cooperative radiative emission from a collection of excited atoms has received considerable attention since the original analysis by Dicke [11] was published in 1954. Some of the recent developments in the theory of superradiance have been reviewed by Bonifacio and Haake, and papers on this subject can be found in this volume. Of particular interest to the present work is the superradiant master-equation derived independently by Agarwal [6,7] and by Bonifacio et al, [8,9].

Bonifacio, Schwendimann and Haake, in particular, have analyzed the dynamics of superradiant emission from a collection of identical two-level systems in a low-Q, pencil-shaped cavity. Under precise conditions (the so-called superradiance limit), and in the Markoff approximation, the superradiant master-equation has been shown to be of the form

$$\dot{W}(t) = 1/2 \; I_1 \left\{ [J^-, W(t)J^+] + [J^-W(t), J^+] \right\} . \qquad (2.1)$$

The operators J^+ and J^- are collective atomic raising and lowering operators which, together with the energy operator J_3, satisfy the angular momentum algebra

$$[J^+, J^-] = 2J_3 \quad , \quad [J_3, J^{\pm}] = \pm J^{\pm} \quad , \tag{2.2}$$

and W is the reduced atomic density operator. The parameter I_1 represents the radiated intensity of a single atom into the diffraction solid angle of the end-fire mode. In the following development this constant will be eliminated by a suitable scaling of the time variable.

It is a simple matter to verify that the arbitrary solution of the superradiant master-equation (2.1) satisfies the conditions

i) $\mathrm{tr}W(t) = 1$ (conservation of probability),

ii) $\mathrm{tr}[J^2\, W(t)] = J^2(0)$ (conservation of the cooperation number).

The conserved nature of the cooperation number is of particular importance for our calculation since it allows the consideration of a continuous representation in the angular momentum subspace of fixed cooperation number.

We are interested in the evolution of the quasi-probability density $P(\Omega,t)$ defined by the diagonal representation

$$W(t) = \int d\Omega P(\Omega,t)\,|\Omega\rangle\langle\Omega| \quad . \tag{2.3}$$

A differential equation for $P(\Omega,t)$ can be constructed as follows. We first substitute Eq. (2.3) in Eq. (2.1) with result

$$\int d\Omega\, \frac{\partial P}{\partial t}\, \Lambda(\Omega) = \int d\Omega\, P(\Omega,t)\Big[J^-\Lambda J^+ - \frac{1}{2}\Lambda(\Omega)J^+J^- - \frac{1}{2}J^+J^-\Lambda(\Omega)\Big] \quad . \tag{2.4}$$

Next, we replace the operators appearing on the right hand side of Eq. (2.4) by the equivalent expressions in terms of the $\mathcal{D}$ operators as shown in Section 1. After a few minor manipulations we find

$$\int d\theta d\phi\, \frac{\partial Q}{\partial t}\, \Lambda(\Omega) = \int d\theta d\phi\, Q(\theta,\phi,t)\Big[\Big(-J\sin\theta - \frac{1-\cos\theta}{2\sin\theta}\Big)\frac{\partial\Lambda}{\partial\theta} + \frac{1}{2}(1-\cos\theta)\frac{\partial^2\Lambda}{\partial\theta^2} - \frac{\cos\theta}{2(1+\cos\theta)}\frac{\partial^2\Lambda}{\partial\phi^2}\Big] \quad , \tag{2.5}$$

where we have defined the new function $Q(\theta,\phi,t)$ as

$$Q(\theta\ ,\phi,t) = \sin\theta\ P(\theta,\phi,t) \quad . \tag{2.6}$$

If we now integrate the left hand side of Eq. (2.5) by parts, and observe that the sum of the surface terms vanishes identically, we obtain the following equation of motion for $Q(\theta,\phi,t)$

$$\frac{\partial Q}{\partial t} = \frac{\partial}{\partial\theta}\left[\left(J\sin\theta + \frac{1-\cos\theta}{2\sin\theta}\right)Q\right] + \frac{\partial^2}{\partial\theta^2}\left[\frac{1-\cos\theta}{2}Q\right] - \frac{\partial^2}{\partial\phi^2}\left[\frac{\cos\theta}{2(1+\cos\theta)}Q\right] \quad . \tag{2.7}$$

For most situations of interest, it will be sufficient to analyze the behavior of the reduced density function

$$Q(\theta,t) = \int_0^{2\pi} d\phi\ Q(\theta,\phi,t)$$

which is solution of the equation of motion

$$\frac{\partial Q}{\partial t} = \frac{\partial}{\partial\theta}\left[\left(J\sin\theta + \frac{1-\cos\theta}{2\sin\theta}\right)Q\right] + \frac{\partial^2}{\partial\theta^2}\left[\frac{1-\cos\theta}{2}Q\right] \quad . \tag{2.8}$$

Equation (2.8) is a Fokker-Planck equation describing the evolution of the atomic density function on the surface of the Bloch sphere. The most significant qualitative features of the evolution of $Q(\theta,t)$ can be derived by inspection of the drift coefficient

$$A(\theta) = -J\sin\theta - \frac{1-\cos\theta}{2\sin\theta} \quad . \tag{2.9}$$

We notice that $A(\theta)$ is made up of two contributions: the first term is proportional to the cooperation number J and represents the dominant contribution over most of the range of variation of θ; the second is of the order of unity everywhere, except in the vicinity of $\theta = \pi$. This value of the angular variable corresponds to an atomic configuration of complete inversion.

Since the first term of the drift coefficient becomes vanishingly small, while the second diverges in the vicinity of $\theta = \pi$, we interpret the second contribution as the source of spontaneous decay. It becomes especially important whenever the initial atomic configuration is described by a density function which is non-zero only in a small region about the north pole of the Bloch sphere between the angle $\theta = \pi$ and some angle θ_1 such that

$$-J \sin\theta_1 \simeq \frac{1-\cos\theta_1}{2\sin\theta_1} \quad .$$

The drift coefficient is responsible for the overall displacement of the probability density toward the south pole of the sphere, as well as for a complicated dispersion. The latter effect results from the explicit θ-dependence of $A(\theta)$ (i.e. different sections of the density function $Q(\theta,t)$ are driven away from the initial configuration at different rates).

In addition, the density function is affected by an angle-dependent diffusion coefficient. The relative importance of the drift and diffusion contributions can be assessed qualitatively from the analysis of the short time behavior of $Q(\theta,t)$ (which can be obtained by replacing $\partial Q/\partial t$ with the incremental ratio $[Q(\theta,t)-Q(\theta,0)]/t$).

The conclusion is that, for initial density functions peaked about $\theta_0 \not\approx \pi$, both the diffusion term and the spontaneous emission contribution to the drift term are negligible.

On the contrary, if $\theta_0 \approx \pi$ both terms contribute significantly to the evolution.

In Section VII of Ref. [10] an exact reduction of the Fokker-Planck equation to quadratures has been reported. Unfortunately, we have not been able to carry out a satisfactory asymptotic expansion of the integral representation based on the large parameter J, so that, at present, the solution appears to have only a formal value.

Here we analyze the simpler case when the initial density function $Q(\theta,0)$ is distributed about a value of θ_0 which is sufficiently removed from π. In view of our comments on the relative importance of the various terms in Eq. (2.8), the Fokker-Planck equation is, to order $1/J$,

$$\frac{\partial Q}{\partial t} = \frac{\partial}{\partial \theta}\left[J \sin\theta \, Q(\theta,t)\right] \tag{2.10}$$

A convenient procedure for solving Eq. (2.10) with the arbitrary initial condition

$$Q(\theta,0) = f(\theta) \tag{2.11}$$

is the method of characteristics. The result of the simple calculation is

$$Q(\theta,t) = [\cosh Jt - \cos\theta \sinh Jt]^{-1} \, f[2tg^{-1} \, (e^{Jt} \, tg\theta/2)] \quad , \tag{2.12}$$

which satisfies the normalization condition

$$\int_0^\pi d\theta\, \varrho(\theta,t) = 1$$

for all time. Notice that if $f(\theta) = \varrho(\theta,0)$ is zero in a small neighborhood of $\theta = \pi$ at time $t = 0$, then Eq. (2.12) shows that this remains true for all time (this excludes the possibility of inconsistencies arising from the solution "spilling over" in the critical region around the north pole of the Bloch sphere).

For initial states of excitation corresponding to a given value $\theta = \theta_0$ we can conveniently choose for $f(\theta)$ the following distribution

$$f(\theta) = \sin\theta\, \delta\, (\cos\theta - \cos\theta_0) \quad , \tag{2.13}$$

whence the solution given by Eq. (2.13) can be cast into the convient form

$$\varrho(\theta,t) = \sin\theta\, \delta \left(\cos\theta - \frac{\cos\theta_0\ \cosh Jt + \sinh Jt}{\cos\theta_0\ \sinh Jt + \cosh Jt}\right) \quad . \tag{2.14}$$

The time dependent expectation values of the collective atomic operators can be readily calculated from the integral representation

$$\langle (J^+)^\ell (J_3)^n (J^-)^{\ell'} \rangle = \int d\Omega P(\Omega,t)\ \langle\Omega|\ (J^+)^\ell (J_3)^n (J^-)^{\ell'}\ |\Omega\rangle \tag{2.15}$$

which is a consequence of the diagonal expansion (2.3). The matrix elements in the integral of Eq. (2.15) are obtained by repeated differentiation of the moment generating function [2]

$$\begin{aligned} X(\alpha,\beta,\gamma) &= \langle\Omega|\ e^{\alpha J^+} e^{\beta J_3} e^{\gamma J^-}\ |\Omega\rangle \\ &= \Big[e^{\beta/2} \sin^2\theta/2 + e^{-\beta/2}\ (\alpha e^{i\phi} \sin\theta/2 + \cos\theta/2) \\ &\qquad (\gamma e^{-i\phi} \sin\theta/2 + \cos\theta/2)\Big]^{2J} \quad . \end{aligned} \tag{2.16}$$

For large values of J and not too large values of ℓ,ℓ', and n, the diagonal matrix elements can be accurately approximated by their leading term. To order $J^{2\ell+n-1}$, we have

$$\langle\Omega|\ (J^+)^\ell (J_3)^n (J^-)^\ell\ |\Omega\rangle \simeq (J\sin\theta)^{2\ell}\ (-J\cos\theta)^n \tag{2.17}$$

(the moments corresponding to $\ell \neq \ell'$ vanish identically if the quasi-probability deunity is ϕ-independent). The integral (2.15) can be easily carried out using the time dependent solution (2.14). The arbitrary moments of the collective atomic operators are given by

$$\langle (J^+)^\ell (J_3)^n (J^-)^\ell \rangle = [J \text{ sech } J(t+\hat{t})]^{2\ell} [-J \tanh J(t+\hat{t})]^n \quad , \quad (2.18)$$

where we have set

$$J\hat{t} = \text{arc cosh } [(\sin\theta_0)^{-1}] \quad .$$

We notice that the same result was obtained by Haake and Glauber [12] using an entirely different procedure. (See also Ref. [10] for an alternative procedure based on the hierarchy of coupled equations of motion for the time-dependent moments

$$R_{2\ell,n}(t) \equiv \frac{\langle (J^+)^\ell (J_3)^n (J^-)^\ell \rangle}{J^{2\ell+n}} \;).$$

3. MULTI-TIME CORRELATION FUNCTIONS FOR COLLECTIVE ATOMIC OPERATORS

The statistical behavior of irreversible processes has been the subject of considerable attention in Quantum Optics. Detailed treatments of a variety of damping models exist in the literature and excellent reviews have been prepared (Haken [13], Haake [14]). In these notes we would like to discuss the application of the ACGT representation to a special class of stochastic quantum systems. To be specific we direct our attention to the description of a system S interacting with a reservoir B. The system will be assumed to be characterized by angular momentum observables and to evolve as a Markoff system. The uninitiated reader may find a comprehensive survey of the background information in Ref. [14].

We begin our analysis with the Nakajima [15]-Zwanzig [16] master-equation for the reduced density operator of the system S

$$\dot{W}(t) = -i \, \text{tr}_B(LBW(t)) + \int_0^t ds \; K(t-s)W(s) \quad . \quad (3.1)$$

In Eq. (3.1), L represents the total Liouvillian of S+B, B is the canonical density operator of the reservoir, and K(t) is an integral kernel given by

$$K(t) = -\text{tr}_B \, L \exp(-i(1-P)Lt)(1-P)LB \quad . \quad (3.2)$$

The operator P projects the total density operator into its relevant part according to the usual prescription, and the traces in Eqs. (3.1) and (3.2) are taken over the complete set of reservoir variables.

Our objective is to direct the reader's attention to the key elements appearing in the calculation of the multi-time correlation functions for system operators (a derivation is outside the scope of our presentation and can be found in Ref. [14]), and to provide an alternative procedure for the evaluation of these functions which bears a strong resemblance to the classical theory of Markoff processes.

For this purpose it will not be necessary to consider explicit solutions of the generalized master-equation. We point out, however, that a formal solution of Eq. (3.1) can be provided in terms of a non-unitary time-translation operator V such

$$W(t') = V(t',t)W(t) \quad . \qquad (3.3)$$

The operator V is the formal solution of the equation of motion

$$\dot{V}(t) = -i\ tr_B LBV(t) + \int_0^t ds\ K(s)\ V(t-s) \quad , \qquad (3.4)$$

subject to the constraint

$$V(0) = 1 \quad , \qquad (3.5)$$

where 1 is the appropriate identity operator in the Hilbert space of the system S. Notice that, strictly speaking, $V(t)$ is not an operator in the Hilbert space of the system but, rather, is a dynamical rule of correspondence that translates operators of the system from some initial time to some later time in an irreversible fashion.

We now consider the time-ordered correlation function

$$\begin{aligned} K_n(t',t) &= \langle A_1(t_1')\ldots A_n(t_n')B_n(t_n)\ldots B_1(t_1)\rangle \\ &= tr_{S+B}(B_n(t_n)\ldots B_1(t_1)\rho(t_1)A_1(t_1')\ldots A_n(t_n')) \quad , \end{aligned} \qquad (3.6)$$

where for simplicity we restrict ourselves to an equal number of A and B operators and where the time sequence is indicated below

$$t_n' \geq t_n \geq \cdots \geq t_1' \geq t_1 \quad . \qquad (3.7)$$

Other time sequences can be chosen: the one given above is adequate for our purposes. The trace in Eq. (3.6) is evaluated in the Heisenberg picture where the operators A and B are explicitly time-dependent and the density operator $\rho(t)$ of the whole system S+B is specified at the earliest time in the time sequence (3.7) and, of course, is time-independent.

It would be an easy matter to continue our considerations with the arbitrary correlation functions $K_n(t',t)$; for convenience, however, we confine our calculations to $K_2(t',t)$ from which most correlation functions of interest can be derived.

In the Markoff approximation, $K_2(t',t)$ is explicitly given by

$$K_2(t',t) = \mathrm{tr}_S \left\{ V(t_2',t_2)[B_2\ V(t_2,t_1')[V(t_1',t_1)[B_1\ W(t_1)]A_1]]A_2 \right\} \quad , \tag{3.8}$$

where $W(t_1)$ is the reduced density operator at the earliest time in the chosen time sequence, and V is the time development operator. In Eq. (3.8) it must be understood that each V acts only on the operators contained in the brackets that immediately follow it.

We can provide a qualitative interpretation of Eq. (3.8) as follows: the measurement of B_1 is performed first at time t_1, with the system in a state described by the density operator $W(t_1)$; $V(t_1',t_1)$ then translates the product $B_1W(t_1)$ from t_1 to t_1' according to the irreversible behavior described by the master-equation; the next measurement A_1 is performed, and the system is successively allowed to undergo its irreversible motion from t_1' to t_2, etc. After the entire set of measurements is completed, one calculates the trace over the dynamical variables of the system S.

It should be clear that, once the effect of the irreversible evolution is accounted for, the calculation of an arbitrary multi-time correlation function becomes a matter of algebraic manipulation. The problem, of course, is the implementation of the formal time translation effected by the operator V.

If the operators A and B are such that the rules of the $\mathcal{D}$-operator calculus can be applied, Eq. (3.8) can be transformed into an interesting integral form. First, we express the operator $B_1W(t_1)$ in terms of the diagonal representation

$$B_1W(t_1) = \int d\Omega_1\ P(\Omega_1,t_1)\ B_1\ |\Omega_1\rangle\langle\Omega_1| \quad , \tag{3.9}$$

and replace B_1 by the corresponding differential operator $\mathcal{D}_{B_1}$ as indicated in Section 1:

$$B_1W(t_1) = \int d\Omega_1\ P(\Omega_1,t_1)\,\mathcal{D}_{B_1}(|\Omega_1\rangle\langle\Omega_1|) \quad . \tag{3.10}$$

The parentheses that follow $\mathcal{D}_{B_1}$ are a reminder that $\mathcal{D}_{B_1}$ operates on the angular variables in the expression that immediately follows.

We proceed next to produce a c-number representation for the time translation operator V. From Eqs. (3.8) and 3.10) we have

$$V(t_1',t_1)[B_1 W(t_1)] = \int d\Omega_1 \; P(\Omega_1,t_1)\, \mathcal{D}_{B_1}(\Omega_1)(V(t_1',t_1)[|\Omega_1\rangle\langle\Omega_1|]) \quad . \tag{3.11}$$

Notice that $V(t_1',t_1)\;[|\Omega_1\rangle\langle\Omega_1|]$ is, by definition, a solution of the generalized master-equation (3.1) subject to the initial condition

$$W(t_1) = |\Omega_1\rangle\langle\Omega_1| \quad . \tag{3.12}$$

This allows us to introduce the following representation for $V(t_1',t_1)\;[|\Omega_1\rangle\langle\Omega_1|]$

$$V(t_1',t_1)[|\Omega_1\rangle\langle\Omega_1|] = \int d\Omega_1' \; P(\Omega_1',t_1'|\Omega_1,t_1)|\Omega_1'\rangle\langle\Omega_1'| \quad , \tag{3.13}$$

with

$$P(\Omega_1',t_1|\Omega_1,t_1) = \delta(\Omega_1'-\Omega_1) \quad . \tag{3.14}$$

For the next step, we multiply Eq. (3.11) from the right by the operator A_1 and construct a new differential operator $\mathcal{D}^*_{A_1^+}$ such that

$$|\Omega_1'\rangle\langle\Omega_1'|\; A_1 = \mathcal{D}^*_{A_1^+}(\Omega_1')\; |\Omega_1'\rangle\langle\Omega_1'| \quad . \tag{3.15}$$

Proceeding in this fashion until every multiplication by the operators A and B is replaced by the differential forms $\mathcal{D}_B$ and $\mathcal{D}^*_{A^+}$ respectively, and every time translation operator is represented in terms of a conditional quasi-probability density, we find the following integral representation for the correlation function

$$K_2(t',t) = \int d\Omega_1 \cdots d\Omega_2' \; P(\Omega_1,t_1)\,\mathcal{D}_{B_1}(\Omega_1)\;(P(\Omega_1',t_1'|\Omega_1,t_1))$$

$$\mathcal{D}^*_{A_1^+}(\Omega_1')\; P(\Omega_2,t_2|\Omega_1',t_1')\;\mathcal{D}_{B_2}(\Omega_2)\;(P(\Omega_2',t_2'|\Omega_2,t_2)) \tag{3.16}$$

$$\langle\Omega_2'|A_2|\Omega_2'\rangle \quad .$$

The generalization to arbitrary multi-time correlation functions is straightforward but of little practical use, and will not be presented here.

There are two special cases of Eq. (3.8) which have frequent use in practice:

a) the two-time correlation function $\langle A(t_1')B(t_1)\rangle$:

$$K_1(t_1',t_1) = \int d\Omega_1 d\Omega_1' \, P(\Omega_1,t_1) \, \mathcal{D}_{B_1}(\Omega_1)(P(\Omega_1',t_1'|\Omega_1,t_1)) \times$$

$$\times \langle\Omega_1'|A_1|\Omega_1'\rangle \quad ; \tag{3.17}$$

b) the case of $t_1' = t_1$:

$$K_2(t,t) = \langle A_1(t_1)A_2(t_2)B_2(t_2)B_1(t_1)\rangle$$

$$= \int d\Omega_1 d\Omega_2 \, P(\Omega_1,t_1) \, \mathcal{D}_{B_1}(\Omega_1) \, \mathcal{D}^*_{A_1^+}(\Omega_1) \, P(\Omega_2,t_2|\Omega_1,t_1)\langle\Omega_2|A_2B_2|\Omega_2\rangle \quad . \tag{3.18}$$

In summary, we have found that an arbitrary multi-time correlation function for operators of the angular momentum type can be constructed in terms of the quasi-probability density, at the earliest time in the required time sequence, and the set of conditional probability densities between all consecutive pairs of times at which the measurements are performed. The measurements are represented by differential operators of the type $\mathcal{D}_A$ and $\mathcal{D}_B$ which can be constructed explicitly using the elementary relations given in Section 1 of these notes.

4. AN APPLICATION OF THE PREVIOUS FORMALISM

We are interested in the calculation of the two-time correlation functions given by Eqs. (3.17) and (3.18). For future use we identify the following expectation values

$$\langle A_1(t_1')\rangle_{\Omega_1,t_1} \equiv \int d\Omega_1' \, P(\Omega_1 ' t_1'|\Omega_1,t_1)\langle\Omega_1'|A_1|\Omega_1'\rangle \quad , \tag{4.1}$$

$$\langle A_2B_2(t_2)\rangle_{\Omega_1,t_1} \equiv \int d\Omega_2 \, P(\Omega_2,t_2|\Omega_1,t_1)\langle\Omega_2|A_2B_2|\Omega_2\rangle \quad , \tag{4.2}$$

where the notation $\langle \cdots \rangle_{\Omega_1 t_1}$ implies that the expectation values are to be calculated from the initial condition

$$W(t_1) = |\Omega_1\rangle\langle\Omega_1|$$

It follows that the two-time correlation functions $K_1(t_1', t_1)$ and $K_2(t_1, t_2, t_2, t_1)$ can be expressed in the more compact form

$$K_1(t_1', t_1) = \int d\Omega_1 \; P(\Omega_1, t_1) \, \mathcal{D}_{B_1}(\Omega_1) \left(\langle A_1(t_1') \rangle_{\Omega_1, t_1} \right) , \tag{4.3}$$

and

$$K_2(t_1, t_2, t_2, t_1) = \int d\Omega_1 \; P(\Omega_1, t_1) \, \mathcal{D}_B(\Omega_1) \, \mathcal{D}^*_{A_1^+}(\Omega_1)$$

$$\left(\langle A_2 B_2(t_2) \rangle_{\Omega_1, t_1} \right) . \tag{4.4}$$

In general Eqs. (4.3) and (4.4) are not simpler to evaluate than the original equations (3.17) and (3.18) because they still require knowledge of the conditional probability density.

The special case of the irreversible relaxation of a two-level system is exceptional because:

a) the one-time expectation values of the operators of interest can be calculated directly from the master-equation for arbitrary initial conditions, and

b) the associated Fokker-Planck equation can be solved exactly; this, of course comes as no surprise because the master-equation can also be solved exactly in closed operator form.

Thus, we can take advantage of this favorable circumstance to illustrate the theory of Section 3.

The irreversible master-equation for the reduced atomic density operator of a two level system interacting with a thermal reservoir in the Markoff approximation is [13].

$$\begin{aligned} \dot{W}(t) = & \frac{\gamma_{21}}{2} \left\{ \left[J^-, W(t)J^+ \right] + \left[J^- W(t), J^+ \right] \right\} \\ & + \frac{\gamma_{12}}{2} \left\{ \left[J^+, W(t)J^- \right] + \left[J^+ W(t), J^- \right] \right\} \\ & - \frac{1}{4} \eta W(t) + \eta J_3 W(t) J_3 \quad , \end{aligned} \tag{4.5}$$

where γ_{12} and γ_{21} are the atomic transition rates and η is a parameter that accounts for phase destroying effects.

Consider, as a first illustration, the two-time correlation function

$$K_1(t_1',t_1) = \langle J_3(t_1')J_3(t_1)\rangle \quad . \tag{4.6}$$

According to Eq. (4.3), we need to calculate the expectation value $\langle J_3(t_1')\rangle_{\Omega_1,t_1}$ subject to the constraint that $W(t_1) = |\Omega_1\rangle\langle\Omega_1|$. From the master-equation we can readily derive the following equation of motion

$$\frac{d}{dt_1'}\langle J_3(t_1')\rangle = tr_s\left[J_3\,\dot{W}(t_1')\right]$$

$$= -(\gamma_{12}+\gamma_{21})\langle J_3(t_1')\rangle - \frac{1}{2}(\gamma_{21}-\gamma_{12}) \quad . \tag{4.7}$$

The solution of Eq. (4.7) is

$$\langle J_3(t_1')\rangle = \left(\langle J_3(t_1)\rangle + \frac{1}{2}\frac{\gamma_{21}-\gamma_{12}}{\gamma_{12}+\gamma_{21}}\right)\exp\left[-(\gamma_{12}+\gamma_{21})(t_1'-t_1)\right]$$

$$- \frac{1}{2}\frac{\gamma_{21}-\gamma_{12}}{\gamma_{12}+\gamma_{21}} \quad , \tag{4.8}$$

where the initial value in question is

$$\langle J_3(t_1)\rangle = \langle\Omega_1|J_3|\Omega_1\rangle = -\frac{1}{2}\cos\theta_1 \quad . \tag{4.9}$$

From the explicit expression for the operator $\mathcal{D}_{J_3}$ given by Eq. (1.17), we finally arrive at the desired result

$$\langle J_3(t_1')J_3(t_1)\rangle = \int d\Omega_1\; P(\Omega_1,t_1)\left(-J\cos\theta_1 + \frac{1}{2}\sin\theta_1\frac{\partial}{\partial\theta_1} + \frac{i}{2}\frac{\partial}{\partial\phi_1}\right)\times$$

$$\times\langle J_3(t_1')\rangle \quad , \tag{4.10}$$

which can be evaluated explicitly once the density function $P(\Omega_1,t_1)$ is assigned. In particular, notice that for $t_1'-t_1\rightarrow\infty$, Eq. (4.10) reduces to

$$\langle J_3(t_1')J_3(t_1)\rangle = -\frac{1}{2}\frac{\gamma_{21}-\gamma_{12}}{\gamma_{12}+\gamma_{21}}\langle J_3(t_1)\rangle = \langle J_3(\infty)\rangle\langle J_3(t_1)\rangle \quad , \tag{4.11}$$

as expected when the observations become statistically independent.

As a second example we calculate the correlation function

$$\langle J^+(t_1)J^+(t_2)J^-(t_2)J^-(t_1)\rangle$$

$$= \int d\Omega_1 \; P(\Omega_1,t_1) \; \mathcal{D}_{J^-}(\Omega_1) \; \mathcal{D}^*_{J^-}(\Omega_1) \left(\langle J^+J^-(t_2)\rangle_{\Omega_1,t_1} \right) \quad . \qquad (4.12)$$

From the identity

$$J^+J^- = \frac{1}{2} + J_3 \qquad (4.13)$$

we derive

$$\langle J^+J^-(t_2)\rangle_{\Omega_1,t_1}$$

$$= \frac{\gamma_{12}}{\gamma_{12}+\gamma_{21}} + \left(-\frac{1}{2}\cos\theta_1 + \frac{1}{2}\frac{\gamma_{21}-\gamma_{12}}{\gamma_{12}+\gamma_{21}}\right) \exp\left[-(\gamma_{12}+\gamma_{21})(t_2-t_1)\right] .$$

(4.14)

From the form of the differential operator $\mathcal{D}_{J^-}\mathcal{D}^*_{J^-}$ given by Eq. (1.25) and the result of Eq. (4.14) we arrive at the required correlation function

$$\langle J^+(t_1)J^+(t_2)J^-(t_2)J^-(t_1)\rangle$$

$$= \int d\Omega_1 \; P(\Omega_1,t_1) \; \frac{1}{2} \; (1-\cos\theta_1) \; \frac{\gamma_{12}}{\gamma_{21}+\gamma_{12}} \; (1-\exp\;[-(\gamma_{12}+\gamma_{21})(t_2-t_1)])$$

$$= \langle J^+J^-(t_1)\rangle \; \frac{\gamma_{12}}{\gamma_{12}+\gamma_{21}} \; (1-\exp\;[-(\gamma_{12}+\gamma_{21})(t_2-t_1)]) \quad . \qquad (4.15)$$

Notice that, as expected, the initial value vanishes since $(J^{\pm})^2 = 0$

Acknowledgments

It is a pleasure to acknowledge the contributions made by several colleagues and friends to the work summarized in these lecture notes. The original derivation and analysis of the superradiant Fokker-Planck equation was performed jointly with Dr. Charles M. Bowden and Professor C. Alton Coulter. Much of the follow-up work on the algebra of the operators and the integral representation of the multi-time correlation functions has been submitted for publication jointly with Dr. Charles M. Bowden, Professors Van Bluemel and Richard A. Tuft and Mr. G. Patricio Carrazana.

It is a special pleasure to express my appreciation to Professor Allan E. Parker for his constant guidance and advice. His dedication has continued to be a source of inspiration.

References

[1] R. J. Glauber, in Quantum Optics and Electronics, edited by C. DeWitt, A. Blandin, and C. Cohen-Tannoudji (Gordon and Breach, New York, 1965).

[2] F. T. Arecchi, E. Courtens, R. Gilmore, and H. Thomas, Phys. Rev. A6 (1972) 2211.

[3] M. Lax, and W. H. Louisell, IEEE J. of Quantum Electronics, QE3 (1967) 47. There is a vast literature on the subject of the so called quantum-classical correspondence. Haken's review article [13] contains many references on this subject.

[4] R. Bonifacio, and F. Haake, Z. Phys., 200 (1967) 526.

[5] We consider angular momentum operators in view of the well known equivalence between the algebra of the observables associated with two-level atomic systems and angular momentum operators.

[6] G.S. Agarwal, Phys. Rev. A2 (1970) 2038.

[7] G.S. Agarwal, Phys. Rev. A4 (1971) 1791.

[8] R. Bonifacio, P. Schwendimann, and F. Haake, Phys. Rev. A4 (1971) 302.

[9] R. Bonifacio, P. Schwendimann, and F. Haake, Phys. Rev. A4 (1971) 854.

[10] L.M. Narducci, C.A. Coulter, and C.M. Bowden, Phys. Rev. A9 (1974) 829.

[11] R. H. Dicke, Phys. Rev. 93 (1954) 99.

[12] F. Haake, and R.J. Glauber, Phys. Rev. A5 (1972) 1457.

[13] H. Haken, Handbuch der Physik, Vol. XXV/2c, Springer Verlag (1970).

[14] F. Haake, Springer Tracts in Modern Physics 66 (1973) 98.

[15] S. Nakajima, Prog. Theor. Phys. 20 (1958) 948.

[16] R. Zwanzig, J. Chem. Phys. 33 (1960) 1338.

Cooperative Phenomena, H. Haken, ed.

RENORMALIZATION GROUP TECHNIQUES ON A LATTICE[+]

Leo P. Kadanoff
Barus & Holley Building
Brown University
Providence, R.I. o2912

These lecture notes are intended to provide a brief introduction to the use of renormalization group techniques [1] for the discussion of critical phenomena. In order to have a particularly simple framework for the discussion, we consider spin variable, $\sigma = \pm 1$, on a lattice. In this case [2], many of the transformation properties may be discussed in a particularly explicit manner.

I. Scale Transformations

Consider a lattice, for example the dotted lattice in Figure 1.

```
.     .     .     .
   x        x
.     .     .     .              . = σ_r
      x
.     .     .     .
   x        x
.     .     .     .              x = μ_r
```

Figure 1: Scaling Transformation on a two dimensional lattice

The dots represent positions of the spin variables, σ_r, each of which takes on the values ± 1. The lattice contains N spins. Consider another lattice, e.g. that produced by the crosses in Figure 1. The other lattice is identical in structure to the first except that it contains only

$$N' = N/\ell^d \tag{1}$$

sites. Here ℓ is the ratio of lattice constants in the two systems and d is the dimensionality of the lattice. Each x is considered to be the site of a new variable μ_r which take on the value ± 1.

The variables σ and μ appear as summation variables in the determination of the free energy. If we start from the Hamiltonian $\mathcal{H}\{\sigma\}$ which is a function of all the σ_r's - then the partition function may be written as

$$Z = \sum_{\{\sigma_r = \pm 1\}} e^{\mathcal{F}\{\sigma\}} \tag{2}$$

where

$$\mathcal{F}\{\sigma\} = -\beta \mathcal{H}\{\sigma\} \tag{3}$$

Here β is the inverse temperature, expressed in energy units. In addition, however, it is possible to re-express the partition function in terms of the μ_r variables. To do this, write

[+]This research was supported in part by the National Science Foundation.

$$e^{\mathcal{F}'\{\mu\}} \sum_{\{\sigma_r=\pm 1\}} T\{\mu,\sigma\} e^{\mathcal{F}\{\sigma\}} \tag{4}$$

If the transformation matrix $T\{\mu,\sigma\}$ has the property that its sum over all μ is unity, i.e.

$$\sum_{\{\mu_r=\pm 1\}} T\{\mu,\sigma\} = 1 \tag{5}$$

Then, the partition function can equally well be written in terms of the transformed free energy, $\mathcal{F}'\{\mu\}$, as

$$Z = \sum_{\{\mu_r=\pm 1\}} e^{\mathcal{F}'\{\mu\}} \tag{6}$$

In effect the transformation defined by Eqs.(4) and (5) is a change in length scale which leaves the free-energy invariant. This type of transformation can be used to describe the concept of scale-invariance near critical points.

It is quite easy to define T's which meets criterion (5). A reasonably general T of this type is

$$T\{\mu,\sigma\} = \prod_r \frac{t_r(\mu_r, \{\sigma\})}{\sum_{\mu'_r=\pm 1} t(\mu'_r,\{\sigma\})} \tag{7}$$

where the product covers all sites of the different μ's. Since each term in the product sums to unity, the entire product obeys condition (5).

In order to proceed further, it is useful to define an appropriate representation for $\mathcal{F}\{\sigma\}$. In a translationally invariant system, $\mathcal{F}\{\sigma\}$ may be written as a sum of terms in the form

$$\mathcal{F}\{\sigma\} = \sum_\alpha K_\alpha S_\alpha\{\sigma\} \tag{8}$$

Here the K_α's are a set of parameters which are defined by the Hamiltonian of the system while the S_α's are a set of extensive translationally invariant functions of the σ_r's. For example, the first few S_α's might be:

$$\begin{aligned} S_o &= \sum_r 1 \\ S_1 &= \sum_r \sigma_r \\ S_2 &= \sum_{rr'} \sigma_r\sigma_{r'} \\ &\text{nearest neighbors} \\ S_3 &= \Sigma\, \sigma_r\sigma_{r'} \\ &\text{next neighbors} \\ S_4 &= \sum_r \sigma_{x,y}\,\sigma_{x+a,y}\,\sigma_{x,y+a}\,\sigma_{x+a,y+a} \end{aligned} \tag{9}$$

Given enough S_α's any possible translationally invariant Hamiltonian may be written in the form (8).

Now notice that $\mathcal{F}'$ as defined by Eq.(4) is a translationally invariant extensive quantity. Therefore $\mathcal{F}'$ can also be represented in the same form as Eq.(8) i.e.

$$\mathcal{F}'\{\mu\} = \sum_\alpha K'_\alpha S_\alpha \{\mu\} \tag{1o}$$

Here $S_\alpha \{\mu\}$ are of exactly the form (9) except that σ has been replaced by μ and, of course, the r-sums now cover all the points of the μ-lattice. In virtue of Eq.(1o), the transformation (4) may be considered to be - in effect - a transformation of the K_α's. In particular we write

$$K'_\alpha = B_\alpha(K_o, K_1, K_2, \ldots) \tag{11a}$$

or in vector notation

$$\vec{K}' = \vec{B}\ (\vec{K}) \tag{11b}$$

II Fixed Points

One of Wilson's major contributions to the theory of critical phenomena [1] was the formulation of the concept of a fixed point. A fixed point of the transformation (11b) is a special vector, $\vec{K}^*$, which is invariant under the transformation, i.e.

$$\vec{K}^* = \vec{B}\ (K^*) \tag{12}$$

Every transformation will have a variety of fixed points. For example, there is a trivial fixed point at zero coupling

$$K^*_o = -\frac{1}{1-e^{-d}} \ell n2$$

$$K^*_\alpha = o \text{ for } \alpha > o$$

However, there are other fixed points which we may try to identify as representations of critical points of phase transitions.

To pin down this identification consider a situation in which there is a small deviation from the fixed point K^*. That is, let

$$K_\alpha = K^*_\alpha + h_\alpha \tag{13}$$

with $h_\alpha \ll 1$.

Given this value of $\vec{K}$ we can write the partition function as

$$Z = e^{Nf(\vec{h})} \tag{14a}$$

where f is proportional to the free-energy per site. In addition, according to Eq.(6) Z may equally well be written as a result of the μ-summation in the form

$$Z = e^{N'f(\vec{h}')} \tag{14b}$$

with h' being given by

$$\vec{h}' = \vec{K}' - \vec{K}^{*} = \vec{B}(K^{*}+h) - B(K^{*}) \tag{15}$$

Given the two expressions (14) for Z and the relationship (1), we find that the free energy per site obeys:

$$f(h) = \frac{N'}{N} f(h') = \ell^{-d} f(h') \tag{16}$$

At a critical point, h = o, f(h) is singular because of a set of cooperative phenomena in which spins far apart exhibit weak but long-ranged correlation. The basic assumption of the transformation theory, is that one can choose $t_r(\mu_r, \{\sigma\})$ in such a way that the h = o singularity does not appear in the transformation law (15). In that case we can expand the right hand side of (15) in a power series so that (15) becomes

$$h'_{\alpha} = B^{1}_{\alpha\beta} h_{\beta} + B^{2}_{\alpha\beta\gamma} h_{\beta} h_{\gamma} + \dots \tag{17}$$

For small h, the first order term dominates the behavior. Assume that by choosing a suitable linear combination of h's B' can be diagonalized. In this diagonal representation, Eq.(17) reads

$$h'_i = b_i h_i \tag{18}$$

It is convenient to write the eigenvalue b_i as

$$b_i = \ell^{x_i} \tag{19}$$

F. Wegner has investigated the consequence of Eqs.(16)-(19) in some detail [3] . He has pointed out that one can classify the variables h_i according to the following scheme:

Case 0: The variable h_i does not appear in f. These so-called "scaling variables" are then completely irrelevant to the critical behavior.

Case 1:
$x_i > o$. After several scale transformations these variables grow and grow. They dominate the behavior in the critical region. These variables are described as "thermodynamically relevant".

Case 2: $x_i < o$. In this case h_i gets smaller and smaller after several scale transformations. In the asymptotic limit as criticality is approached, these "thermodynamically irrelevant" variables tend to get less and less important.

Case 3: $x_i = o$. These marginal variables play a special and complex role in near-critical behavior. However, they are believed to only occur a few special problems [4].

III. Scaling and Universality

Equations (16) and (19) together have a solution

$$f(h) = h_1^{d/x_1} f^* \left(\frac{h_\partial^{x_1}}{h_1^{x_\partial}} \right) \qquad (20)$$

This scaling solution is a generalization of the homogeneous function concept of Widom [5] . From this form, one can see that as h_1 goes to zero only the variables with $x_i > 0$ remain in the free energy. The remaining variable drop out. Thus, if two starting Hamiltonia differ only in the values of irrelevant or scaling variables, their eventual critical behavior will be identical. This weak dependence of the answer upon the starting point is described by the words smoothness or universality [6].

IV. Examples

A. One Dimensional Case

The one-dimensional Ising model has a kind of phase transition at T = o. To see this, write the free energy as

$$\mathfrak{F}\{\sigma\} = \sum_\partial \left[K(\sigma_\partial \sigma_{\partial+1} - 1) + \frac{h}{2} (\sigma_\partial + \sigma_{\partial+1}) \right] \qquad (21)$$

and define

$$h_1 = h \qquad h_2 = e^{-K} \qquad (22)$$

In the limit in which h_1 and h_2 are small, the free energy per site is

$$f(h_1, h_2) = \sqrt{h_1^2 + h_2^4} \qquad (23)$$

which is certainly singular at $h_1 = h_2 = 0$.

A partially simple form of the transformation $T\{\mu,\sigma\}$ can serve to generate the critical indices for this case. Let the new spin variable be

$$\mu_K, \; K = 1, 2, \ldots, N/2 \equiv N' \qquad (24)$$

Each μ_K is placed upon an even-numbered site as shown in Figure 2.

```
    x       x       x       x
.   .   .   .   .   .   .   .        x = μi

                                     . = σi
```

Figure 2: Scaling Transform in one dimension

Since N' = N/2, the length change is

$$\ell = 2 \qquad (25)$$

Define $T\{\mu,\sigma\}$ in such a way that $\mu_K = \sigma_{2K}$, i.e. with

$$t_K(\mu_K, \{\sigma\}) = \frac{1 + \mu_K \sigma_{2K}}{2}$$

Then, the new free energy function is given by

$$e^{\mathcal{F}\{\mu\}} \quad \prod_K \sum_{\sigma_{2K+1}=\pm 1} \tag{26}$$

$$e^{K(\mu_K \sigma_{2K+1} - 1) + h(\frac{\mu K}{2} + \sigma_{2K+1} + \frac{\mu K + 1}{2})}$$

$$e^{K(\sigma_{2K+1}\mu_{K+1} - 1)}$$

After the sum over the intermediate spin variables σ_{2K+1} is performed, $\mathcal{F}\{\mu\}$ is exactly of the form (21)

$$\mathcal{F}'\{\mu\} = \sum_K \left[K'(\mu_K \mu_{K+1} - 1) + h'(\mu_K + \mu_{K+1})/2\right]$$

In the limit of small $h_1 = h$ and $h_2 = e^{-K}$, we find

$$h_1' = 2h \qquad \text{and}$$

$$h'_2 = e^{-K'} = \sqrt{2}\; h_2$$

A comparison with Eqs (18) and (19) shows that

$$x_1 = 1$$

$$x_2 = \sqrt{2}$$

Furthermore, the free energy (23) is exactly of the form (2o) with these values of x and

$$f^* \left(\frac{h_2^{\,x_1}}{h_1^{\,x_2}} \right) = \sqrt{1 + (h_2^4/h_1^2)} \tag{27}$$

B. A Failure

The approach just employed may be directly extended to two dimensions by defining $\mu_{j,K}$ to lie on the lattice shown in Figure 3, and letting

$$\mu_{j,K} = \sigma_{j,K} \quad \text{if } j + K \text{ is even}$$

```
.   x   .   x
x   .   x   .
.   x   .   x                        . = σ_r
x   .   x   .                        x = σ_r and μ_r
```

Figure 3: Another transformation in two dimensions

Thus

$$T\{\mu,\sigma\} = \prod_{\substack{j,K \\ j+K \text{ even}}} \frac{1 + \mu_{j,K}\sigma_{j,K}}{2} \tag{28}$$

At first sight this approach looks very promising. One can exactly calculate the sum

$$\sum_{\{\sigma\}} T\{\mu,\sigma\} \quad e^{\mathfrak{F}\{\sigma\}}$$

for the case in which $\{\sigma\}$ includes any nearest neighbor interactions. This sum, however, produces an $\mathfrak{F}'\{\mu\}$ in which there are next nearest neighbor and 4-spin interaction terms. Thus, this situation is different from the one dimensional case in which the nearest neighbor interaction produces only more nearest neighbor interaction.

Nevertheless, the new interactions produced seem small enough so that they can be treated in perturbation theory. In first order perturbation theory, the recursion relations for nearest neighbor, next neighbor and 4-spin interactions is

$$K_1' = K_1^o(K_1) + K_2 + A_{12}K_2 + A_{13}K_4$$

$$K_2' = \frac{1}{2} K_1^o(K_1) + A_{22}K_2 + A_{23}K_4$$

$$K_4' = K_4^o(K_1) + A_{42}K_2 + A_{44}K_4$$

with, for example,

$$K_1^o = \frac{1}{4}\ell n \cosh 4 K_1$$

$$K_4^o = \frac{1}{2}K_1^o - \frac{1}{2}\ell n \cosh 2 K_1$$

$$A_{12} = .5x(\tanh 2K + \frac{1}{2}\tanh 4K) \ x \ (\frac{7}{2} \tanh 4K - \tanh 2K)$$

These equations have a fixed point at

$$K_1 = K_1^* = .37$$

$$K_2 = K_2^* = .13$$

$$K_4 = K_4^* = -.04$$

and an eigenvalue [3] defined by

$$x_2 = 1.03$$

In the next order of perturbation theory

$$x_2 = 1.002$$

Since the exact result from the Onsager solution is

$$x_2 = 1$$

the perturbation theory seems to be working beautifully.

However, third order perturbation theory reveals a disaster, namely

$$x_2 = 1.37$$

What is the trouble? The difficulty can actually be seen in a reasonably simple fashion.

The spin-spin correlation function is given by

$$g^*(r) = \sigma_o \sigma_r = \frac{\sum_{\{\sigma\}} \sigma_o \sigma_r e^{\mathcal{H}^*}}{\sum_{\{\sigma\}} e^{\mathcal{H}^*}} \qquad (29)$$

at the fixed point. However, if o and r are sites for the μ-spins, the recursion relations simply

$$\langle \sigma_o \sigma_r \rangle = \langle \mu_o \mu_r \rangle$$

In the σ-lattice the number of lattice constants between o and r is given by

$$r^2/a^2 = n_x^2 + n_y^2$$

where n_x and n_y is the separation in lattice constants. In the μ- lattice

$$n_x^2 + n_y^2 = r^2/2a^2$$

Thus, the method of calculations directly implies that

$$g^*(r) = g^*(r/\sqrt{2})$$

On the other hand, the Onsager solution implies

$$g(r) \frac{1}{r^{1/4}} \quad \text{for large r.}$$

Thus, the fixed point we have reached in our approximation is either completely spurious or at least different from the critical point of the two dimensional Ising model [7].

D. More Successful Calculations

Other authors have produced apparently more successful fixed point calculations, by using different forms for the transformation matrix. However, one should notice that there is in all cases, a possibility that these other calculations also reach a non-existant or spurious fixed point.

Niemeijer and van Leeuwen [6] start from a $T\{\mu,\sigma\}$ similar to one proposed by LPK in reference [1], namely

$$t_r(\mu_r,\{\sigma\}) = \frac{1 + \mu_r \, S_r/|S_r|}{2}$$

where

$$S_r = \sum_{r'} \sigma_{r'}$$

$$\text{for} \begin{cases} |x - x'| < \ell \\ |y - y'| < \ell \end{cases}$$

In reference k, a square lattice is used and ℓ is chosen to be much larger than 1. This approach is only useful for qualitative arguments. On the other hand, in reference [6] a triangular lattice is used, and ℓ is chosen so that S_r includes exactly three spins. This approach leads to excellent quantitative results, namely

$$x_1 = 1.8760$$

$$x_2 = 1.028$$

In comparison, the exact statements are

$$x_1 = 1.8750$$
$$x_2 = 1.000$$

At this moment, A. Houghton and the author are following a slightly different approach where

$$t_r(\mu,\{\sigma\}) = \frac{e^{\bar{K}\sigma_r S_4}}{\cosh KS_4}$$

where S_4 is the sum of the four neighboring σ's. At first sight, this approach seems to lead to good convergence and slightly simpler calculations than those in reference [6].

References

[1] These techniques were introduced by K. Wilson, Phys.Rev.B4, 1374, 3184 (1971), and K. Wilson and J.Kogut, Physics Reports (to be published). In some sense, this work is an extension of L. Kadanoff, Physics 2, 263 (1966).

2 A general discussion of this case occurs in Kadanoff (Ref. 1). More detailed calculations have also been performed by Th. Neimeijer and J.MJ. van Leeuwen, Phys.Rev. Letters 31, 1412 (1973) and Physica 71, 17 (1974) by K. Wilson (private communication) and by L.Kadanoff, A. Houghton and M.Grover, (Proceedings of the Temple Conference on Critical Phenomena (to be published).

3 F.J. Wegner, Phys.Rev. B5, 4529 (1972)

4 In the Baxter model, [R.J.Baxter, Phys.Rev.Lett. 26, 831 (1971)] Also in most critical phenomena problems there is a stress tensor with this eigenvalue. However, this does not contribute to the free energy.

5 B. Widom, J.Chem.Phys. 43, (1965) 3898, 3892

6 R. Griffiths, Phys.Rev. Lett. 24 (197o) 1479,
D. Jasnow and M.Wortis, Phys.Rev. 176 (1968), 739
L. Kadanoff, Proc. of the 197o Enrico Fermi Summer School on Critical Phenomena, 1971 Academic Press, New York

7 K. Wilson (private communication) has analyzed this situation and finds that the problem comes from two scaling variables with x_i equal to and slightly greater than one.

Cooperative Phenomena, H. Haken, ed.

REVERSIBILITY INVARIANTS AND NONLINEAR THERMODYNAMICS

F. Schlögl, Institut für Theoretische Physik, RWTH Aachen

Introduction

Nonlinear thermodynamics shows very interesting features such as instabilities and the building up of dissipative structures [1]. A special kind of phenomena in the field of dissipative structures is the surprisingly closed analogy of certain steady state changes to phase transitions. These phenomena therefore are called "nonequilibrium phase transitions" [2-8]. It seems suitable to regard the equilibrium phase transitions as a special case of a general class which contains the nonequilibrium phase transitions as well. In the special case the steady states become equilibrium states.

In search of quantities which show characteristic behaviour in nonlinear thermodynamics in a most general way, independent of individual features of the special system, and which therefore can be called thermodynamic quantities, a certain class of statistical quantities shall be discussed in the following. These quantities are invariant with respect to reversible changes of thermodynamic states and therefore will be called "reversibility invariants". They are introduced as cumulants of bit-numbers of a probability distribution or as bit-number differences of two distributions. A particular one of these quantities is the entropy. Two more are discussed, information gain, and bit-number variance. The first one is an essential quantity in open systems. It is closely related to entropy prodcution and to stability criteria for steady states. The second one is related to the heat capacity in thermostatics and seems to be a quantity which behaves dramatically in nonequilibrium at the critical point, in an analogous way as specific heat in thermostatics.

1. The cumulants of bit-numbers

If b is a random variable, the generating function

$$\Psi(\alpha) = \sum_{s=o}^{\infty} \frac{\alpha^s}{s!} < b^s >_c \qquad (1.1)$$

of the cumulants of b

$$< b^s >_c = \left(\frac{d^2\Psi}{d\alpha^s}\right)_{\alpha=o} \qquad (1.2)$$

is defined by means of the expectation value

$$< e^{\alpha b} > = e^{\Psi} \quad . \qquad (1.3)$$

The subscript c of the bracket distinguishes the cumulants from the ordinary expectation values. Let b be a random variable which is defined in the microstates of a physical system which can be divided into two subsystems. Let b, moreover, be additive for the two subsystems, then, in general, Ψ is not additive. Ψ becomes, however, additive for the two subsystems if these are not correlated. Therefore the cumulants, unlike the moments $<b^s>$, become additive as well if the subsystems are uncorrelated. This fact is a main reason to introduce the cumulants. The cumulants are polynomials of the moments. In particular:

$$< b >_c = < b > \qquad (1.4)$$

$$< b^2 >_c = < b^2 > - < b >^2 = <(\Delta b)^2> \quad . \qquad (1.5)$$

The last quantity is the mean square of the fluctuations of b, also called the "variance" of b.

The possible microstates of a physical system may be denumerated by a subscript i. If the whole set p of the probabilities p_i of the microstates is known, then

$$b_i = - \ln p_i \qquad (1.6)$$

is a measure for the number of bits which are necessary to give the communication that the special microstate i is accepted by the system. We call the random variable b with the possible values b_i the "bit-number" and we can form it's cumulants.

The first cumulant

$$\langle b \rangle = - \sum_i p_i \ln p_i = S(p) \qquad (1.7)$$

is the information entropy, the negative of Shannon's information measure [9]. It is the thermodynamic entropy if p is a thermal equilibrium distribution or an adequate nonequilibrium distribution. It shall be shown later that the second cumulant, the variance of b is closely related to the heat capacity in thermostatics and it seems that it is an interesting quantity in nonequilibrium statistics.

We can compare two probability distributions p, p' over the same set of microstates and introduce the relative bit-number

$$(b_r)_i = b'_i - b_i = \ln \frac{p_i}{p_i'}\,. \qquad (1.8)$$

If b_i is small, then the expectation that i will occur is large. That means, if i occurs, the knowledge about this fact given already before by p was large. Thus, with respect to i, the relative bit-number (1.8) gives the surplus knowledge of p compared with p'. If p is the true distribution which shall be compared with a reference distribution p', then b_r is a random variable in the distribution p. We can form the mean value of (1.1) for b_r instead of b in the distribution p and get relative measures for p with respect to p' by the cumulants. In particular

$$\langle b_r \rangle_c = \langle b_r \rangle = K(p,p') \qquad (1.9)$$

$$\langle b_r^2 \rangle_c = \langle (\Delta b_r)^2 \rangle = Q_r(p,p')\,. \qquad (1.10)$$

The first one

$$K(p,p') = \sum_i p_i \ln \frac{p_i}{p_i'} \qquad (1.11)$$

is called Kullback's information or "information gain" [10,11] .

The change to continuous microstates, which are characterized by a set of continuous parameters

$$x = (x_1, x_2, ..),$$ (1.12)

requires a change of the definition of bit-number. If w(x) is the probability density in x-space, we define the bit-number by

$$b(x) = - \ln w(x).$$ (1.13)

This definition becomes obvious by dividing the x-space into small cells and omitting the contribution from the interior of a cell to the total bit-number. b(x) and $\langle b \rangle_c$ then are dependent on the metric of the x-space. Therefore $\langle b \rangle_c$ becomes thermodynamic entropy only if x is the canonical phase space of the system. In contrast, the cumulants (1,5), (1,9), (1.10) are independent of the x-metric. Thus K(w,w'), Q(w), Q_r(w,w') are metrical invariants.

All the notions can be extended to quantum mechanical density matrices ρ by introducing the operator "bit-number" [12]

$$b = - \ln\rho \ .$$ (1.14)

The expectation values are formed with ρ in the usual way

$$\langle A \rangle = \mathrm{tr} \ (\rho A) \ .$$ (1.15)

An important feature of the cumulants of the bit-number b und b_r is that they don't change by a reversible motion of the system. This follows from the invariance of the generating functions $\Psi(\alpha)$ of these cumulants with respect to a reversible motion. In the case of discrete microstates i these functions remain unchanged by a permutation of the microstates and a reversible motion is a one to one mapping of the microstates; that is a permutation. The corresponding transformation in continuous space is a canonical transformation, in particular the Liouville motion. A canonical transformation from phase space variables x to x' does not change the phase volume. Therefore probability density transforms like a scalar

$$w'(x') = w(x). \qquad (1.16)$$

The same is valid for b. The functions $\Psi(\alpha)$ are unchanged. In quantum mechanics a reversible motion is a unitary transformation in pure state space and the considered functions $\Psi(\alpha)$ are unitary invariants.

Thus we have the result that all the cumulants of b and b_r can change by irreversible motion only.

2. Open systems and information gain.

The first cumulant of b_r, that is the information gain K(p,p'), gives the surplus knowledge which is contained in the true probability distribution p in comparison to the reference distribution p'.

If for instance p' is the thermal equilibrium distribution of an open system with its environment, p' is determined by the thermal states of the environment. A measurement in the system itself can lead to another distribution p, say a frozen equilibrium. Then K(p,p') gives a measure for the knowledge gained by this measurement as surplus knowledge to the knowledge about the environment. This concept can be extended to nonequilibrium states if p' is determined by the environment and p by the environment plus an additive measurement in the system. Thus K is a useful quantity for open systems.

K(p,p') has some remarkable features. It is positive definite and vanishes only in case that p and p' are identical. It is a convex and monotonous function of p. The latter means that any surface with constant values of K contains nowhere larger values of K. As due to normalization the p_i are not independent, the p-space is a space of independent parameters, which uniquely describe a distribution p. And now it is valid, that these parameters always can be found so that in this parameter space K is convex and monotonous [13] . That K does not change by a reversible motion is already said. A special class of irreversible motions is given by linear master equations

$$\dot{p} = \Lambda p \,, \tag{2.1}$$

where Λp is a linear expression in p taken at the same t as $\dot{p}$. A remarkable fact is now that K does not increase if p,p' are solutions of the same equation (2.1). This can be shown generally in the following way. The transformation from t to t + τ changes p,p' into p, p' in the following way

$$\hat{p}_j = \sum_i r_{ji}\, p_i \tag{2.2}$$

$$\hat{p}'_j = \sum_i r_{ji}\, p'_i \,, \tag{2.3}$$

where

$$\sum_j r_{ji} = 1 \,. \tag{2.4}$$

With the inequality for all positive x

$$\ln x \geqq 1 - \frac{1}{x} \tag{2.5}$$

we get

$$K - \hat{K} = \sum_{ji} r_{ji} p_i \left(\ln \frac{p_i}{p'_i} - \ln \frac{\hat{p}_j}{\hat{p}'_j} \right) \tag{2.6}$$

$$\geqq \sum_{ji} r_{ji} p_i \left(1 - \frac{p'_i \hat{p}_j}{p_i \hat{p}'_j} \right) = 0 \,. \tag{2.7}$$

If p' is an equilibrium state of the system with its environment and p(t) is the changing true state of the system which goes into p', then

$$P = - \dot{K}(p,p') \qquad (2.8)$$

is the entropy production in the interior of the system [14] . It looks reasonable to describe such a process by a linear master equation. At least, if such a description is assumed, it is garanteed that p is positive definite.

As $K(p,p')$ is a monotonous function of p, it can be used as test function in Liapounoff's stability theory. Thus we get the following stability criterion for a steady state p' [13,15] : Any region of states p round a given steady state p' given by an inequality

$$K(p,p') \leqq \text{const} \qquad (2.9)$$

is a stability region, so that the momentary nonequilibrium state p(t) of the system never will move out of this region, if everywhere in its interior

$$\dot{K}(p,p') \leqq 0. \qquad (2.10)$$

is always valid.

This criterion gets a macroscopic form when K is expressed by macroscopic variables A^{σ} . These variables can be local thermal variables which are already used in thermostatics. They can moreover by typical nonequilibrium variables such as rate changes and transport flows. The variables A^{σ} shall be called "direct" variables when they are expectation values of quantities which are defined already in the microstates. Under very general conditions -K is the nonlinear part of the deviation δS of the entropy S from the steady state in an expansion in powers of the deviations δA^{σ} of the direct variables [13] :

$$-K(p,p') = \delta S - \left(\frac{\delta S}{\delta A}\right)' \delta A \equiv \delta_{NL} S. \qquad (2.11)$$

Here and in the following equal greek sub- and superscripts, such as σ, are summation indices.

The restriction to infinitesimal deviations from the steady state gives the stability criterion of Glansdorff and Prigogine [1,16]:

$$\delta^2 \dot{S} \geqq 0. \tag{2.12}$$

K(p,p') ist not only a convex and monotonous function of p but it has this features also as a function of the macroscopic observables A^σ because (2.11) is valid also if p' is not a steady state. Therefore under the same general conditions [13]:

$$K(p+\delta p,p) = -\delta_{NL} S \geqq 0 \tag{2.13}$$

and

$$\delta K \equiv K(p+\delta p,p') - K(p,p') \tag{2.14}$$

$$= \sum_i \delta p_i \, (\ln p_i - \ln p_i') - \delta_{NL} S \tag{2.15}$$

$$\delta_{NL} K = -\delta_{NL} S \geqq 0 \tag{2.16}$$

This however, means that the nonlinear part of δK in powers of δA^σ is always positive definite. K is convex and monotonous in A-space as well.

3. Hydrodynamical systems

Now systems in nonequilibrium states may be considered for which a description by local thermodynamical variables is possible. These variables $m^\nu(\underline{r})$ may be densities of such direct variables M^ν which are used also in thermostatics and therefore may be called local thermostatic variables.

Let us first speak of systems in equilibrium. The macroscopic state may be given by a set of macrovariables M^ν which will be interpreted as expectation values $\langle M^\nu \rangle$ of direct variables M_i^ν. The probability distribution

$$p_i = \exp \left[\emptyset - \lambda_\nu M_i^\nu \right] , \qquad (3.1)$$

which gives an adequate description of this equilibrium state, may be called a generalized canonical distribution. λ_ν are extensive quantities, $\emptyset$ is given by the normalization. Here M^ν are global quantities.

If now a nonequilibrium state can be described by local macroscopic densities m^ν $(\underline{r},t)$, we can construct a generalized canonical distribution in the same way:

$$p_i^a (t) = \exp \left[\emptyset \ (t) - \int d\underline{r} \ \ \lambda_\nu (\underline{r},t) \ m_i^\nu \ \ (\underline{r}) \right] . \qquad (3.2)$$

It may be called the accompanying canonical distribution p^a. It is a useful approximation for certain purposes. It is lacking, however, in other respects. For instance in this approximation expectation values of irreversible transport flows will vanish.

A substantially better description can be given for so called hydrodynamical systems by a distribution which we shall call a Mori distribution [17,18,19] . These distributions are constructed under the assumptions that there exists a distinct separation of short time and long time behaviour. Let τ be a time which is very large on the short time scale, yet small on the long time scale. During the time τ the systems behaves practically like a closed system with a Liouville operator L. Then the Mori distribution is constructed as

$$p(t) \ = e^{-iL\tau} \qquad p^a(t- \tau). \qquad (3.3)$$

Due to the distinct separation of short and long time behaviour this distribution is independent of the special value of τ .

The cumulants of the bit-numbers b and b_r are invariant with respect to reversible motion. Therefore they are equal for p(t) and $p^a(t-\tau)$. As τ is small on the macroscopic time scale, these cumulants as macroscopic quantities are equal for p and p^a at any time. Thus we are not restricted to the approximation by the accompanying distribution p^a when we express these cumulants in terms of p^a. We get the same expressions also for the substantially better description by Mori states. In particular one finds:

$$S(p) = \int d\underline{r} \; \lambda_\nu m^\nu - \emptyset \tag{3.4}$$

$$K(p,p') = \delta\emptyset - \int d\underline{r} \; m^\nu \, \delta\lambda_\nu \tag{3.5}$$

$$= - \delta S + \int d\underline{r} \; \lambda_\nu \, \delta m^\nu . \tag{3.6}$$

δ always symbolizes the difference between the values for p^a and p'^a.

These results give a justification of the well known approximation in macroscopic thermodynamics by local equilibria for hydrodynamical systems if we define this approximation by the validity of the thermostatic relations between the macroscopic local quantities. It should, however, be stressed that S and K remain integrals over $\underline{r}$ space and its important inequality features are not garanteed for local densities of S and K by this considerations.

4. Nonequilibrium phase transitions

Before we consider the bit-number cumulants of second order, it seems useful to discuss a nonequilibrium phase transition because it seems that these cumulants lead to an interesting question in connection with nonequilibrium phase transitions.

A special type of chemical reaction systems shows two competing steady states [7]. The concentration of all components with exception of only one shall be held constant by appropriate feeding of a reactor. The phenomenon of a nonequilibrium phase transition occurs already in a deterministic theory in which the change of the concentration n of the only one variable component is already given by the value of n itself:

$$\dot{n} = \varphi(n) . \tag{4.1}$$

The function $\varphi(n)$ can have three zero points. Two of them belong to stable steady states. These two states behave like two phases in a phase transition of first order when different boundary conditions are compared. The boundary conditions are given by the concentrations of the constant components. The diagram of the steady states is very similar to the equilibrium state diagram of the fluid gas system. Inclusion of diffusion of the variable component leads to a Maxwell construction for the coexistence states of the two phases in a similar way as for a Van der Waals gas. The two phases have the tendency to form domains with minimal surface. The Coexistence values depend in the same way on the curvature of the surface layer as vapor pressure does. A further analogy to equilibrium phase transitions is the critical slowing down of fluctuations in a steady state.

The same chemical reaction system can be described also in a stochastic theory in which n does not uniquely determine its change $\dot{n}$ with time. In the stochastic theory a master equation is assumed for the probability P_N that the number of molecules of the variable component has a certain value N in the homogeneous reactor:

$$\dot{P}_N = \sum_{N'} R_{NN'} P_{N'} . \tag{4.2}$$

It should be stressed that P is not a probability distribution over microstates but over the random variable N. To one value of N belongs a very large number of microstates.

Such a stochastic theory for the considered reaction system was developed and discussed by H.-K. Janssen [20]. The Matrix R in (4.2) is constructed with the reaction rates of ideal reaction kinetics of rarified gases. It has nonvanishing elements only if N' is different from N by one. This fact allows to find exactly the steady solutions of (4.2). In case that the deterministic theory predicts three steady states, two of them n_1, n_2 are stable and the third n_3 lying between n_1, n_2 is unstable. In this case the stochastic theory gives two peaks of P_N at the corresponding numbers N_1, N_2 and a relative minimum between them at N_3. In contrast to the deterministic theory the system can pass through the value N_3 with a small yet final probability.

The stochastic theory is interesting also in methodical respects. For instance a Fokker-Planck approximation would fail totally because the stochastic process is far away from a Gaussian process.

The thermodynamic limit means that volume and N go to infinity whereas n remains finite. By this limit the higher peak of P_N increases, the lower decreases and vanishes. Only in the case that the two peaks are equal height they both remain in the limit. That gives a coexistence condition of the two phases. By passing the critical point the two peaks merge into one. The phase transition becomes sharp in the thermodynamic limit. That shows another important analogy to equilibrium phase transitions. They also occur only in the thermodynamic limit.

On the search for more analogies between phase transitions in equilibrium and nonequilibrium it seems desirable to find a quantity which could play a similar role in nonequilibrium to the specific heat in equilibrium. The specific heat in general behaves dramatically in the vicinity of a critical point. One would expect a dramatical behaviour of the analogous quantity in nonequilibrium. In the following chapter it shall be shown that such an analogous quantity is given by the second cumulant of the bit-number.

5. The variance of bit-number

The second cumulant of the bit-number b is its variance, that is the mean square of its fluctuations [21] :

$$Q = \langle (\Delta b)^2 \rangle . \tag{5.1}$$

For a generalized canonical distribution (3.1) we get

$$\Delta b_i = - \lambda_\nu \quad \Delta M_i^\nu \tag{5.2}$$

$$Q = \lambda_\mu \lambda_\nu < \Delta M^\mu \quad \Delta M^\nu > . \tag{5.3}$$

The correlation matrix of the fluctuations of M^ν is given by

$$< \Delta M^\mu \quad \Delta M^\nu > = - \frac{\partial}{\partial \lambda_\nu} < M^\mu > \equiv - \frac{\partial M^\mu}{\partial \lambda_\nu} \tag{5.4}$$

where M^μ are the mean values.

In thermostatics the temperature

$$T = \lambda_o^{-1} \tag{5.5}$$

always occurs as one of the λ_ν. Then

$$\xi_\nu = - T\lambda_\nu \tag{5.6}$$

are used as the common intensive paramters (eg. negative pressure, chemical potentials, magnetic field). For any thermostatic function A we get

$$-T \left(\frac{\partial A}{\partial T}\right)_\xi = \lambda_\nu \frac{\partial A}{\partial \lambda_\nu} . \tag{5.7}$$

The subscript ξ indicates that all ξ_k, $(k \neq O)$, shall be fixed. With

$$\lambda_\mu = \frac{\partial S}{\partial M^\mu} \tag{5.8}$$

we get

$$Q = -\lambda_{\nu} \frac{\partial S}{\partial \lambda_{\nu}} = T \left(\frac{\partial S}{\partial T}\right)_{\xi} . \tag{5.9}$$

If all occuring M^{ν} are of the type that their changes do not change the material composition of the system, that means the particle numbers, then

$$C_{\xi} = T \left(\frac{\partial S}{\partial T}\right)_{\xi} \tag{5.10}$$

is the heat capacity of the system for fixed intensive parameters ξ_k. We can call C_{ξ} a generalized heat capacity in all cases, and get the result that the variance Q of the bit-number of a generalized canonical distribution is the corresponding heat capacity:

$$Q = C_{\xi} . \tag{5.11}$$

It should be stressed that, with respect to this relation, we have to distinguish between different generalized canonical distributions also in the thermodynamic limit.

Now it becomes plausible that C_{ξ} and in particular the specific heat behaves dramatically when passing a critical point of a phase transition. If the degrees of freedom of a many particle system are uncorrelated, then their contributions to Q are additive. Yet if they become correlated, this gives rise to an additive correlation term in Q. The region of the critical point always is characterized by the building up of long range correlations and one has to expect a dramatical change of Q there.

To better understand what features of a probability distribution are described by the quantity Q, some special and typical cases shall be discussed.

If we look first for one peak distributions w(x) over one continuous parameter x, we see that Q, unlike to the information entropy S, is independent of the width. Yet Q is very susceptible to the shape of w(x).

For the exponentials

$$w(x) \sim \exp(-\alpha|x|^n) \tag{5.12}$$

of various degrees n one finds

$$Q = \frac{1}{n} . \tag{5.13}$$

Q is independent of α and therefore of the width. For fixed α the quantity Q decreases with increasing steepness of the flanks. The special case n=2 leads to the equipartition theorem of classical statistical mechanics. The case n = 1/3 occurs for a highly relativistic gas of particles with negligible rest mass.

The polynomial distribution

$$w(x) \sim 1 - |x/\alpha|^n \tag{5.14}$$

for $|x| < \alpha$ and vanishing for $|x| > \alpha$. The quantity Q can be expressed in series and be calculated numerically.

The same is true for Student's distribution

$$w(x) \sim (x^2 + \alpha^2)^{-(n+1)/2} . \tag{5.15}$$

In both cases we find that Q is independent of α and decreases with increasing n. This again means that for fixed α the quantity Q decreases with increasing steepness of the flanks of the distribution.

If the steepness becomes infinite, that means for a square wall distribution, Q vanishes.

The probability distribution over two microstates

$$p_{\pm} = \frac{1}{2}(1 \pm \gamma) \tag{5.16}$$

can be characterized by one parameter γ which lays between +1 and -1. It leads to

$$Q = \frac{1}{4}(1-\gamma^2)\left(\ln\frac{1+\gamma}{1-\gamma}\right)^2. \tag{5.17}$$

In its dependence on γ this quantity has two peaks at $\pm 0{,}84$ and vanishes at the points 0 and $\pm$ 1.

The power distribution over discrete states n = 0,1,2...

$$p_n \sim \gamma^n \tag{5.18}$$

leads to

$$Q = \frac{\gamma}{1-\gamma^2}(\ln\gamma)^2. \tag{5.19}$$

As already explained, an interesting question is whether there exists an analogy between the dramatic behaviour of the specific heat at a critical point in equilibrium and corresponding phenomena in nonequilibrium as well. As the measure Q is attached to any probability distribution and becomes heat capacity for generalized canonical distributions, it seems reasonable to examine Q of probability distributions in a stochastic theory of a non-equilibrium phase transition. Such a theory is that of H.K. Janssen [20] cited in chapter 4. The distribution P_N in it is not a distribution over numerous degrees of freedom of microstates but over only one random variable N. Therefore one would not expect that Q goes to infinity at the critical point but that it behaves nevertheless remarkably. The solutions of the stochastic equations are very complicated for the calculation of Q. Therefore only simulations of the solutions were discussed which describe the merging of the two peaks in the probability distribution P_N to one peak when passing the critical point from below, that means from the region of two coexistent phases to the region of one only. As N goes to infinity in the thermodynamic limit, it is to be replaced by a continuous concentration variable. In the following this one is called x.

The first model is a sum of two equally shaped square walls

$$w(x) = f\left(x - \frac{\sigma}{2}\right) + f\left(x + \frac{\sigma}{2}\right) \tag{5.20}$$

$$f(x) = \begin{cases} C & \text{for } |x| < \frac{1}{2} \quad (5.21) \\ 0 & \text{for } |x| > \frac{1}{2} \quad (5.22) \end{cases}$$

which merge to one if the parameter τ in

$$\sigma = \frac{1}{2} e^{-\tau} \tag{5.23}$$

passes through zero from below. One finds

$$Q = (\ln 2)^2 (1 - \sigma)\sigma \tag{5.24}$$

which is plotted in Fig. 1. The result indeed shows a dramatic behaviour near the critical point $\tau = 0$.

The second model is an analytic function

$$w(x) \sim \exp -(x^2 + \tau)^2 \tag{5.25}$$

which describes the merging of two peaks to one when τ passes the zero point form below. Q can be found as function of τ by series expansions and finally by numerical calculation. The result is plotted in Fig. 2. It again shows a dramatical behaviour at the critical point $\tau = 0$.

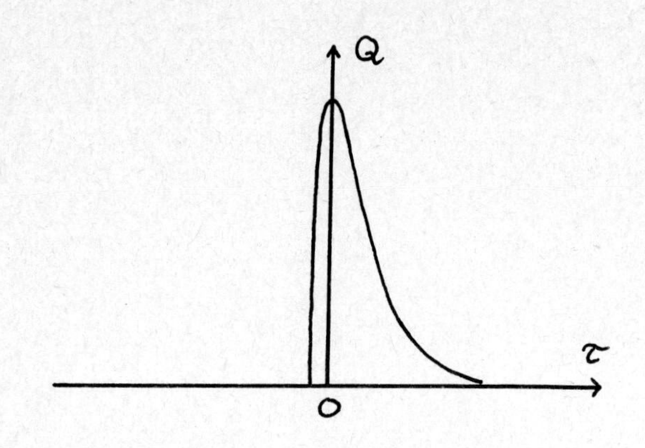

Fig. 1: Bit-number variance Q for two square walls merging to one at $\tau = 0$.

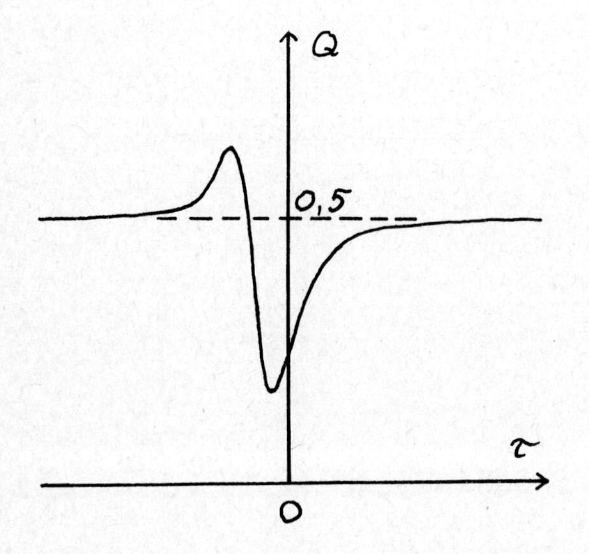

Fig. 2: Bit-number variance Q for an analytical probability distribution with two peaks merging to one at $\tau = 0$.

These results show indeed a new remarkable analogy between phase transitions in equilibrium and nonequilibrium. So far the analogy is shown on a phenomenological level. The deeper question remains open whether similar reasons in the nonequilibrium case can be found for the dramatic behaviour of Q as in the equilibrium case. Such reasons might be critical correlations, for instance.

About the relative bit-number variance Q_r we cannot say very much. It is

$$Q_r = \langle (\ln \frac{p}{p'})^2 \rangle - \langle \ln \frac{p}{p'} \rangle^2 \tag{5.26}$$

and becomes equal to 2K(p,p') if the deviations

$$\delta p = p - p' \tag{5.27}$$

are infinitesimal up to the quadratic terms in δp:

$$Q_r = \sum_i \frac{\delta p_i^2}{p_i'} + .. \approx 2\, K(p,p') \approx -\delta^2 S. \tag{5.28}$$

The positivity of this expression gives the stability of an equilibrium state p' in the same way as the positivity of K(p,p') does.

For infinitesimal deviations from a steady state p' the expression Q_r in (5.28) is again the Liapounoff function of the stability criterion of Glansdorff and Prigogine. Unlike to K, however, Q_r is in general not monotonous for final deviations δp. Therefore it has not the same features of a test function in Liapounoff's stability theory as K has.

To prove that Q_r in general is not a monotonous function of direct variables, it is sufficient to give one example for counter evidence. Such an example is any chemical reaction with variable concentration n of only one component. The general expression for generalized canonical distributions p,p' namely

$$Q_r = -(\lambda_\mu - \lambda'_\mu) \frac{\partial M}{\partial \lambda_\mu} (\lambda_\nu - \lambda'_\nu), \tag{5.29}$$

reduces to

$$Q = \frac{1}{RT} (\mu - \mu') \frac{\partial n}{\partial \mu} \tag{5.30}$$

in the isothermic case.

If the component is an ideal gas, the chemical potential is

$$\mu(T,n) = \mu_o (T) + RT \ln n \tag{5.31}$$

$$Q = n \left(\ln \frac{n}{n'}\right)^2, \tag{5.32}$$

which indeed is not a monotonous function of n.

References:

1 Glansdorff, P. and I. Prigogine: Thermodynamic Theory of Structure, Stability and Fluctuations. New York: Wiley (1971)

2 Haken, H. In: Festkörperprobleme X (Ed. O. Madelung). Braunschweig: Vieweg 1970.

3 Pytte, E. Thomas, H.: Phys. Rev. A 179, 431 (1969).

4 De Giorgio, V., Scully, M.O.: Phys. Rev. A2, 1170 (1970).

5 Grossmann, S., Richter, P.H.: Z. Physik 242, 458 (1971).

6 Woo, J.W.F., Landauer, R.: IEEE J. Qu. El. 7, 435 (1971).

7 Schlögl, F.: Z. Physik 253, 147 (1972).

8 Synergetics (ed. H. Haken) Stuttgart: Teubner 1973

9 Shannon, C., Weaver, W.:
The Mathematical Theory of Communication.
Urbana: University of Illinois Press 1949.

10 Kullback, S. Ann. Math. Statistics 22, 79 (1951).

11 Kullback, S.: Information theory and statistics.
New York: Wiley 1951.

12 Schlögl, F.: Z. Physik 249, 1 (1971).

13 Schlögl, F.: Z. Physik 243, 303 (1971).

14 Schlögl, F.: Z. Physik 191, 81 (1966).

15 Schlögl, F.: Z. Physik 248, 446 (1971).

16 Glansdorff, P., Prigogine, I.: Physica 46, 344 (1970).

17 Mori, H.: Phys. Rev. 115, 298 (1959).

18 Mc Lennan, J.A.: Advances in Chemical Physics
(ed. I. Prigogine). New York:
Interscience Publishers, Inc. (1963).

19 Zubarev, D.N.: Soviet Phys. Doklady 10, 526 (1965).

20 Janssen, H.K. to be published in Z. f. Physik

21 Schlögl, F.: Z. Physik 267, 77 (1974).

Cooperative Phenomena, H. Haken, ed.

CURRENT INSTABILITIES

H. Thomas
University of Basel, Switzerland

Instabilities of current distributions in conducting media are cooperative phenomena which bear a certain resemblance to phase transitions occurring in equilibrium systems. The dynamic behaviour of an equilibrium system undergoing a phase transition is characterized by the existence of a soft mode the frequency of which goes to zero at the stability limit, and of critical fluctuations of low frequency and large amplitude. It is of interest to determine whether analogous phenomena are associated with current instabilities, which occur in dissipative systems far from equilibrium.

As an introduction to the problem, we discuss in Section 1 briefly the simple case of instabilities in electric circuits consisting of isolated macroscopic elements without internal degrees of freedom. Section 2 introduces the basic concepts for the description of current instabilities in uniform continuous media. The electromagnetic interactions between the carriers, which give rise to the cooperative behaviour of the system, are studied in Section 3. In the mean-field approximation, they can be described in terms of a feedback mechanism, which relates the cooperative behaviour of the interacting system to the behaviour of a system of non-interacting carriers, and one obtains simple stability criteria. In Section 4, we review the dynamic behaviour for the two specific cases of the Gunn instability and the Erlbach instability.

1. INSTABILITIES IN ELECTRIC CIRCUITS

1.1. Negative-resistance elements

In many applications, an electric circuit can be considered to consist of isolated elements, the states of which are described by macroscopic variables like the current I through the element or the

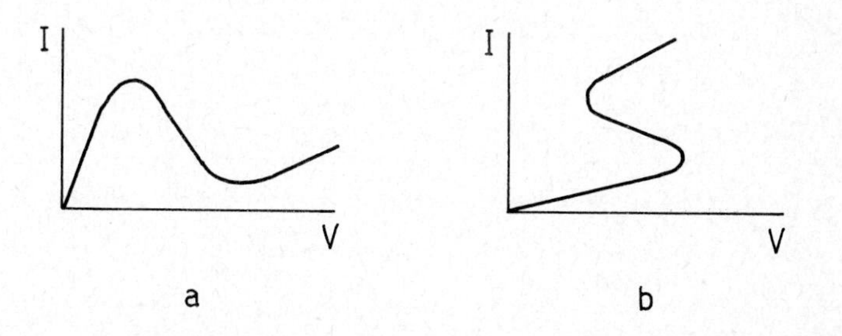

Fig.1. Nonlinear characteristics giving rise to instabilities.

a. N-shaped characteristic, voltage-controlled instability (example: tunnel diode)

b. S-shaped characteristic, current-controlled instability (example: gaseous discharge) voltage V across it. All internal degrees of freedom of the elements are disregarded. Instabilities can occur in circuits containing elements with negative differential resistance, i.e. with a nonlinear characteristic $I=g(V)$ of one of the two types displayed in Fig.1.

In order to analyse the stability of a steady state $\{V_o, I_o=g(V_o)\}$ of the system, we study its linear response

$$\delta I(\omega) = K(\omega)\,\delta V^{ext}(\omega) \tag{1.1}$$

to a small dynamic external perturbation $\delta V^{ext}(\omega)\exp(-i\omega t)$ of frequency ω. It is important to take into account the resistance R of the source,

$$V = V^{ext} - R^{-1} I \tag{1.2}$$

("load line"), and the reactances of the system (including any frequency dependence in the response of the element) which can be represented by a series inductance L and a parallel capacitance C. We thus have to consider the nonlinear element in the circuit shown in Fig.2. It is an elementary problem to calculate the current response function $K(\omega)$ in terms of R,L,C, and the differential conductance $G=g'(V_o)$ of the element. One obtains

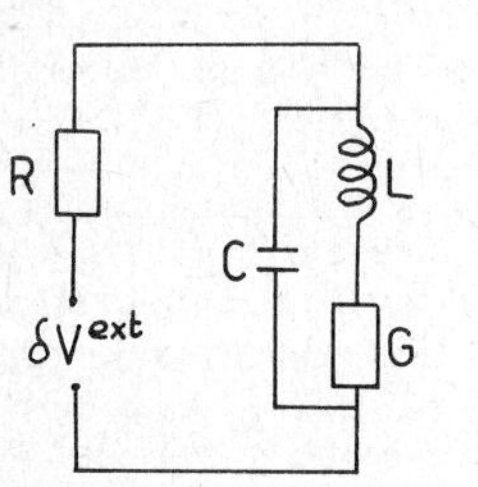

Fig.2. Circuit containing nonlinear element G

$$K(\omega) = \frac{G - i\omega C - \omega^2 GLC}{1 + GR - i\omega(GL+RC) - \omega^2 GRLC}\,. \tag{1.3}$$

The current I(t) in the circuit after switching off an initial perturbation can be represented as a superposition of normal modes $I\exp(-i\Omega t)$ the frequencies Ω of which are the poles of the response function $K(\omega)$ in the complex ω-plane:

$$1/K(\Omega) = 0. \tag{1.4}$$

The steady state is stable if and only if any initial perturbation decays in time, i.e. if all normal modes are represented by poles in the lower half plane. We thus have the stability criterion

$$\begin{aligned} \text{stable:}\quad & \mathrm{Im}\,\Omega < 0 \text{ for all normal modes} \\ \text{unstable:}\quad & \mathrm{Im}\,\Omega > 0 \text{ for at least one normal mode.} \end{aligned} \tag{1.5}$$

By applying this criterion, we find that the steady state becomes unstable for

$$G = -1/R \quad \text{for } R < \sqrt{L/C} \tag{1.6a}$$

$$G = -RC/L \quad \text{for } R > \sqrt{L/C}\,. \tag{1.6b}$$

In the case of Eq. (1.6a), the instability occurs via a "soft" relaxation-type mode which moves through the origin at the stability limit. In the case of Eq. (1.6b), on the other hand, the system becomes unstable by the undamping of a pair of oscillating modes which cross the real axis at finite frequencies.

It should be noted that in the above treatment we have tacitly made an assumption equivalent to the mean-field approximation (MFA): The replacement of the nonlinear element by the differential conductance is justified if the system responds linearly not only to the external perturbation δV^{ext} but also to the fluctuations occurring in the circuit. We represent the mechanism giving rise to the fluctuations by a noise source δV^{fluct}. If the above condition is satisfied, then the current fluctuations are given by

$$\delta I^{fluct} = K(\omega)\, \delta V^{fluct} \tag{1.7}$$

and can be linearly superimposed on the response to the external perturbation. However, if the fluctuations are large, they drive the element into the nonlinear regime, and the approximation breaks down.

Under ordinary circumstances, the fluctuations occurring in macroscopic elements will be very small, and the MFA is well justified. Near a stability limit, on the other hand, the fluctuations get enhanced by the same mechanism which gives rise to the instability, and one may anticipate the MFA to become invalid in a critical region around the stability limit. However, one might expect that fluctuations have a less pronounced effect in the "zero-dimensional" systems considered here where the instability is due to a single isolated mode, as compared to the case of a continuous medium where a whole branch of modes is affected by the instability.

1.2. Temperature-dependent conductors

In conductors with a strongly temperature-dependent conductance, Joule heating produces important effects and can give rise to instabilities. We discuss here the case that the sample is at a uniform temperature $T=T_o+\Theta$ determined by the balance between power input and heat transfer to the environment at temperature T_o, such that its state can be described by two variables, the voltage V across the element, and the temperature difference Θ. For simplicity, we consider a conductance which depends on Θ only,

$$I = G_o(\Theta)V. \tag{1.8}$$

We assume a linear law for the heat transfer, such that the temperature satisfies the equation

$$\dot{\Theta} = -\gamma\Theta + \frac{1}{c}P \tag{1.9}$$

where $P=VI$ is the power input and c is the heat capacity of the element. Linearization of these equations with respect to a deviation of frequency ω from a steady state yields

$$\delta I = G_o\, \delta V + b\, I_o\, \delta\Theta \tag{1.10}$$

$$\delta\Theta = (I_o\, \delta V + V_o\, \delta I)/c(\gamma - i\omega) \tag{1.11}$$

where $b=(1/G_o)dG_o/d\Theta$. By eliminating $\delta\Theta$ one finds the effective frequency-dependent conductance

$$G(\omega) = G_o \frac{c(\gamma - i\omega) + bP_o}{c(\gamma - i\omega) - bP_o} . \qquad (1.12)$$

Clearly, the static conductance may become negative for either sign of b, and one obtains an S-shaped or an N-shaped characteristic for $b > 0$ and $b < 0$, respectively. The possible instabilities of such an element can now be investigated with the method outlined in Section 1.1, where the frequency dependence of $G(\omega)$ has to be incorporated in the reactances L and C.

In reality, the temperature will not be uniform within the sample, because power dissipation occurs in the volume, but heat transfer only through the surface. The treatment of such a problem with a non-uniform steady state is outside the scope of this contribution. A critical review of the various physical effects, their interpretation, and the pertinant literature has recently been given by Landauer [1] .

2. CURRENT INSTABILITIES IN UNIFORM MEDIA

We consider electrical conduction in a continuous medium. For the long wavelengths considered, it is well justified to disregard the atomic structure and use a continuum description. A current state is described by the current density field $\tilde{j}(\tilde{x},t)$, which couples to an externally controlled electric field $\tilde{E}^{ext}(\tilde{x},t)$. Experimentally, only uniform or slightly non-uniform fields are available, but a field the space and time-dependence of which can be specified arbitrarily, is a useful theoretical concept. In general, the current state will depend in a very complex way on the history of the external field. In these notes, we shall restrict the discussion to uniform steady states and small perturbations thereof.

2.1. Linear response and normal modes

We assume that in the presence of a static uniform external field $\tilde{E}^{ext}$, there exists a uniform steady current state $\tilde{j}(\tilde{E}^{ext})$. We are interested in materials with a nonlinear characteristic $\tilde{j}(\tilde{E}^{ext})$ showing a negative differential conductivity. Such a characteristic may have one of the two forms shown in Fig.1, with V and I replaced by E^{ext} and j. The Gunn instability (Gunn [2]; for a review see Mc Groddy [3]) in certain single-crystal semiconductors like GaAs and Ge is an example for a behaviour according to Fig.1a. Certain glassy semiconductors (Ovshinsky [4]; for a review and references to other work see Fritzsche [5] and Ref. [6]) behave as in Fig.1b. On the other hand, an instability may also occur with respect to a direction perpendicular to the applied field. The Erlbach instability (Erlbach [7], Shyam [8]) in Ge is of this type. Finally, instabilities can also occur with respect to a finite wavelength. This would be reflected in the static conductivity for the corresponding wavevector q.

A small dynamic perturbation $\delta\tilde{E}^{ext}(\tilde{x},t)$ gives rise to a linear response

$$\delta\tilde{j}^{ind}(\tilde{x},t) = \int d^3x' \int_{-\infty}^{t} dt \;\; \tilde{\tilde{K}}(\tilde{x}-\tilde{x}'; t-t') \cdot \delta\tilde{E}^{ext}(\tilde{x}',t'). \qquad (2.1)$$

It is convenient to decompose the perturbation into Fourier components

$$\delta \underset{\sim}{j}^{ind}(\underset{\sim}{x},t) = \frac{1}{\sqrt{2\pi}}\sum_{q}\int_{-\infty}^{+\infty} d\omega \; \delta\underset{\sim}{j}^{ind}(\underset{\sim}{q},\omega)e^{i(\underset{\sim}{q}\cdot\underset{\sim}{x}-\omega t)}$$

$$\delta \underset{\sim}{E}^{ext}(\underset{\sim}{x},t) = \frac{1}{\sqrt{2\pi}}\sum_{q}\int_{-\infty}^{+\infty} d\omega \; \delta\underset{\sim}{E}^{ext}(\underset{\sim}{q},\omega)e^{i(\underset{\sim}{q}\cdot\underset{\sim}{x}-\omega t)} \tag{2.2}$$

The Fourier amplitudes are then related by

$$\delta\underset{\sim}{j}^{ind}(\underset{\sim}{q},\omega) = \underset{\approx}{\kappa}(\underset{\sim}{q},\omega)\cdot\delta\underset{\sim}{E}^{ext}(\underset{\sim}{q},\omega) \tag{2.3}$$

where

$$\underset{\approx}{\kappa}(\underset{\sim}{q},\omega) = \int d^3\xi \int_0^{\infty} d\tau \; \underset{\approx}{\kappa}(\underset{\sim}{\xi},\tau)e^{-i(\underset{\sim}{q}\cdot\underset{\sim}{\xi}-\omega\tau)} \tag{2.4}$$

is the current-response tensor, which depends on the state of the system. It contains information about the dynamics of the system, and its calculation is therefore an important problem. - It should be noted that $\underset{\approx}{\kappa}$ is different from the conventional conductivity tensor, because the latter is defined with respect to the local field $\underset{\sim}{E}$ which differs form $\underset{\sim}{E}^{ext}$ by the field induced by the response of the medium (see Section 3).

A problem closely related to linear response is the natural motion of the unperturbed system after switching off an initial perturbation. It is assumed that for small deviations from the steady state this can be written as a superposition of normal modes

$$\underset{\sim}{j}(\underset{\sim}{q},\Omega_q)e^{i(\underset{\sim}{q}\cdot\underset{\sim}{x}-\Omega_q t)} \tag{2.5}$$

The normal-mode frequencies Ω_q are the poles of the current response tensor $\underset{\approx}{\kappa}(\underset{\sim}{q},\omega)$ in the complex ω-plane, and the $\underset{\sim}{j}(\underset{\sim}{q},\Omega_q)$ are the corresponding eigenvectors. They can thus be found as the solutions of

$$\underset{\approx}{\kappa}^{-1}(\underset{\sim}{q},\Omega_q)\cdot\underset{\sim}{j}(\underset{\sim}{q},\Omega_q) = 0\,. \tag{2.6}$$

In order for the steady state to be stable, all deviations must decay in time, i.e. all Ω_q must lie in the lower half of the complex ω-plane. At the stability limit of the uniform state, at least one of the normal modes becomes undamped, and the corresponding pole crosses the real axis. For the dynamics of the instability, it is of prime importance to determine whether it is caused by a "soft mode" (crossing the real axis at $\omega=0$) as in equilibrium phase transitions, or by a finite-frequency mode.

2.2. Fluctuations and correlations

Even in the absence of external perturbations, the system is not truly static, but there occur spontaneous fluctuations $\delta\underset{\sim}{j}^{fluct}(\underset{\sim}{x},t)$ about the steady state. The properties of the fluctuations can be characterized by the two-point correlation tensor

$$\underset{\approx}{S}(\underset{\sim}{x}-\underset{\sim}{x}',\, t-t') = \langle\delta\underset{\sim}{j}^{fluct}(\underset{\sim}{x},t)\;\delta\underset{\sim}{j}^{fluct}(\underset{\sim}{x}',t')\rangle. \tag{2.7}$$

Its Fourier transform, the spectral tensor

$$\underset{\approx}{S}(\underset{\sim}{q},\omega) = \int d^3\xi \int_{-\infty}^{+\infty} d\tau \; \underset{\approx}{S}(\underset{\sim}{\xi},\tau)e^{-i(\underset{\sim}{q}\cdot\underset{\sim}{\xi}-\omega\tau)} \tag{2.8}$$

is related to the correlations of the Fourier amplitudes by

$$\langle \delta \underset{\sim}{j}^{fl}(\underset{\sim}{q}, \omega)\, \delta \underset{\sim}{j}^{fl}(\underset{\sim}{q}', \omega')\rangle = \underset{\approx}{S}(\underset{\sim}{q}, \omega)\, \delta(\underset{\sim}{q}+\underset{\sim}{q}')\, \delta(\omega+\omega') . \qquad (2.9)$$

The spectral tensor contains information about the dynamics of the current fluctuations, and its calculation is therefore another important problem.

It should be noted that the close connection which the Nyquist theorem provides between linear response and the fluctuation spectrum of an equilibrium system, does not exist for the type of system far from thermodynamic equilibrium considered here. Therefore, current response tensor and spectral tensor require separate calculation.

3. ELECTROMAGNETIC INTERACTIONS

Current instabilities are cooperative phenomena produced by the electromagnetic interactions between the carriers. For a system of non-interacting carriers, the differential conductivity could become negative, without causing an instability. In this Section, we therefore study the effect of the interactions on the dynamic behaviour of the system, following the presentation given in Ref. [9].

3.1. Interaction fields

The electromagnetic interactions give rise to an interaction field $\underset{\sim}{E}^{int}(\underset{\sim}{x},t)$ which acts on the carriers in addition to the external field $\underset{\sim}{E}^{ext}(\underset{\sim}{x},t)$. According to Maxwell's equations the Fourier component $\underset{\sim}{E}^{int}(\underset{\sim}{q}, \omega)$ is related to the Fourier component $\underset{\sim}{j}(q, \omega)$ of the total current by the equation

$$\omega^2 \underset{\sim}{E}^{int} + c^2 \underset{\sim}{q} \times (\underset{\sim}{q} \times \underset{\sim}{E}^{int}) = -4\pi \underset{\sim}{j} \qquad (3.1)$$

which can be solved for $\underset{\sim}{E}^{int}$. The result can be expressed in terms of the dynamical screening tensor

$$\underset{\approx}{s}(\underset{\sim}{q}, \omega) = 4\pi\, (\omega^2 \underset{\approx}{1} - c^2 \underset{\sim}{q}\underset{\sim}{q})/(\omega^2 - c^2 q^2) \qquad (3.2)$$

in the form

$$\underset{\sim}{E}^{int}(\underset{\sim}{q}, \omega) = -\frac{i}{\omega}\, \underset{\approx}{s}(\underset{\sim}{q}, \omega) \cdot \underset{\sim}{j}(\underset{\sim}{q}, \omega) . \qquad (3.3)$$

The principal axes of the screening tensor are the directions parallel and perpendicular to the wavevector $\underset{\sim}{q}$, corresponding to purely longitudinal and purely transverse waves, respectively. The corresponding eigenvalues are

$$s_{\ell}(q, \omega) = 4\pi \qquad (3.4a)$$

$$s_t(q, \omega) = 4\pi\, \omega^2/(\omega^2 - c^2 q^2) . \qquad (3.4b)$$

In the longitudinal case, the interaction field is produced by the space charge layers originating in div $\underset{\sim}{j}$. In the transverse case, it is generated by electromagnetic induction of the magnetic flux sheets originating in curl $\underset{\sim}{j}$.

It should be noted that on account of the linearity of Maxwell's equations, the relation (3.3) between interaction field and current density is strictly linear. Therefore, Eq. (3.3) holds separately for the field $\delta \tilde{E}^{ind}$ induced by the current response $\delta \tilde{j}^{ind}$ to an external perturbation, and for the field $\delta \tilde{E}^{fluct}$ induced by the current fluctuations $\delta \tilde{j}^{fluct}$, no matter how large the fluctuations.

3.2. Mean-field approximation

The behaviour of a system of interacting carriers under the action of the external field $\tilde{E}^{ext}$ is thus the same is the behaviour of a system of non-interacting carriers under the action of the total field $\tilde{E}^{ext} + \tilde{E}^{int}$. The mean-field approximation (MFA) is obtained by assuming that the system responds linearly not only to the external perturbation $\delta \tilde{E}^{ext}$ but also to the interaction field $\delta \tilde{E}^{fluct}$ generated by the fluctuations $\delta \tilde{j}^{fluct}$. Then, we obtain separately for the two contributions

$$\delta \tilde{j}^{ind}(\tilde{q}, \omega) = \underset{\approx}{\kappa}^{free}(\tilde{q}, \omega) \cdot [\delta \tilde{E}^{ext}(\tilde{q}, \omega) + \delta \tilde{E}^{ind}(\tilde{q}, \omega)] \tag{3.5}$$

and

$$\delta \tilde{j}^{fl}(\tilde{q}, \omega) = \delta \tilde{j}^{fl,free}(\tilde{q}, \omega) + \underset{\approx}{\kappa}^{free}(\tilde{q}, \omega) \cdot \delta \tilde{E}^{fl}(\tilde{q}, \omega) \tag{3.6}$$

where $\delta \tilde{E}^{ind}$ and $\delta \tilde{E}^{fluct}$ are given in terms of $\delta \tilde{j}^{ind}$ and $\delta \tilde{j}^{fluct}$, respectively, by Eq. (3.3). In MFA, the interactions are thus represented by a linear feedback mechanism as shown in Fig.3. This is quite similar to the method discussed in Ref.[10] for equilibrium systems.

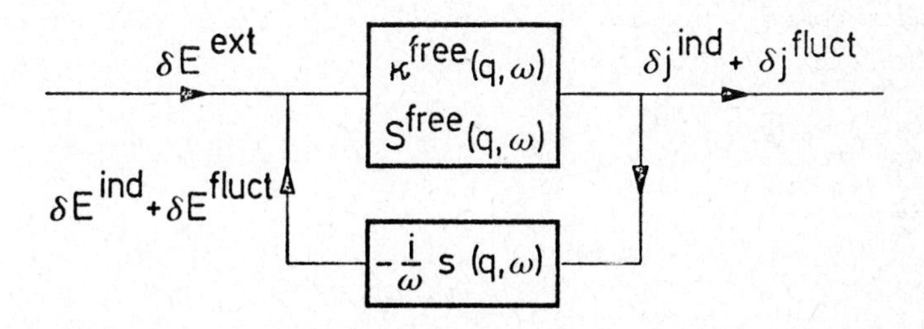

Fig.3. Representation of the electromagnetic interactions by a feedback mechanism.

The conductivity tensor $\underset{\approx}{\sigma}(\tilde{q}, \omega)$ is defined by the current response to the total non-fluctuating field $\delta \tilde{E}^{ext} + \delta \tilde{E}^{ind}$. Equation (3.5) shows that in MFA it is given by

$$\underset{\approx}{\sigma}(\tilde{q}, \omega) = \underset{\approx}{\kappa}^{free}(\tilde{q}, \omega) . \tag{3.7}$$

By making use of Eq. (3.3), we solve the feedback equations (3.5) and (3.6) for the resulting currents $\delta \tilde{j}^{ind}$ and $\delta \tilde{j}^{fluct}$ in terms of $\delta \tilde{E}^{ext}$ and $\delta \tilde{j}^{fluct,free}$, respectively. We can thus express the current response tensor and, by making use of Eq. (2.9), the fluctuation tensor of the interacting system

$$\underset{\approx}{K} = (\underset{\approx}{1}+(i/\omega)\,\underset{\approx}{\sigma}\cdot\underset{\approx}{s})^{-1}\cdot\underset{\approx}{\sigma} \tag{3.8}$$

and

$$\underset{\approx}{S} = (\underset{\approx}{1}+(i/\omega)\,\underset{\approx}{\sigma}\cdot\underset{\approx}{s})^{-1}\cdot\underset{\approx}{S}^{free}\cdot(\underset{\approx}{1}+(i/\omega)\,\underset{\approx}{\sigma}\cdot\underset{\approx}{s})^{\dagger\,-1}\cdot\underset{\approx}{\sigma} \tag{3.9}$$

in terms of the corresponding quantities $\underset{\approx}{\sigma} = \underset{\approx}{K}^{free}$ and $\underset{\approx}{S}^{free}$ of the non-interacting system.

3.3. Collective modes and fluctuations

We restrict the discussion to the case that $\underset{\approx}{\sigma}$ and $\underset{\approx}{S}^{free}$ have principal axes parallel and perpendicular to $\underset{\sim}{q}$, such that the normal modes are purely longitudinal or purely transverse.

In the longitudinal case one finds

$$K_\ell(\underset{\sim}{q},\omega) = \sigma_\ell(\underset{\sim}{q},\omega)/\varepsilon_\ell(\underset{\sim}{q},\omega) \tag{3.10}$$

and

$$S_\ell(\underset{\sim}{q},\omega) = S_\ell^{free}(\underset{\sim}{q},\omega)/|\varepsilon_\ell(\underset{\sim}{q},\omega)|^2 \tag{3.11}$$

where ε_ℓ is the longitudinal component of the dielectric tensor

$$\underset{\approx}{\varepsilon}(\underset{\sim}{q},\omega) = 1+\frac{4\pi i}{\omega}\underset{\approx}{\sigma}(\underset{\sim}{q},\omega). \tag{3.12}$$

The frequencies of the longitudinal collective modes (plasma waves) are thus obtained as the solutions of

$$\varepsilon_\ell(\underset{\sim}{q},\Omega_{\ell,q}) = 0 \tag{3.13}$$

in the complex ω-plane.

In the transverse case, one finds

$$K_t(\underset{\sim}{q},\omega) = \frac{\omega^2-c^2q^2}{\varepsilon_t(\underset{\sim}{q},\omega)\,\omega^2-c^2q^2}\,\sigma_t(\underset{\sim}{q},\omega) \tag{3.14}$$

and

$$S_t(\underset{\sim}{q},\omega) = \left|\frac{\omega^2-c^2q^2}{\varepsilon_t(\underset{\sim}{q},\omega)\,\omega^2-c^2q^2}\right|^2 S_t^{free}(q,\omega) \tag{3.15}$$

and the frequencies of the transverse collective modes are obtained as the solutions of

$$\varepsilon_t(q,\Omega_{t,q})\Omega_{t,q}^2-c^2q^2 = 0 \tag{3.16}$$

in the complex ω-plane.

Once the collective-mode frequencies have been found, the stability of the given uniform state can be tested. The state becomes unstable with respect to a longitudinal or to a transverse mode, when a root of Eq. (3.13) or (3.16), respectively, crosses the real ω-axis into the upper half-plane.

Equations(3.11) and (3.15) show that near a stability limit the fluctuations get enhanced by the same cooperative mechanism which gives rise to the instability. The MFA will therefore become invalid in a critical region around the stability limit.

4. SPECIFIC CASES

The above considerations have shown that in the mean-field approximation the linear response and the fluctuation spectrum of a particular system can be found from the behaviour of the corresponding system in the absence of electromagnetic interactions between the carriers according to Eqs. (3.8) and (3.9) in connection with Eq. (3.7). Thus, for the systems considered here, we have to investigate the transport properties of an electron gas in a semiconductor interacting only with the phonons and imperfections of the crystal. The treatment of this single-electron transport problem on a microscopic level, e.g. by means of the Boltzmann equation, is still quite complicated and requires powerful numerical methods. We have therefore studied the transport problem on a more macroscopic level by means of balance equations for the particle number and momentum densities, which describe the scattering processes in terms of phenomenologically introduced scattering rates. We briefly discuss here the two cases of the Gunn instability (Pytte [11]) and the Erlbach instability (Pfundtner [12]).

4.1. Gunn Instability

This instability which is associated with a characteristic of the type shown in Fig.1a, can occur in n-type semiconductors with a conduction band consisting of a high-mobility valley (1) which forms the bottom of the band, and a set of low-mobility valleys (2) which lie a few tenths of an electron volt higher in energy. At small values of the applied field, all electrons are in valley 1, and one finds ohmic conduction corresponding to its high mobility. As the field increases, the electrons are heated up, and when their energy becomes sufficiently high, a further increment of the field will give rise to competing contributions to the current: The drift velocity will continue to increase, but the carrier concentration in valley 1 will decrease due to electron transfer to the upper valleys. At the critical field, the two contribution cancel each other.

We treat the upper valleys formally as a single valley, and describe the state of the system by the particle densities n_1 and n_2 and the average drift velocities $\underset{\sim}{v}_1$ and $\underset{\sim}{v}_2$ in the lower and upper valleys, respectively. For simplicity, we restrict the discussion to uniform states (q=0). Particle conservation requires $n_1+n_2 = n_c$, where n_c is the total carrier concentration in the conduction band. The change of the state with time is determined by the balance equations for the particle and crystal-momentum densities (Fig.4). For uniform states, we write

$$dn_1/\,dt = -n_1 r_1 + n_2 r_2 = -\,dn_2/\,dt \tag{4.1}$$

and

$$\begin{aligned} d(n_1 m_1 v_1)/dt &= n_1 eE - (\gamma_1 + r_1) n_1 m_1 v_1 \\ d(n_2 m_2 v_2)/dt &= n_2 eE - (\gamma_2 + r_2) n_2 m_2 v_2 \end{aligned} \tag{4.2}$$

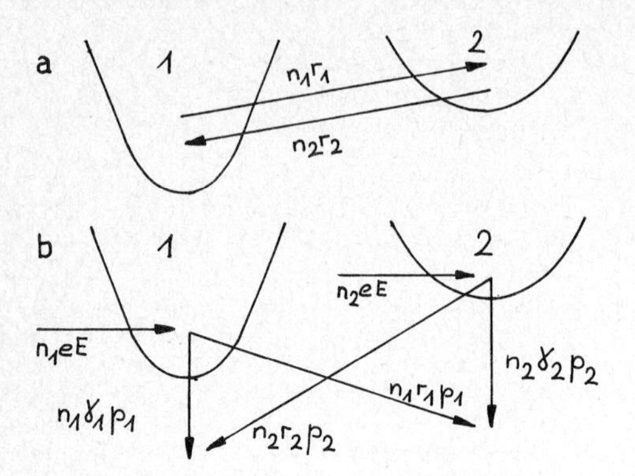

Fig.4. Mechanism of the Gunn instability for two-valley model
a. Particle density balance
b. Momentum density balance

Here, we have introduced phenomenological scattering rates γ_1, γ_2 for intravalley scattering and r_1, r_2 for intervalley scattering. Their dependence on drift velocity has to be obtained from a microscopic theory; in Ref. [11] we have made use of results based on approximate solutions of the Boltzmann equation (Conwell [13]).

The conditions for a stationary state in the presence of a constant field E are obtained by setting the left hand sides of Eqs. (4.1) and (4.2) equal to zero. Solving these conditions for n_1, n_2 and $\underset{\sim}{v}_1, \underset{\sim}{v}_2$ in terms of E and substituting into the expression for the current

$$\underset{\sim}{j} = e(n_1\underset{\sim}{v}_1 + n_2\underset{\sim}{v}_2) \tag{4.3}$$

gives the static characteristic. It is found that the steep rise of r_1 with v_1 when transfer from valley 1 to valley 2 becomes energetically possible does indeed lead to a negative differential conductivity (Pytte [11]).

The linear response to a dynamic perturbation

$$\delta\underset{\sim}{E}(t) = \delta\underset{\sim}{E}\, e^{-i\omega t} \tag{4.4}$$

is obtained by solving Eqs. (4.1) and (4.2) for small deviations $\delta n_1 = -\delta n_2$, $\delta\underset{\sim}{v}_1$, $\delta\underset{\sim}{v}_2$ from the stationary state in terms of $\delta\underset{\sim}{E}$. The frequency-dependent conductivity $\sigma_c(\omega;E)$ is then found from

$$\begin{aligned}\delta\underset{\sim}{j} &= e\,[(n_1\delta\underset{\sim}{v}_1 + n_2\delta\underset{\sim}{v}_2) + (\underset{\sim}{v}_1 - \underset{\sim}{v}_2)\delta n_1] \\ &= \sigma_c(\omega;E)\,\delta\underset{\sim}{E}.\end{aligned} \tag{4.5}$$

One recognizes easily the two competing contributions to the current due to drift and transfer. - The response of the bound electrons is described by the dielectric constant ε_L of the lattice, such that the total response of the semiconductor is given by the dielectric function

$$\varepsilon(\omega;E) = \varepsilon_L + \frac{4\pi i}{\omega}\sigma_c(\omega;E). \tag{4.6}$$

So far, we have calculated the response of non-interacting carriers. Instabilities can occur only by the electromagnetic feedback mechanism discussed in Section 3. According to Eq. (3.13), the zeros of $\varepsilon(\omega;E)$ are the eigenfrequencies of the longitudinal collective modes. Their dependence on the applied field is studied in detail in Ref. [11].

For a discussion of the dynamic properties connected with the instability, we disregard for simplicity the contribution of the low mobility valleys to the current, and assume that the intervalley transfer rates are small compared to the momentum relaxation rates (Drude cutoff). Then, interesting phenomena occur for frequencies small compared to the Drude cutoff frequency. In this frequency regime, we find for the real parts $\sigma'(\omega)$ and $\varepsilon'(\omega)$ the approximate expressions

$$\sigma'(\omega) = \sigma_0[1 - \Theta R/(\omega^2 + R^2)] \tag{4.7}$$

and

$$\varepsilon'(\omega) = \varepsilon_0 + 4\pi\, \sigma_0 \Theta/(\omega^2+R^2) \tag{4.8}$$

where

$$R = r_1+r_2; \qquad \Theta = v_1 dr_1/dv_1 \tag{4.9}$$

As the field increases, Θ rises steeply; at the critical field, Θ=R. The behaviour of $\sigma'(\omega)$ and $\varepsilon'(\omega)$ is shown in Fig.5. Parallel

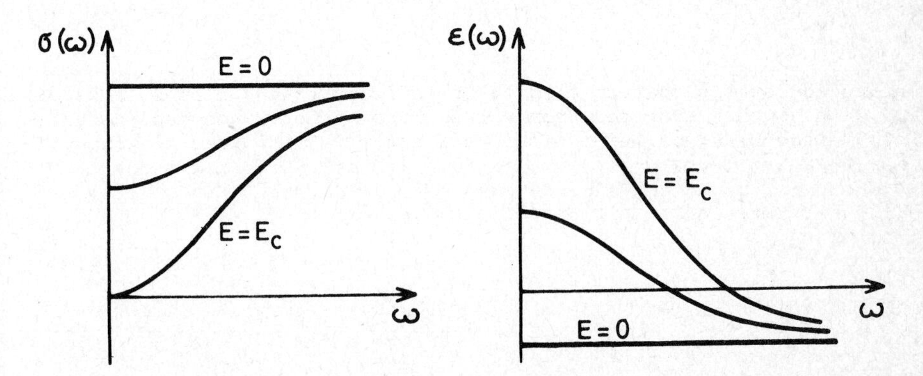

Fig.5. Variation of the real parts of conductivity $\sigma'(\omega)$ and dielectric constant $\varepsilon'(\omega)$ in the frequency range $\omega < R$ with applied field.

with the decrease of the static conductivity $\sigma'(0)=\sigma_0(1-\Theta/R)$, the static dielectric constant $\varepsilon'(0)=\varepsilon_0+4\pi\,\sigma_0\Theta/R^2$ rises from the value ε_0 which may be positive or negative, depending on the carrier concentration, to the large positive value $\varepsilon_0+4\pi\,\sigma_0/R$. Significant changes occur in the whole frequency range $\omega \lesssim R$.

From Eq. (3.13), one finds for fields close to the critical field a longitudinal mode with a purely imaginary frequency

$$\Omega_\ell = -\,i\Gamma_d \tag{4.10}$$

where

$$\Gamma_d = 4\pi\, \sigma'(0)/\varepsilon'(0) \tag{4.11}$$

is the dielectric relaxation rate. Since $_d$ goes to zero proportional to $\sigma'(0)$ as $E \to E_c$, we have thus found a soft dielectric relaxation mode. - The longitudinal fluctuation spectrum is given by

$$S_\ell(\omega) = \frac{1}{[\varepsilon'(0)]^2} \frac{\omega}{\omega^2 + \Gamma_d^2} S_\ell^{free}(\omega). \tag{4.12}$$

For the transverse modes, one finds from Eq. (3.15) close to the critical field a frequency

$$\Omega_t = \pm v_t q - \frac{i}{2}\Gamma_d \quad , \tag{4.13}$$

with a phase velocity

$$v_t = c/\sqrt{\varepsilon'(0)} \tag{4.14}$$

and a damping proportional to Γ_d. Thus, also this transverse mode becomes undamped for $E \to E_c$ in the long-wave-length limit. - The transverse fluctuation spectrum is given by

$$S_t(\underset{\sim}{q},\omega) = \frac{1}{[\varepsilon'(0)]^2} \frac{(\omega^2 - c^2 q^2)^2}{(\omega^2 - v_t^2 q^2)^2 + (\omega\Gamma_d)^2} S_t^{free}(\underset{\sim}{q},\omega). \tag{4.15}$$

4.2. Erlbach Instability

This instability which is associated with the current response in a direction perpendicular to the uniform field, can occur in n-type semiconductors with a conduction band consisting of a set of equivalent anisotropic valleys. The simplest model consists of two equivalent valleys 1 and 2 with anisotropic axes pointing into different directions (Fig.6). The static field E_0 is applied along the symmetry axis (x-axis). It induces drift velocity components

$$v_{1x}^0 = v_{2x}^0 \; ; \; v_{1y}^0 = -v_{2y}^0 \tag{4.16}$$

in the two valleys, and leaves the carrier concentrations unaltered:

$$n_1^0 = n_2^0 = n_0 \; . \tag{4.17}$$

Thus, because of the anisotropy, the drift velocity is tilted in

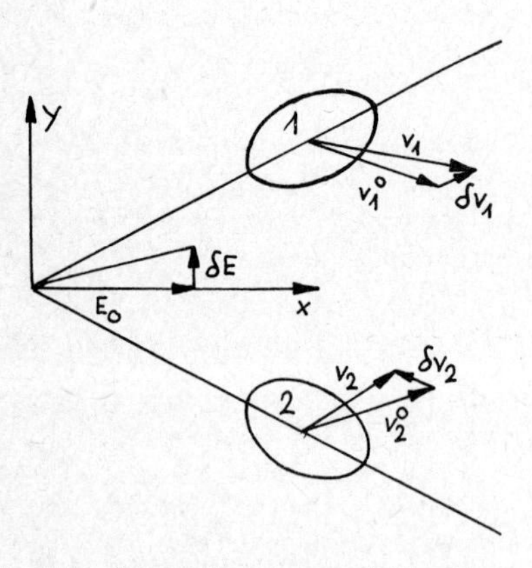

Fig.6. Mechanism of the Erlbach instability

each valley with respect to the field direction, but the perpendicular contributions to the current cancel on account of symmetry. A small field δE_y perpendicular to E_O induces a linear change in the drift velocities

$$\delta v_{1x} = - \delta v_{2x} \; ; \qquad \delta v_{1y} = \delta v_{1y} \tag{4.18}$$

leading to a positive contribution $2en_O \delta v_{1y}$ to the current δj_y. However, due to this change the total drift velocity in valley 1 has become larger than the total drift velocity in valley 2, and the electrons are scattered more effectively from valley 1 to valley 2 than in the opposite direction. The resulting unbalance

$$\delta n_1 = - \delta n_2 \tag{4.19}$$

in the carrier concentration leads to a negative contribution $2ev_{1y}^O \delta n_1$ to the current δj_y which is therefore given by

$$\delta j_y = 2e[n_O \delta v_{1y} + v_{1y}^O \delta n_1] \quad . \tag{4.20}$$

Thus, we find again a competition between contributions due to drift and transfer. At the critical field, the two contributions cancel.

The dynamic behaviour of this system has been studied in Ref. 12 by means of balance equations for particle and crystal momentum densities similar to Eqs. (4.1, 4.2). In contrast to model discussed for the Gunn instability, the two valleys are equivalent in this case, and the anisotropy of the effective masses is of crucial importance and has to be taken into account. - For the intravalley and intervalley scattering rates, use has been made of the results of an approximate microscopic calculation (Reik [14]). It is found that the scattering rates increase sufficiently strongly with the applied field for an instability to occur. The behaviour for fields close to the critical field is qualitatively very similar to the behaviour discussed above for the case of the Gunn instability: One finds again a soft longitudinal dielectric relaxation mode, and undamping of transverse modes.

It should be noted that the considerations presented here apply only to the states of uniform field distribution present for fields below the critical field. In the case of the Gunn instability, in the non-uniform state a high-field domain moves through the material with a velocity corresponding to the drift velocity of the carriers. Such a domain can in principle be obtained as the solution of the (nonlinear) balance equations for non-uniform states. Actually, the assumption of a uniform field distribution is an idealization even below the critical field, because the electrodes always give rise to field inhomogeneities. As a consequence, a Gunn domain is ordinarily nucleated in the non-uniform field region near the cathode before the bulk stability limit is reached. - In the case of the Erlbach instability, the structure of the non-uniform state is not known.

REFERENCES AND FOOTNOTES

[1] R. Landauer, J.W.F. Woo, Comments on Solid State Physics A2 (1974) 139.

[2] J.B. Gunn, Solid State Comm. 1 (1963) 88; IBM J. Res. Develop. 8 (1964) 141.

[3] J.C. McGroddy, M.I. Nathan, J.E. Smith, Jr., IBM J. Res. Develop. 13 (1969) 543; J.E. Smith, Jr., M.I. Nathan, J.C. McGroddy, IBM J. Res. Develop. 13 (1969) 554.

[4] S.R. Ovshinsky, Phys. Rev. Letters 21 (1968) 1450.

[5] H. Fritzsche, IBM J. Res. Develop. 13 (1969) 515.

[6] Report of the Ad Hoc Committee on the Fundamentals of Amorphous Semiconductors, National Academy of Sciences, Washington, D.C. 1972.

[7] E. Erlbach, Phys. Rev. 132 (1963) 1976.

[8] M. Shyam, H. Kroemer, Appl. Phys. Letters 12 (1968) 283.

[9] H. Thomas, Dynamics of Current Instabilities, in Synergetics, Edited by H. Haken, B.G. Teubner, Stuttgart 1973, p. 87. See also: H. Thomas, Current Instabilities: Critical undamping of Normal Modes, and Critical Fluctuations, Lectures presented at the Conference on Fluctuation Phenomena, Chania, Crete, August 1969. (To be published)

[10] H. Thomas, Mean-Field Theory of Phase Transitions, Lectures presented at the course "Local Properties at Phase Transitions" of the International School of Physics "Enrico Fermi", Varenna, July 1973 (to be published).

[11] E. Pytte, H. Thomas, Phys. Rev. 179 (1969) 431.

[12] K. Pfundtner, H. Thomas, Phys. kondens. Materie 16 (1973) 245.

[13] E.M. Conwell, M.O. Vassell, Phys. Rev. 166 (1968) 797.

[14] H.G. Reik, H. Risken, Phys. Rev. 124 (1961) 777.

Cooperative Phenomena, H. Haken, ed.

GRAVITATIONAL INSTABILITIES OF LIQUID CRYSTALS

P. G. de Gennes
Collège de France, 75231 Paris Cedex 05, France

Abstract
After a short review of the main types of liquid crystals, we discuss the instabilities which occur when a heavy fluid is placed on top of a lighter fluid, one (or both) of them being a liquid crystal. The most interesting case is found with twisted nematics (assumed immiscible). Here a relatively small twist angle may be enough to stabilize the structure.

I - GENERAL FEATURES OF LIQUID CRYSTALS

1) Classification

Liquid crystals are phases of intermediate symmetry between a true crystal and an isotropic liquid [1]. They are found mainly with long organic molecules. The chemical aspects are reviewed in a classic book by G. Gray [2]. A list of general references on the physical aspects is given under [3].

The two main types are called nematics and smectics. In the nematic phase (fig 1) the centers of the molecules are disordered as in a liquid, but the long axis of the molecules is alined along one direction : the liquid is optically uniaxial. The direction of the optical axis is usually described by a unit vector, or director $\underset{\sim}{n}$.

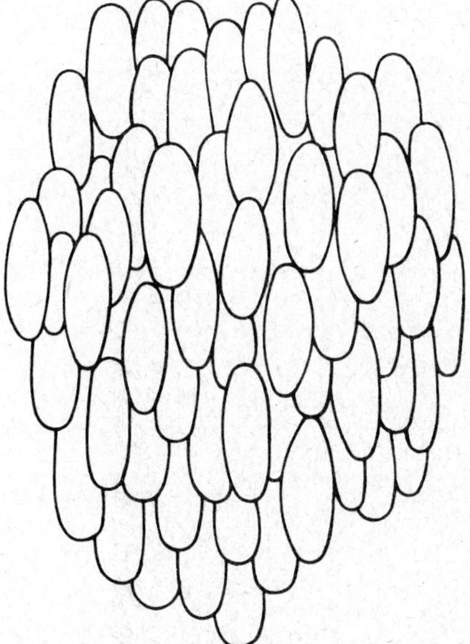

Fig. 1 - Molecular organisation in the nematic phase
(After R. Steinstrasser, L. Pohl, Ang. Chem. Inter. Ed. 12 (1973) 617 .

The smectic phases are more complex, but they are all characterized by a layer structure. The simplest smectic phase is called the smectic A, and is shown on fig 2 . Here each layer is a two dimensional fluid. The thickness of the layers is usually comparable to the molecular length, and of order 20 Å .

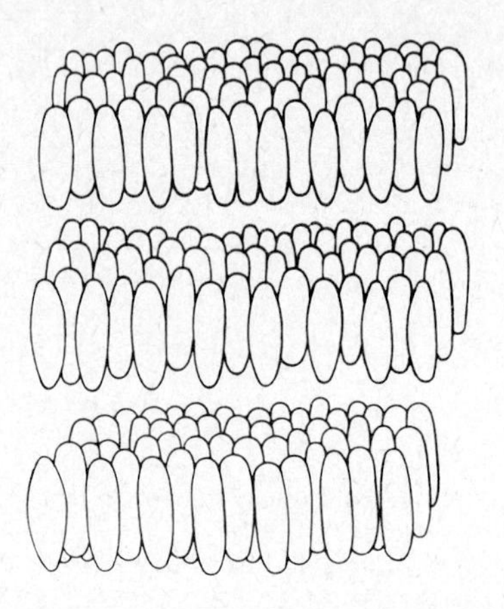

Fig. 2 - Molecular organisation in the smectic A phase
(After R. Steinstrasser, L. Pohl, Ang. Chem. Inter. Ed. 12 (1973) 617 .

Finally we should mention a variant of the nematic phase, which is obtained when the constituent molecules do not have mirror symmetry : then locally we still have nematic order, but on a large distance scale the director field is spontaneously twisted : the structure is helical as shown on fig 3 .

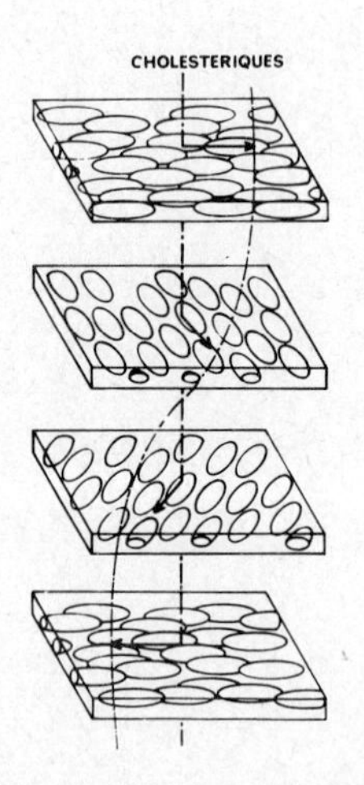

Fig. 3 - Molecular organisation in the cholesteric phase
(Courtesy J. Billard)

This variant is usually called cholesteric (since it is obtained most conspicuously with cholesterol esters). On a molecular scale it is very close to a nematic. But on a larger scale (for distances much larger than the helical pitch P) it is essentially a structure with one dimensional periodicity, and thus similar to a smectic.

2) Continuous elasticity

The main interest of liquid crystals is their flexibility, i.e. their strong large scale response to external perturbations. To discuss this one must first construct an elastic theory describing the energies stored in a weak distortion of the structure. For the nematics, this has been carried out most completely by Frank [4]. A simplified version of his distortion energy (per cm^3) may be written as

$$F_d = \frac{1}{2} K (\nabla n(\underset{\sim}{r}))^2 \equiv \frac{1}{2} K \; \partial_i n_j \; \partial_i n_j \tag{I.1}$$

and is then very similar to the Landau-Lifschitz distortion energy for a ferromagnet. The constant K has the dimensions of energy per unit length, and is of order 10^{-6} dynes.

In a smectic A the director is locked normal to the layers, and it is enough, to describe the distorted state, to specify the displacement of the layers u. Note that u is a one dimensional parameter : since each layer is fluid, displacements inside the layer are essentially irrelevant. In terms of u the elastic energy may be reduced to [5] :

$$F_d = \frac{1}{2} B \left(\frac{\partial u}{\partial z}\right)^2 + \frac{1}{2} K \left(\frac{\partial^2 u}{\partial x^2} + \frac{\partial^2 u}{\partial y^2}\right)^2 \tag{I.2}$$

Here z is the normal to the unperturbed layers. B is an elastic constant associated with changes in the interlayer thickness, and has the dimensions energy /cm^3. K describes the curvature energy of the layers, and is very similar to the constant K of eq. (I.1). To eq. (I.2) is associated a characteristic length

$$\lambda = (K/B)^{1/2} \tag{I.3}$$

This length is usually comparable to the repeat period, and is thus of order 20 Å. Note that eq (I.2) does not contain any term proportionnal to $\left(\frac{\partial u}{\partial x}\right)^2$ or $\left(\frac{\partial u}{\partial y}\right)^2$: for instance a small, constant, $\frac{\partial u}{\partial x}$ represents a simple rotation of the layers along y and not a real distortion.

Finally, the deformations of cholesterics may also be described by an energy of the form (II.2), and the constant K retains the same order of magnitude. But λ is now in the range of 1 micron.

3) Typical instabilities

In nematic fluids some very spectacular convective instabilities have been found : the principle of one of them is shown on fig 4, and a more detailed discussion can be found in ref [6] : This is a thermal instability induced by a temperature gradient. In an isotropic fluid, thermal instabilities occur only when the fluid is heated from below. With a nematic fluid, however, there are cases (such as the one shown on fig 4) where the instability is driven by heating from above! Also the gradients required at threshold are typically a hundred times smaller in nematics than in normal liquids.

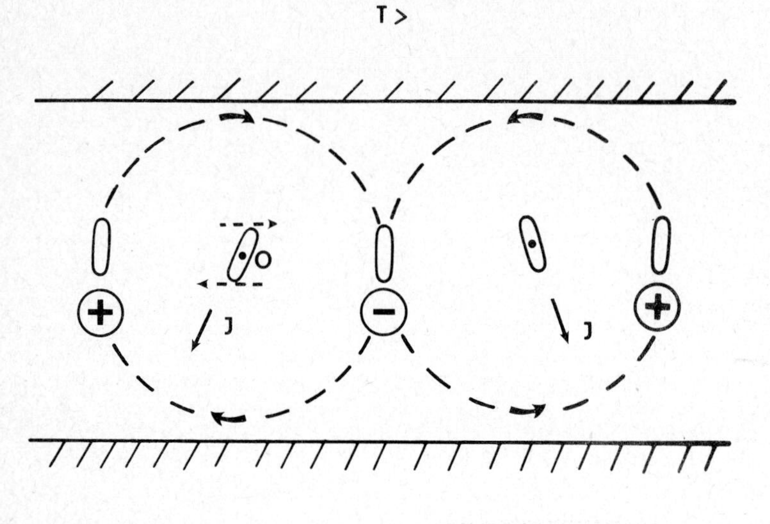

Fig. 4 - A typical instability in a nematic slab. The slab is prepared with its optical axis normal to the walls. A temperature gradient is applied, the upper plate being warmer. A small distortion of the molecular orientation (induced by thermal noise) is shown. The heat conductivity along the molecules $K_{/\!/}$ is larger than the transverse conductivity $K_{\perp}$. Thus the heat current J is deflected as shown, and the temperature is raised in the regions marked +, and decreased in the region marked -. The buoyancy forces then induce a flow pattern (dotted lines). This in turn tends to rotate the molecules as shown at point O, and thus to increase the original fluctuation.

Similar instabilities are found when the driving force is an electric field (through the very weak electric currents flowing inside these organic materials) [7]. It is thus possible to modify the optical properties of a thin (20 micron) slab by applying weak voltages ($\sim$ 10 volts) : this has found some interesting display applications.

Another class of instabilities has been found recently with purely mechanical driving forces : for instance in simple shear flow with the unperturbed optical axis normal to the velocity and to the velocity gradient [8]. Here it is found that above a critical shear rate S_c, the molecular alignment is distorted. The value of S_c is not large, and the corresponding Reynolds numbers are very weak ($\sim 10^{-3}$). Thus these purely mechanical instabilities are completely different from those of isotropic fluids [9].

In smectics and cholesterics one can also have convective instabilities. But the most spectacular effect is purely mechanical : it is illustrated on fig 5 . Relevant experiments are described in ref [10]. Here a very small traction on the upper plate (leading to displacements of order 200 Å) is enough to induce a static distortion of the smectic layers : the smectic fills the available space with undulated sheets rather than with flat sheets . This effect may also lead to some interesting devices [11] .

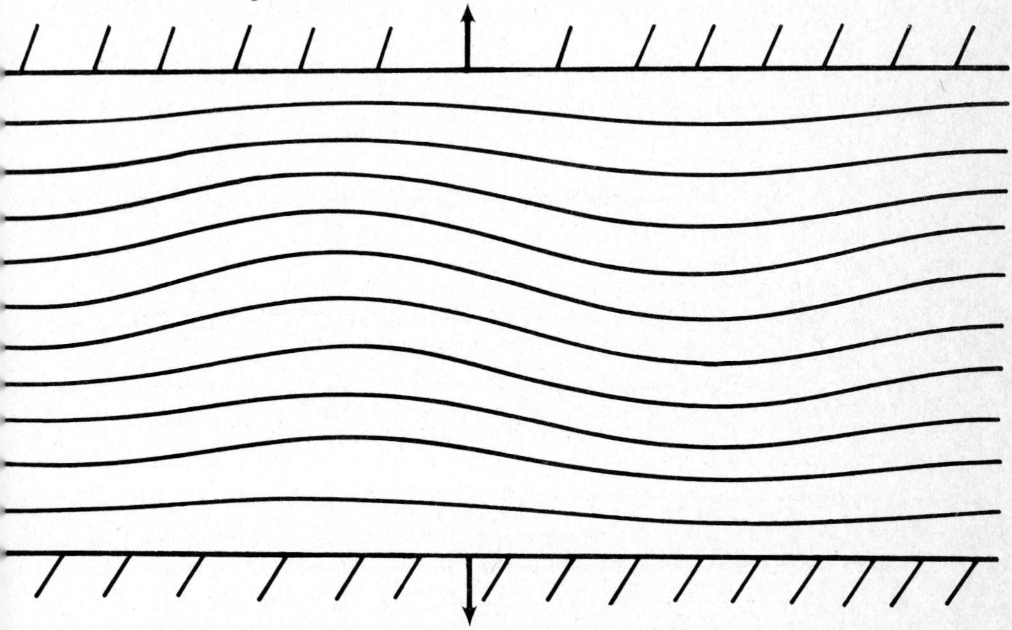

Fig. 5 - Static undulation mode induced in a smectic A by mechanical traction.

II - GRAVITATIONAL INSTABILITIES

1) Conventional liquids

Let us now consider a system of two immiscible liquids, separable by an interface which is horizontal at rest . If the upper liquid is denser, there is the possibility of an unstable situation . For conventional liquids, if we look at a small vertical displacement $\zeta(x)$ of the interface, we find a change of free energy (per unit area of the interface)

$$F_s = -\frac{1}{2}\,\Delta\rho g\,\zeta^2 + \frac{1}{2}\,A\left(\frac{d\zeta}{dx}\right)^2 \qquad \text{(II. 1)}$$

where $\Delta\rho$ is the density difference, g is the gravitational constant, and A the interfacial tension . Looking for modes of an infinite interface, of the form

$$\zeta = \zeta_\circ\; e^{iqx} \qquad \text{(II. 2)}$$

we find $F_s < 0$ (i.e. : instability) whenever $q < q_c$ where

$$q_c = \left(\frac{\Delta\rho\, g}{A}\right)^{1/2} \sim 10\ \mathrm{cm}^{-1} \qquad \text{(II. 3)}$$

Does this have a counterpart if we now replace one (or both) fluids by liquid crystals ? We shall now discuss this and show that with smectics the instability is suppressed, but that with nematics some interesting effects could occur.

2) Smectic-isotropic interface

The situation is represented on fig 6a .

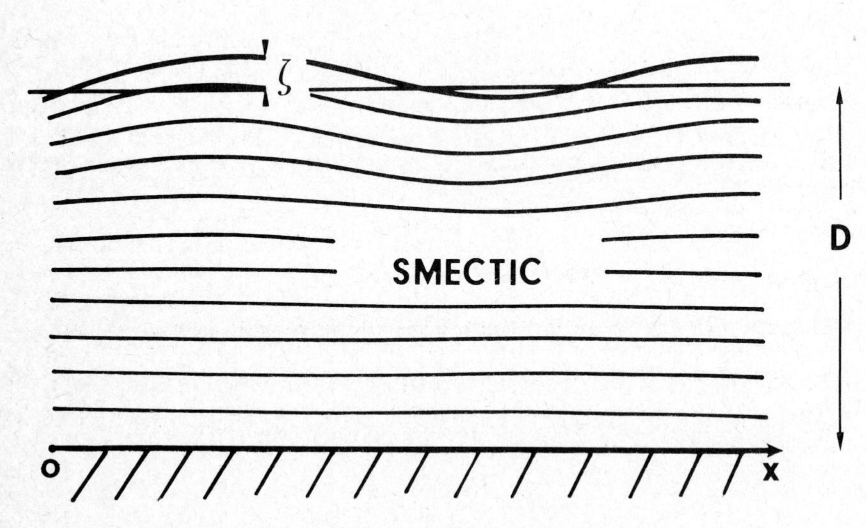

Fig. 6a - Distortion in a capillary mode of a smectic A .

The smectic slab, of thickness D, has its layers originally horizontal. It is lying on a solid surface, and is covered by a heavy liquid . We considered the distorted state associated with finite displacements ζ of the interface , and make the following assumptions .

a) The layers are anchored to both interfaces : this means that the displacement u satisfies

$$\begin{aligned} u(x, z = 0) &\equiv 0 && \text{(bottom plate)} \\ u(x, z = D) &\equiv \zeta(x) && \text{(smectic / isotropic interface)} \end{aligned} \qquad \text{(II. 4)}$$

b) The u field is continuous (no dislocations or other defects)

We then look at modes where the displacement is of the form

$$u(x, z) = u_\circ(z)\, e^{iqx} \tag{II. 5}$$

The energy F_D is then derived from eq (I. 2) and takes the form

$$F_D = \frac{1}{2} K \left[\left(\frac{d\, u_\circ}{d z}\right)^2 + \lambda^2 q^4 u_\circ^2 \right] \tag{II. 6}$$

The Lagrange Euler eq. for $u_\circ$ is then [12]

$$- \frac{d^2 u_\circ}{d z} + \lambda^2 q^4 u_\circ = 0 \tag{II. 7}$$

leading to the following form (compatible with the boundary conditions II. 4)

$$u_\circ = \zeta \frac{\sin h\, \lambda q^2 z}{\sin h\, \lambda q^2 D} \tag{II. 8}$$

Knowing the displacements we can integrate the distortion energy over the sample thickness and we obtain, instead of (II. 1) :

$$F_s = - \frac{1}{2} \Delta \rho g \zeta^2 + \frac{1}{2} A_{eff}\, q^2 \zeta^2 \tag{II. 9}$$

where A_{eff} is an effective, q dependent, surface tension, given by

$$A_{eff}(q) = A + B\lambda \;\; \cot\!an h \;\; (q^2 \lambda D) \tag{II. 10}$$

For large q , A_{eff} goes to a finite limit. But for small q , A_{eff} is very large : this expresses the fact that a finite ζ imposes a change of the interlayer distance : we than have to fight against the strong (solid type) elastic constant B . Note that $q^2 A_{eff}(q)$ is a monotonously increasing function of q . Thus instability occurs first at $q = 0$. In this region we have :

$$F_s \;(q \longrightarrow 0) = \frac{1}{2} \zeta^2 \left[- \Delta\rho g + \frac{B}{D} \right] \tag{II. 11}$$

In principle, for very large D ($D > D_c$) this can still be negative, and lead to an instability . However the critical thickness

$$D_c = \frac{B}{\Delta \rho g} \tag{II. 12}$$

is enormous . (B is of order 10^8 . Even putting a very heavy liquid such as mercury on top, we would have $D_c \sim 100$ meters !) . Thus in practice a smectic will be stable under those conditions . (*)

We may think of decreasing D_c by decreasing B . Since $B = K/\lambda^2$ K being roughly the same for all liquid crystals, this means going to large λ , i. e. substituting a cholesteric to the smectic. However, the boundary conditions(II.4)

(*) Instabilities due to large defects at the interface could still occur, but at very slow rates .

are then not adequate at the interface, and a more complicated system with two cholesteric slabs would be required . We shall dismiss this case, and go directly to a slightly more convenient situation .

3) Nematics

We consider now two superimposed nematic slabs (fig 6b) of thicknesses D_1 and D_2, and (for simplicity) restrict our attention to modes of very small q. We assume that the director $\tilde{n}$ is everywhere horizontal

$$n_x = \cos \varphi(z)$$

$$n_y = \sin \varphi(z)$$

$$n_z = 0$$

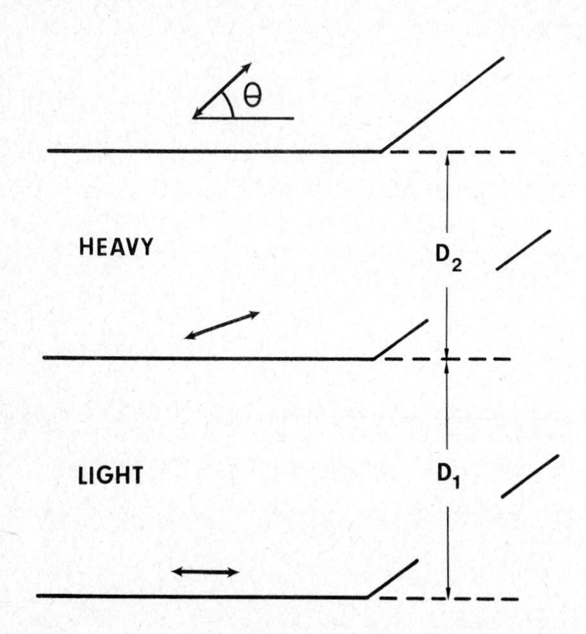

Fig. 6b - A system of two sumperimposed nematic films with twist .

The boundary conditions are the following

a) φ is continuous at the nematic nematic interface

b) $\varphi = 0$ at the bottom plate

c) $\varphi = \Theta$ at the upper plate

Conditions (b, c) can be realised in practice with suitably treated plates [3]. Note that for $\Theta \neq 0$ there is an overall twist imposed on the director field. Values of $|\Theta|$ smaller than $\frac{\pi}{2}$ can be achieved without difficulty (*) . We

---

(*) Larger values of Θ lead to instabilities through the emission of lines of discontinuities ("disclinations") .

now consider a displacement ζ of the interface and write the elastic energy (per cm^2 of interface) from eq (I. 1)

$$F_{el} = \int_0^{D_1+\zeta} \frac{1}{2} K_1 \left(\frac{\partial \varphi}{\partial z}\right)^2 dz + \int_{D_1+\zeta}^{D_2} \frac{1}{2} K_2 \left(\frac{\partial \varphi}{\partial z}\right)^2 dz \qquad (II.13)$$

Here K_1 and K_2 are the elastic constants of the two nematics. It is essential to have $K_1 \neq K_2$: If the two constants were equal, we would have a uniform helical distortion completely independent of the displacement ζ . The Lagrange Euler equations derived from (II. 13) at fixed ζ lead to constant slopes $\frac{\partial \varphi}{\partial z}$ in each medium, and more specifically to

$$K_1 \left.\frac{\partial \varphi}{\partial z}\right|_1 = K_2 \left.\frac{\partial \varphi}{\partial z}\right|_2 = C \qquad (II.14)$$

Physically, C is the torque exerted by one plate on the other, and transmitted by the nematic. The energy is then easily integrated to

$$F_{el} = \frac{1}{2} C \Theta \qquad (II.15)$$

We may split the total torsion Θ into its components Θ_1 and Θ_2 in the two fluids, and write (II. 14) in the form

$$\frac{K_1 \ \Theta_1}{\tilde{D}_1} = \frac{K_2 \ \Theta_2}{\tilde{D}_2} = C \qquad (II.16)$$

$$\text{where} \quad \left.\begin{aligned} \tilde{D}_1 &= D_1 + \zeta \\ \tilde{D}_2 &= D_2 + \zeta \end{aligned}\right\} \qquad (II.17)$$

Thus

$$\Theta = \Theta_1 + \Theta_2 = C \left(\frac{\tilde{D}_1}{K_1} + \frac{\tilde{D}_2}{K_2}\right) \equiv \frac{C}{G(\zeta)} \qquad (II.18)$$

where we have introduced

$$G(\zeta) = \left(\frac{\tilde{D}_1}{K_1} + \frac{\tilde{D}_2}{K_2}\right)^{-1} \qquad (II.19)$$

In terms of this function we may write

$$F_{el} = \frac{1}{2} \Theta^2 G(\zeta)$$

$$= \frac{1}{2} \Theta^2 \left[G(0) + \zeta G'(0) + \frac{1}{2} \zeta^2 G''(0) + \ldots \right] \qquad (II.20)$$

The first order term is not interesting for our purposes. The first order form will not contribute (for an infinite interface) when integrated over x with $\zeta = \zeta_0 \cos qx$. The third term is the interesting one for small motion instabilities. When added to the gravitational contribution it gives

$$F_s = \frac{1}{2} \zeta^2 \left[-\Delta\rho g + M\theta^2 \right] \tag{II.21}$$

where

$$M = G^3(0) \left(\frac{1}{K_1} - \frac{1}{K_2} \right)^2 \tag{II.22}$$

General stability conditions impose that $G(0)$ be positive. Thus M is positive and a finite twist θ may stabilize the structure. The threshold value of θ is

$$\theta_c = \left(\frac{\Delta\rho g}{M} \right)^{1/2} \tag{II.23}$$

Qualitatively we may write that M is of order

$$M \sim \frac{K}{D^3}$$

where K is an average elastic constant, and D is the total nematic thickness. Putting $K = 10^{-6}$, $D = 20$ microns, and $\Delta\rho = 0.01$ we find $\theta_c \sim 20°$ a very reasonable value. Going from $\theta > \theta_c$ to $\theta < \theta_c$ by a rotation of the upper plate, one should see the instability set in.

Experiments of this type have not been attempted yet. There are of course various difficulties -some of which are listed below :

a) One must find a couple of nematics which do not mix completely, and which build up an interface where the director is tangential (or nearly so). Also the elastic constants of the two materials must be significantly different.

b) We have discussed here only the instabilities of an infinite interface. For finite extensions along the x direction, the problem is much more complex, and has not been entirely solved (to the author's knowledge) even for isotropic fluids. We might fear an "instability wave" nucleates at the edges of the sample, even for $\theta > \theta_c$. However, this appears as a rather remote possibility ; in any case it could be distinguished optically from the general instability, occurring when $\theta < \theta_c$.

Finally, we must emphasize that our discussion has been strictly limited to static stability. For isotropic liquids and an infinite interface, the dynamics (at small ζ) are easily analysed [13]. For nematics the capillary waves are well understood [14] [15] ; dynamical calculations on gravitational instabilities would be rather complex, but are not unfeasible.

[1] G. Friedel, Annales de Phys. 18 (1922) 273.
[2] G. Gray "Molecular structure and the properties of liquid crystals" Acad. Press (1962).
[3] a) A. Saupe, Angewandte Chemie (Int. edition) 7 (1968) 97.
b) G. B. Brown, J. W. Doane, V. Neff, C. R. C. Critical rev. Solid State Sc. 1 (1970) 303.
c) P. G. de Gennes "The physics of liquid crystals" Oxford Un. Press (1974).
[4] F. C. Frank, Disc. Faraday Soc. 25 (1958) 19.
[5] P. G. de Gennes, Journal de Phys. 30 Colloq. C4 (Suppt to N°11-12) (1969) p. 65.
[6] P. Pieranski, E. Dubois Violette, E. Guyon, Phys. Rev. Lett. 30 (1973) 736.
[7] W. Helfrich, Molecular Crystals 21 (1973) 187.
[8] E. Guyon, P. Pieranski, Physical Review A, 9 (1974) 404.
[9] For a comparison between the hydrodynamics of ordinary fluids and of nematics, see P. G. de Gennes, Les Houches, lecture notes (1973) to be published by Gordon and Breach).
[10] M. Delaye, R. Ribotta, G. Durand, Phys. Lett. A44 (1973) 139.
[11] F. Kahn, Appl. Phys. Lett. 22 (1973) 111.
[12] G. Durand, C. R. Acad. Sci., Paris B275 (1972) 629.
[13] See for instance Chia Sun Yih "Fluid mechanics" Mc Graw Hill (1969)
[14] M. Papoular, A. Rapini, J. Phys. 30 (1969) 406
J. Phys. 31 (1970) C1-27.
[15] D. Langevin, M. Bouchiat, Mol. Crystals 22 (1973) 317
D. Langevin, Thèse, Paris (1974).

Cooperative Phenomena, H. Haken, ed.

DISSIPATIVE STRUCTURES WITH APPLICATIONS TO CHEMICAL REACTIONS

G. Nicolis,
Faculté des Sciences,
Université Libre de Bruxelles, Belgium.

Introduction

Our principal goal in these notes will be to analyze the origin of certain types of self-organization processes. More specifically, we shall try to understand the physical basis and the mathematical mechanisms responsible for the spontaneous emergence of ordered patterns in physico-chemical systems. The most striking example of such self-organizing systems are, of course, biological systems.

We first develop, in Part I, a purely phenomenological description of self-organization processes aiming to interpret the macroscopic properties of the observed patterns and to provide a preliminary classification of the various possible types of organized states. As we shall see, the transition to an organized configuration can best be understood as a phenomenon of branching of solutions of a certain set of nonlinear partial differential equations. Moreover, this phenomenon will be possible only if the system is driven beyond a critical distance from the state of thermodynamic equilibrium. In these lectures we deal primarily with chemical reacting and diffusing systems. Most of the phenomena will be illustrated on simple models, although the conclusions we shall draw from our analysis will be quite general and applicable to large classes of nonlinear differential systems of the parabolic type.

In Part II we attempt to provide a more fundamental explanation of the origin of transitions leading to ordered patterns by means of a stochastic analysis of fluctuations around non equilibrium states in the neighborhood of the branching points.

Throughout theses notes we focus on the qualitative aspects of the various questions and on the relation between our treatment and other methods encountered in the analysis of related problems. Although we shall attempt to make our presentation self-contained, those readers interested in technical details could also consult references [1] to [4] and [22] to [25].

PART I. MACROSCOPIC DESCRIPTION

I.1. Systems described by reaction-diffusion equations

It is almost a tautology to note that cooperative phenomena like the spontaneous appearance of ordered structures are only possible in macroscopic systems involving a large number of strongly coupled subunits. In many situations of interest these systems communicate with the external environment through the exchange of energy, matter or information. Fig. 1 represents such an open system containing n chemical species $1, \dots, n$ within a volume V .

We shall assume that the system is isothermal and at mechanical equilibrium. The latter restriction rules out convection. In physical chemistry and biology both conditions are very frequently obeyed to a good approximation. Within the framework of a macroscopic description the only state variables which are still allowed to evolve in time are the composition variables $X_1, \dots, X_n$ denoting e.g. the densities of the n chemical substances. Two kinds of processes may be responsible for the evolution of these variables :

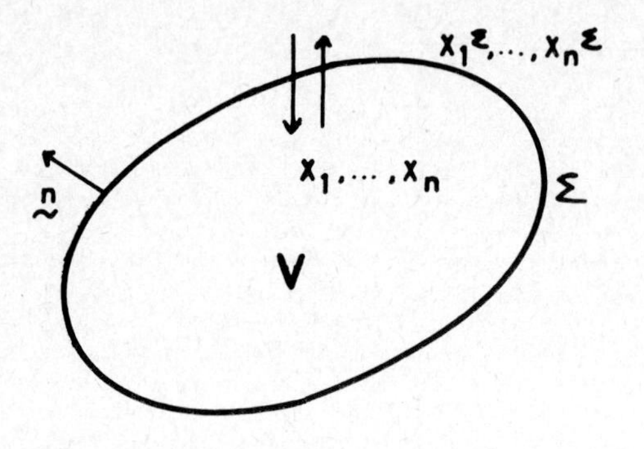

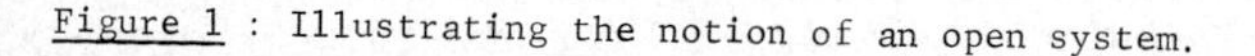
Figure 1 : Illustrating the notion of an open system.

- A local relaxation process, like e.g. a chemical reaction.
- The diffusion of 1, ..., n which will tend to damp local inhomogeneities. Diffusion can also be thought of as a coupling between neighbor spatial regions. Analytically, it can be approximated by Fick's law as long a the spatial gradients are appreciable only for distances much larger than the mean free path.

The equations describing the evolution of $X_1, \ldots, X_n$ take the form :

$$\frac{\partial X_i}{\partial t} = v_i(\{X_j\}) + D_i \nabla^2 X_i \quad (i=1,\ldots,n) \tag{I.1}$$

where v_i are the rates of production of X_i from the chemical reactions and D_i the diffusion coefficients of species i. For a physico-chemical system obeying the law of mass action v_i will be nonlinear (usually polynomial) functions of $\{X_j\}$'s. Thus, eq. (I.1) constitute a set of coupled nonlinear partial differential equations of the parabolic type. In order to have a well-posed problem, we will have to supplement these equations with the appropriate boundary conditions on the surface Σ of the volume V. Two types of condition will be envisaged :

- Dirichlet conditions :

$$\{X_1^{\Sigma}, \ldots, X_n^{\Sigma}\} = \{\text{const.}\} \tag{I.2 a}$$

- Neumann conditions :

$$\{ \underset{\sim}{n} \cdot \underset{\sim}{\nabla} X_1^{\Sigma}, \ldots, \underset{\sim}{n} \cdot \underset{\sim}{\nabla} X_n^{\Sigma} \} = \{ \text{const.} \} \tag{I.2 b}$$

Whenever some of the constants in (I.2 b) are equal to zero, then the system will not exchange the corresponding chemical substances with the external environment.

An alternative way to understand conditions (I.2) is to realize that they represent the constraints the system is experiencing from the outside world. In some exceptional situations these constraints will guarantee the equality of chemical potentials inside the system, on the surface and in the outside world. The system will then be in the state of thermodynamic equilibrium. In general however, it will be subject to gradients of chemical potentials which will drive it far from equilibrium.

Although the above relations have been set up for a reacting mixture, it is quite clear that they describe other types of situation as well. For instance, for a set of populations in an ecosystem, the term v_i will describe the competition for food or other types of interaction between the populations, whereas a suitably modified diffusion term may stand for motion within the ecosystem, or for migration. Moreover, with slight modifications and adjustments system (I.1) describes the dynamics of a laser, the activity of a population of excitatory and inhibitory neurons, etc...

I.2. A thermodynamic Theorem

Owing to the nonlinearity of the reaction rates in eq. (I.1), the dynamics of a purely dissipative system will in general not be expressed in terms of a potential function. Exceptions to this rule are systems near thermodynamic equilibrium, where the validity of Onsager's reciprocity relations guarantees the existence of certain state functions like free energy or entropy production playing the role of a potential. This ensures automatically, the asymptotic stability of the steady states of the system with respect to small perturbations. The proof of this statement is straightforward. Indeed, if a potential exists, eq. (I.1) will take the form :

$$\frac{\partial X_i}{\partial t} = -\frac{\delta \mathcal{L}(\{X_j\})}{\delta X_i} \tag{I.3}$$

and

$$\frac{\partial \mathcal{L}}{\partial t} = \sum_i \frac{\delta \mathcal{L}}{\delta X_i} \frac{\partial X_i}{\partial t} = -\sum_i \left(\frac{\delta \mathcal{L}}{\delta X_i} \right)^2 \leq 0 \tag{I.4}$$

If in addition $\mathcal{L}$ is chosen to be non negative everywhere in the space of $\{X_j\}$'s and zero only at the reference steady state, then according to Lyapounov's theorems [5], inequality (I.4) will imply asymptotic stability.

Conversely, the non existence of a potential in far from equilibrium situations leaves open the problem of stability of steady states. The following theorem, due to Glansdorff and Prigogine [6], describes the possibilities that may arise in this case.

Theorem : Consider a single phase, open, nonlinear system, subject to time-independent non equilibrium boundary conditions. Then steady states belonging to a finite neighborhood of the state of thermodynamic equilibrium, hereafter referred to as the thermodynamic branch, are asymptotically stable. Beyond a critical distance from equilibrium, they may become unstable.

By continuity the states on the thermodynamic branch will show equilibrium like behavior. In particular, in the limit of long times the system will be expected to attain a quasi-homogeneous steady solution. Thus, a self-organization process leading to ordered patterns will necessarily be accompanied by an instability of the solutions on the thermodynamic branch. The term dissipative structures [6] will be used to denote these ordered patterns, whose appearance and maintainance requires a minimum level of dissipation inside the system. For further remarks on thermodynamic stability theory we refer to Schlögl's paper in this Volume.

I.3. Examples of self-organization processes

Experimental examples for dissipative structure formation are now available both for non-biological and for biological reactions. Among the former, the Belousov-Zhabotinski reaction is the best documented example [7]. This process involves the oxidation of analogs of malonic acid by potassium bromate in the presence of Ce (or Fe, or Mn) ions. Among the most striking patterns observed [8] we mention the appearance of rotating propagating fronts of chemical activity which, in 3-dimensional media, can give rise to quite complex geometrical configurations.

The biological implications of the spontaneous emergence of order in a previously structureless medium are also quite obvious. Several biochemical networks at the cellular level have been shownto give rise to dissipative structures ([6], [9]). Even more spectacular phenomena occur in large scale processes such as embryonic development, morphogenesis and, even, systems of interacting populations, involving the emergence of spatial ordering in previously homogeneous media. In these cases one starts with a space (e.g. a "morphogenetic field") where some entities are distributed uniformly. Then, as a result of a distrubance, this homogenity is broken. The result is that the system evolves subsequently to states where the constituting entities are distributed regularly in space ; in addition they may sometimes exhibit, locally, periodic behavior. In the remaining of Part I we will attempt to understand the mechanism for breaking this homogeneity and to classify the kinds of structure one could evolve to subsequently.

I.4. Bifurcation Analysis of Dissipative Structures

The thermodynamic stability theory outlined in Section I.2 suggests that the natural approach to a mathematical analysis of dissipative structures would be in terms of the bifurcation and stability theory of partial differential equations [10]. The purpose of this theory is, precisely, to study the possible branchings of solutions which may arise under certain conditions. Suppose, for instance, that the right hand side of eq. (I.1) contains a parameter, ϵ , and that the family of (uniform) steady states on the thermodynamic branch depends smoothly on ϵ . It will always be possible to choose ϵ such that for ϵ small enough, the system (I.1) admits, for $t \to \infty$ a unique positive solution. But, there might exist critical values of ϵ at which new steady states arise, possibly exhibiting some kind of spatial organization, or at which time periodic solutions emerge. At these values the stability properties of thermodynamic branch will change and the uniqueness of solutions will fail.

In order to illustrate these phenomena we briefly compile here the results obtained recently ([1] - [4]) from the analysis of a simple chemical network [6], in fact the simplest one capable to exhibit cooperative behavior ([11], [12]) :

$$A \rightarrow X$$
$$B + X \rightarrow Y + D$$
$$2X + Y \rightarrow 3X$$
$$X \rightarrow E$$

The equations describing this mechanism are :

$$\frac{\partial X}{\partial t} = A - (B+1)X + X^2 Y + D_1 \frac{\partial^2 X}{\partial r^2}$$
$$\frac{\partial Y}{\partial t} = BX - X^2 Y + D_2 \frac{\partial^2 Y}{\partial r^2} \qquad (I.5)$$

We consider a single space dimension, with $0 \leq r \leq 1$. Note the presence of the cubic term in the right hand side, describing the "three-body collision" associated with the autocatalytic step in the reaction mechanism. It can be shown ([11], [12]) that for a system involving two variables this nonlinearity is the first nontrivial one compatible with the laws of chemical kinetics and allowing an instability of the thermodynamic branch.

Suppose the concentrations of substances A, B are held constant throughout the system and the concentrations of X, Y obey the following boundary conditions :

$$X(0) = X(1) = A$$
$$Y(0) = Y(1) = \frac{B}{A} \qquad (I.6\ a)$$

or

$$\frac{\partial X(0)}{\partial r} = \frac{\partial X(1)}{\partial r} = \frac{\partial Y(0)}{\partial r} = \frac{\partial Y(1)}{\partial r} = 0 \qquad (I.6\ b)$$

Then, eq. (I.5) admit a unique uniform steady state solution on the thermodynamic branch :

$$X_0 = A, \qquad Y_0 = \frac{B}{A} \qquad (I.7)$$

A linear stability analysis ([1] - [4], [6]) reveals that this state may become unstable and that, under appropriate conditions, the system may evolve to a steady state dissipative structure. Under other conditions one has bifurcation of a standing or propagating concentration wave. In the limit of very fast diffusion the latter solutions reduce to a homogeneous temporal oscillation of the limit cycle type. The quantity playing here the role of the bifurcation parameter ϵ is the fixed concentration of B.
The surprising aspect of these transitions is that one obtains quite different behavior depending on the symmetry properties of the critical mode. For a steady state solution bifurcating from (I.7) the latter has the form :

$$\begin{pmatrix} X \\ Y \end{pmatrix} = \begin{pmatrix} A \\ B/A \end{pmatrix} + \begin{pmatrix} x_0 \\ y_0 \end{pmatrix} \sin n_c \pi r \qquad (I.8\ a)$$

for the boundary conditions (I.6 a) and

$$\begin{pmatrix} X \\ Y \end{pmatrix} = \begin{pmatrix} A \\ B/A \end{pmatrix} + \begin{pmatrix} x_0 \\ y_0 \end{pmatrix} \cos n_c \pi r \qquad \text{(I.8 b)}$$

for the boundary conditions (I.6 b), where n_c is the critical wavenumber. When n_c is even, the system exhibits a symmetry-breaking transition to two possible new states, both of which are stable. Fig. 2 describes the concentration profile of X in these states.

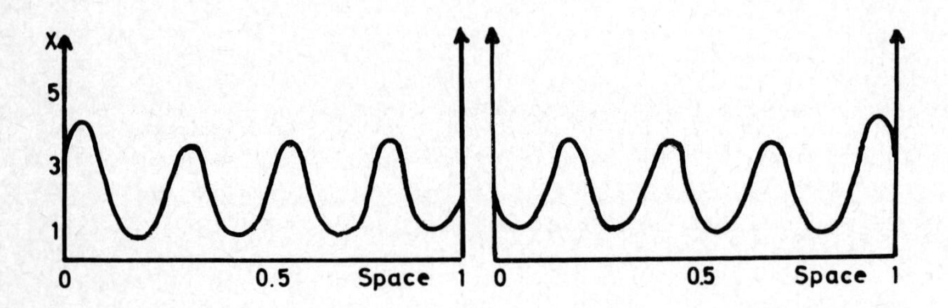

Figure 2 : Steady state dissipative structures for X in the case of an even critical wave number. $A = 2$, $B = 4.6$, $D_1 = 0.0016$, $D_2 = 0.0080$.

The bifurcation mechanism leading to these states is illustrated on the bifurcation diagram of Fig. 3.

In contrast, when the critical wave number is odd one observes both symmetry-breaking as well as bistable behavior combined with hysteresis, as shown on the bifurcation diagram of Fig. 4.

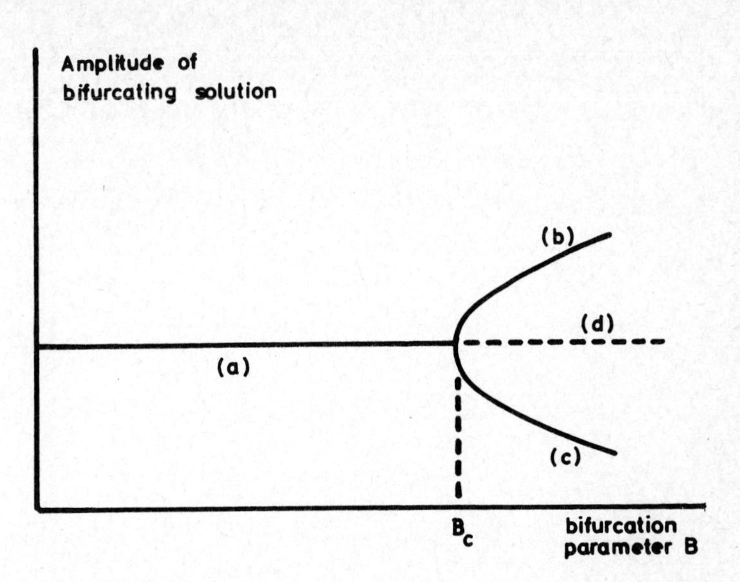

Figure 3 : Bifurcation diagram for an even critical wavenumber. (b), (c) = stable dissipative structures emerging beyond a symmetry-breaking instability.

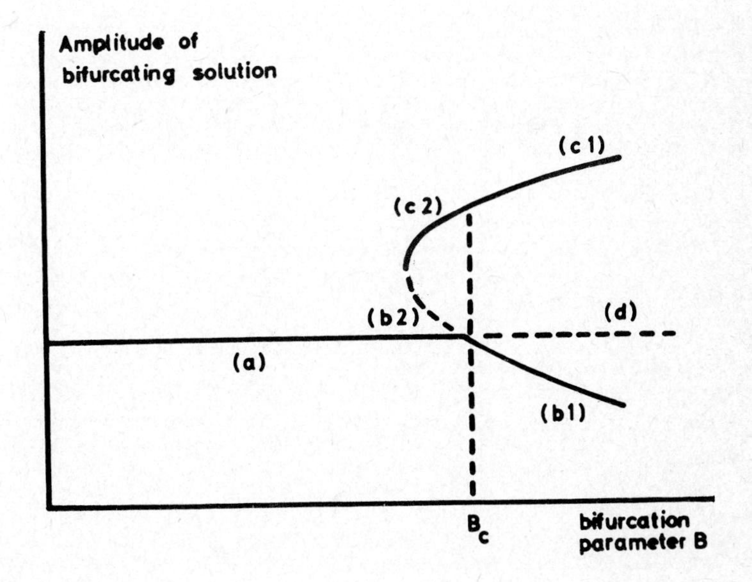

Figure 4 : Bifurcation diagram for an odd critical wave number. (C1), (C2) = branches arising by a finite jump from the thermodynamic branch (a) or (d).

For $B < B_c$ there may be two stable solutions, the uniform one and a dissipative structure. For $B \geq B_c$ there are two dissipative structures, one of which requires an abrupt transition from the uniform solution. For B sufficiently large these two solutions will be reached with equal probability by perturbing slightly the thermodynamic branch and one would have, again, the phenomenon of symmetry-breaking.

The spontaneous character of the symmetry-breaking transitions should be pointed out. One can really speak of a self-organization process whereby the system evolves by itself to a solution representing an endogenous spatial distribution or rhythmicity. Another striking point is that when the boundary conditions (I.6 b) apply, then the solution will be dominated by the critical mode (I.8 b). The spatial distribution of X and Y will exhibit, in this case, a macroscopic concentration gradient along the system, as the values of X or Y will be different at r= 0 and r = 1. We have therefore a spontaneous onset of polarity in the system, an effect which was long considered to be a puzzle in developmental biology.

The bifurcation of time-periodic solutions can also be handled straightforwardly with the same techniques. Particularly striking in this case is the appearance of propagating fronts of chemical activity [2], which bear many similarities with those observed in the Belousov-Zhabotinski reaction [8]. The biological implications of this behavior seem to be far-reaching ([8], [9]). Finally, when the concentration of the initial product A in the reaction scheme is no longer imposed throughout but is left, instead, to adjust by diffusion, then the dissipative structures may become localized inside "natural" boundaries within the reaction space ([2] - [4]). These boundaries seem to be related to the turning points of the differential system and are therefore determined intrinsically by the system itself, independently of the external environment.

One fascinating point which remains practically unexplored is the multiplicity of dissipative structures. In principle, each time the bifurcation parameter crosses one of the denumerably infinite critical values corresponding to a marginal stability of the thermodynamic branch, a new structure becomes available to the system. These various structures are dominated by different wavelengths and have different amplitudes. This whole hierarchy of transitions has tempting analogies with what one expects to happen in large scale biological phenomena like development and morphogenesis. However, the stability of these patterns is still an open question. Computer simulations indicate that those structures which emerge after the first few bifurcations, remain stable even for much higher values of the bifurcation parameter.

Several of the phenomena described in this section present striking similarities with what is observed in quite different areas like fluid dynamics, laser theory or the activity of neural masses. The lecture notes by Haken and Wilson in this Volume should be consulted for these topics. This frequent appearance of a limited number of of systematic patterns prevailing in self-organization processes suggests that most of the results described in these notes are not dependent heavily on the choice of models but rather, that they reflect some deep properties of nonlinear parabolic systems which one only begins to suspect and to explore.

I.5. Comparison with the theory of catastrophes

In the preceding sections we outlined a mechanism of spontaneous pattern formation based on the phenomenon of bifurcation and branching of solutions of nonlinear partial differential equations of the reaction-diffusion type. In recent years, Thom [13] and his followers elaborated a detailed theory aiming to analyze this phenomenon, based on an analysis of the structural stability of ordinary differential equations describing potential systems with a finite number of degrees of freedom. The mathematical structure of these equations is closely related to eq. (I.3) :

$$\frac{dx_i}{dt} = - \frac{\partial \mathcal{L}(\{x_j\},\{\mu\})}{\partial x_i} \quad (i=1,\dots,n<\infty) \tag{I.9}$$

The difference is that $\mathcal{L}$ is now an ordinary function and not a functional like in eq. (I.3). Moreover, the set of parameters μ in eq. (I.9) determines the points of structural instability of the steady state solutions. The latter points are manifolds of the parameter space on which two or more such solutions coalesce. Taking μ to stand for the space coordinates, one obtains then the shape of spatial domains corresponding to different kinds of regimes described by eq. (I.9) ; the boundaries separating these domains being the loci of points of structural instability.

Two major differences can immediately be established between this approach and the one outlined in the previous sections.

a. One is limited to potential systems which, from the point of view of irreversible thermodynamics, seem to arise in exceptional situations corresponding to systems near thermodynamic equilibrium.
b. The explicit influence of diffusion is not taken into consideration. Instead, the "reaction rates" in eq. (I.9) are taken to depend explicitly on space.

Point (a) rules out, automatically, the possibility of temporal organization in the form of sustained oscillations. Indeed (see also relation (I.4)) :

$$\frac{d\mathcal{L}}{dt} = - \sum_i \left(\frac{dx_i}{dt}\right)^2 \tag{I.10}$$

Integrating over a time interval T one obtains :

$$\Delta \mathcal{L} \equiv \mathcal{L}(t_1+T) - \mathcal{L}(t_1) = - \int_{t_1}^{t_1+T} d\tau \sum_i \left(\frac{dx_i}{d\tau}\right)^2 \leq 0 \tag{I.11}$$

If a periodic solution could exist, then $dx_i/d\tau$ would only vanish on a set of points of measure zero in the interval (t_1 , $t_1 + T$). Thus, after one complete period :

$$\mathcal{L}(t_1 + T) < \mathcal{L}(t_1) \tag{I.12}$$

which is in contradiction with the existence of a (regular) potential function $\mathcal{L}$ satisfying the condition $\mathcal{L}(t_1+T) \equiv \mathcal{L}(t_1)$. We may note, however, that these arguments do not exclude the possibility of almost-periodic solutions.

Coming now to point (b), it may be observed that imposing on the "reaction rates" an external space dependence implies that the initial spatial homogeneity is broken by some unspecified external action. The spontaneous character of self-organization processes is therefore not accounted for. Moreover it is natural to expect that those quantities which describe a biological function will depend both on space and time and thus will be governed by partial differential equations. This will result in a mathematical problem involving an infinity of coupled degrees of freedom for which Thom's classification of the regions structural instabilities, or catastrophes according to his terminology, will no longer be applicable. At this time there exists no classification of the solutions of partial differential equations comparable in generality to Thom's theory for ordinary differential equations. However, by performing bifurcation analyses as in the previous section, one can construct explicit forms of the solutions in the neighborhood of the bifurcation points and obtain in this way a preliminary classification of these solutions.

PART II. THE ONSET OF A SELF-ORGANIZATION PROCESS

II.1. The role of fluctuations

We want to understand now, on a microscopic basis, the spontaneous emergence of patterns of the dissipative structure type. To simplify the analysis we restrict ourselves to systems which remain macroscopically homogeneous. Let $\{\bar{X}_j\}$ be a set of macroscopic state variables, which are assumed to evolve according to the equations (I.1) (with diffusion terms neglected) :

$$\frac{d\bar{X}_i}{dt} = v_i(\{\bar{X}_j\}) \quad (i = 1, \cdots, n) \tag{II.1}$$

From the point of view of mechanics a macroscopic variable is a statistical average, over a suitable probability distribution function, of a mechanical quantity. Thus :

$$\bar{X}_i \longrightarrow \langle X_i \rangle \tag{II.2}$$

For instance, if X_i stands for the number of particles of a constituent i within the reaction volume, $\langle X_i \rangle$ in (II.2) will be given by :

$$\langle X_i \rangle = \sum_{X_i=0}^{\infty} X_i \, P(\{X_j\}, t) \tag{II.3}$$

where P is the probability function. Starting from the equation of evolution for P an equation for $\langle X_i \rangle$ can be obtained by taking the first moment of the equation for P. This relation will have the form :

$$\frac{d\langle X_i \rangle}{dt} = \langle v_i(\{X_j\}) \rangle \tag{II.4}$$

In a nonlinear system $\langle v_i(\{X_j\}) \rangle$ will in general differ from $v_i(\{\langle X_j \rangle\})$ by terms related to the mean square deviation of the variables or to higher order terms. Thus :

$$\frac{d\langle X_i \rangle}{dt} = v_i(\{\langle X_j \rangle\}) + \text{terms containing } \{\langle (\delta X_i)^2 \rangle - \langle X_i \rangle\} + \cdots \tag{II.5}$$

where

$$\langle (\delta X_i)^2 \rangle = \sum_{X_i} (X_i - \langle X_i \rangle)^2 \, P(\{X_j\}, t) \tag{II.6}$$

We notice that the first term in the right hand side of (II.5) is identical to the one appearing in the phenomenological equation (II.1). The corrections to the phenomenological description arise from the statistical deviations from the average behavior, that is, from the fluctuations. Suppose now that fluctuations are small and (II.1) is valid to a good approximation. Let $\{\bar{X}_j^0\}$ be a steady state solution. By varying slowly the bifurcation parameter denoted by ϵ in Section I.4 we drive the system to the threshold for an instability. Obviously, if the description based on (II.1) remains valid the system is not going to evolve spontaneously to a new state, but is going to remain on the unstable branch of solutions. According to eq. (II.5) however, we see that $d\langle X_i \rangle/dt$ is going to **increase** - and thus the system will be able to evolve spontaneously from the unstable solution - provided the fluctuations may increase in time and take appreciable values. The purpose of the following sections will be to analyze the conditions under which this evolution through fluctuations is possible.

II.2. The birth and death description of fluctuations

The behavior of fluctuations can be analyzed by means of a master equation based on the recognition that eq. (II.1) define a stochastic process in the space of certain appropriate variables. The most natural assumption is that the latter are of the same nature as the macroscopic variables appearing in eq. (II.1), and that the stochastic process is a stationary Markov process. Then one obtains a master equation of the Kolmogorov type [14] :

$$\frac{dP(\{X_i\},t)}{dt} = \sum_{\{X'_i\}} w(\{X'_i\}|\{X_i\})\, P(\{X'_i\},t) \qquad \text{(II.7)}$$

where w is the transition probability per unit time between states $\{X'_i\}$ and $\{X_i\}$. Equations of this type have been used extensively in chemical kinetics, among others by Mc Quarrie [15] (for equilibrium situations), Nicolis and Babloyantz [16] (for linear networks far from equilibrium), Nicolis and Prigogine ([17], [18]), Schlögl [19], Walls et al [20] (for nonlinear systems). In these investigations, the additional assumption is made that the stochastic process is of the birth and death type. As a result, for dilute mixtures, the transition probabilities w become proportional to products of the stochastic variables $\{X_i\}$. For instance, for the reaction

$$X + Y \xrightarrow{k} E$$

one would have :

$$w(X', Y'|X, Y) \equiv w(X+1, Y+1|X, Y) = k\,(X+1)(Y+1) \qquad \text{(II.8)}$$

where k is the chemical rate constant.

Let us now summarize briefly the results obtained from a birth and death analysis of fluctuations.

1. Small fluctuations in systems near equilibrium, or in linear systems arbitrarily far from equilibrium, are described by an extension of the Einstein formula [21], well known from equilibrium thermodynamics :

$$P(\{\delta X_i\}) \propto \exp\left(\frac{1}{2k_B}(\delta^2 S)_0\right) \qquad \text{(II.9)}$$

Here k_B is Boltzmann's constant and $(\delta^2 S)_o$ is the second order excess entropy of fluctuations, evaluated around the nonequilibrium reference state. For a dilute mixture, (II.9) implies

$$\langle (\delta X_i)^2 \rangle = \langle X_i \rangle \simeq \bar{X}_i$$
$$\langle \delta X_i \, \delta X_j \rangle = 0 \, , \quad i \neq j \qquad \text{(II.10)}$$

Thus, fluctuations are always of the Poisson type [16].

2. Nonlinear systems possessing a single asymptotically stable steady state, attain a steady state regime for the fluctuations. The probability distribution however does not reduce to the generalized Einsetin formula (II.9). Instead, one obtains ([17], [18]) :

$$\langle (\delta X_i)^2 \rangle = \bar{X}_i \, (1 + \mu_i) \, , \quad \mu_i = O(1) \qquad \text{(II.11)}$$

where the numerical factor μ_i depends on the detailed properties of the chemical steps. Note however, that near thermodynamic equilibrium (II.11) reduces always to (II.10).

3. Systems for which the steady state is not asymptotically stable cannot attain, in the neighborhood of marginal stability, a steady state regime for the fluctuations. In particular the mean quadratic deviations $\langle (\delta X_i)^2 \rangle$ etc increase in time. According to eq. (II.5) this influences, eventually, the averages themselves which are driven in this way to a new regime, e.g. to a limit cycle. Obviously, the occurrence of such abnormal, critical fluctuations implies that the description of the system in terms of the macroscopic evolution equations (II.1) is insufficient ([17], [18]).

A closer inspection of the birth and death formulation outlined in this section reveals several deficiencies. In the first place, the fact that the transition probabilities are computed in terms of macroscopic, "collective" variables referring to the entire system, obviously overestimates these quantities. For instance, in a chemical mixture only those particles which are sufficiently close will be able to undergo a reactive collision. An even more crucial omission is the fact that one discards the size and the range of the fluctuations, and argues as if the whole system fluctuated coherently within the (macroscopic) reaction space. Now, in their overwhelming majority, fluctuations are local events, with a coherence length determined by the dynamics of the system itself, independently of the size of the box confining the system. Related to this critique is also the remark that most physico-chemical systems including biological ones, are, locally, quite close to thermodynamic equilibrium, although on the whole they may be subject to strongly nonequilibrium constraints. Thus, it is not clear why, at least locally, the behavior of fluctuations should deviate from the equilibrium like behavior described by eq. (II.9) and (II.10).

In the next section we outline a local theory of fluctuations in nonequilibrium systems ([17], [18], [22] - [25]), which will enable us to relate their coherence length to the stability properties of the reference state.

II.3. Nonlinear Master Equation Description of local Fluctuations

The main feature of our analysis is to adopt again a stochastic description in the space of macroscopic variables, but to take into account, in an average sense, the interaction between a "small" subvolume and the surrounding "big" system.

Let ΔV be the size of this subvolume (see Fig. 5), which is taken to be sufficiently small to permit a stochastic description of the dynamical processes therein in terms of the macroscopic variables X only. In addition to the local processes, like chemical reactions, ΔV will be coupled to the remaining part of the system, $V - \Delta V$. As we are interested in chemical instabilities we neglect all processes responsible for this coupling other than the transport of

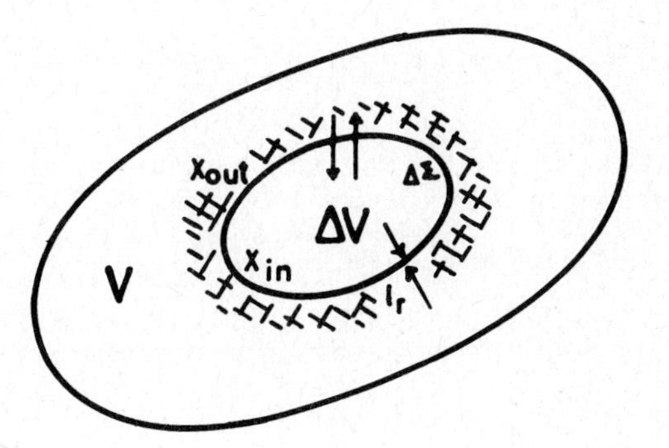

Figure 5 : Illustrating the coupling through exchange of matter between a small subvolume ΔV and its local environment (indicated by the dashed area around the surface $\Delta\Sigma$).

matter across the $V - \Delta V$ interface, $\Delta\Sigma$. This transport will arise from exchanges between ΔV and a layer surrounding $\Delta\Sigma$ whose width is of the order of the mean free math, l_r of the chemical species in the mixture. Let $\mathcal{D}$ be the corresponding transport coefficient. One may derive a master equation for the probability $P(X,t)$ within the subvolume ΔV, by averaging the master equation for the entire system over the "external" environment of ΔV. One obtains ([22], [23]) :

$$\begin{aligned}\frac{dP(x,t)}{dt} = {} & R(x, P(x,t)) + \\ & + \mathcal{D}\langle x\rangle [P(x-1,t) - P(x,t) + \\ & + \mathcal{D}(x+1) P(x+1,t) - \\ & - \mathcal{D}\, x \quad P(x,t)\end{aligned} \qquad \text{(II.12)}$$

with

$$\langle x\rangle = \sum_{x=0}^{\infty} x \; P(x,t) \qquad \text{(II.13)}$$

$R(x, P(x,t))$ stands for the contributions coming from the chemical reactions inside ΔV described on the average by eq. (II.1) or (II.5). The system surrounding ΔV has been taken into account in an average sense, and the requirement of macroscopic homogeneity has permitted to relate this average to the mean value $\langle x \rangle$ of the subsystem itself . An important consequence of this is that the master equation (II.12) is nonlinear. The formal similarity of eq. (II.12) to the kinetic equations, of statistical mecahnics like Boltzmann's equation should be pointed out.

The moment equations generated by (II.12) can be derived straightforwardly. The first moment equation turns out to be independent of $\mathcal{D}$ (and approximately identical to (II.1)), in agreement with the requirement of spatial homogeneity. The second moment equation reads

$$\frac{d}{dt}\langle x^2 \rangle = \sum_{x=0}^{\infty} x^2 R(x, P(x,t)) + 2\mathcal{D}\left[\langle x \rangle - \langle (\delta x)^2 \rangle\right] \quad \text{(II.14)}$$

We note that the role of $\mathcal{D}$ in the evolution of fluctuations becomes more important as one deviates from the Poisson regime. In the limit $\mathcal{D} \to \infty$ the steady state solution of (II.12) - (II.14) reduces to a Poisson distribution and the Einstein-like results (II.9), (II.10) are recovered.

The formulation outlined above shares some common features with previous theories dealing with transitions to instability in the context of equilibrium phase transitions. In particular, the average way the local environment of the subvolume ΔV is taken into account reminds strongly the Becker-Döring theory of nucleation in a liquid-vapor transition, where the evolution of the nucleating embryo is determined entirely by macroscopic factors [26].

Suppose now that a system described macroscopically by eq. (II.1) is driven to the threshold of an instability. The possibility of an amplification of fluctuations which, according to eq. (II.5), could then drive the averages to a new regime, depends on whether the second moment equations (II.14) admit a critical point separating a steady state regime from a regime where fluctuations grow in time. At this critical point, the transport parameter $\mathcal{D}$ will be related to the chemical parameters, e.g. a rate constant k . We shall call k_c the critical value of k corresponding to the onset of an instability in the limit $\mathcal{D} \to 0$ On the other hand $\mathcal{D}$ is related to the size of the volume ΔV . A qualitative estimate yields ([23], [25]) :

$$\mathcal{D} = \frac{D}{\ell^2} \quad \text{(II.15)}$$

where D is Fick's diffusion coefficient and ℓ a characteristic size parameter of ΔV . Introducing then eq. (II.15) into the critical point condition one will find a relation between size of the subsystem and rate of growth of fluctuation. This calculation has been carried out on a number of chemical examples [23]. The results are shown qualitatively on Fig. 6.

The meaning of this diagram is as follows. If the range of a disturbance, i.e. the length over which it preserves a coherent character, acting at the vicinity of a point inside the subvolume ΔV is in the dashed region of Fig. 6, then the decay processes (roughly measured by $\mathcal{D}^{-1}$) will take over and the disturbance will die out. For $k < k_c$ even if an infinite coherence length is imposed, the disturbance will decay. For $k > k_c$ only those disturbances whose range exceeds ℓ_c will be amplified and spread throughout the system.

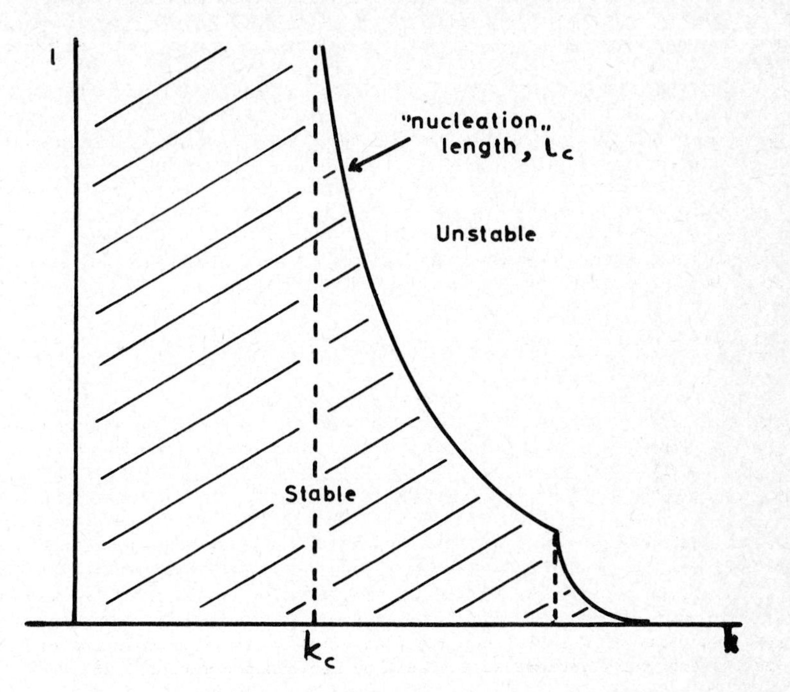

Figure 6 : Coherence length of a fluctuation at the critical point versus a rate constant.

The similarity between this picture and first order phase transitions should be pointed out. In essence, Fig. 6 gives the values of an entropy of activation necessary to form an unstable nucleus under nonequilibrium conditions. The ideas developed in this section have some implications in the understanding of the stability properties of several types of complex systems, including systems of interacting populations or economics. A first qualitative formulation of these points is given in ref. [24].

II.4. Relation with the Langevin force method

One of the most widely spread methods for studying fluctuations is the so-called random, or Langevin force approach. The evolution equations for the state variables $\{X_i\}$ are regarded as stochastic equations differing from the phenomenological ones (II.1) by the addition of random forces $f_i(t)$ taking into account the effects of "noise" :

$$\frac{dX_i(t)}{dt} = v_i(\{X_j\}) + f_i(\underset{\sim}{r},t) \tag{II.16}$$

Moreover, in most applications (see also the lecture notes by Haken in this Volume) one imposes on f_i the following additional conditions :

$$\langle f_i(\mathbf{r},t) \rangle = 0$$

$$\langle f_i(\mathbf{r}_1,t_1)\, f_j(\mathbf{r}_2,t_2) \rangle = 2Q\, \delta(t_1-t_2)\, \delta(\mathbf{r}_1-\mathbf{r}_2)\, \delta^{kr}_{ij} \qquad \text{(II.17)}$$

where the brackets denote a statistical average over the distribution of f's. From (II.16) to (II.17) one can deduce a Fokker-Planck equation for the probability distribution of X_j's which presumably will give all desired information about fluctuations.

The use of Langevin's method in nonlinear systems far from equilibrium has been criticized by Van Kampen [27], who pointed out that the very fact that, in general,

$$\langle v_i(\{X_j\}) \rangle \neq v_i(\{\langle X_j \rangle\})$$

compromises the validity of the first of the two conditions (II.17). Here we would like to make a few comments on the very meaning of the Langevin method as compared to the master equation approach used in the preceding sections.

Implicit in the master equation approach is the idea that fluctuations are internal processes generated by the system itself and due to the same mechanisms as those giving rise to the macroscopic evolution. A particularly striking illustration of this idea is the result derived in the previous section, according to which the range of the critical fluctuations is related to the system's parameters such as the rate constants of the chemical reactions.

On the other hand relations (II.17) impose in fact the statistics of the random forces, as they define a Gaussian Markov process. Once this noise source is imposed on the system, the latter responds according to eq. (II.16) or according to the related Fokker-Planck equation. Thus, in the Langevin picture, fluctuations are regarded as a reponse to an externally imposed (stochastic) disturbance, and not as internal events.

Remarkably, near equilibrium or in linear systems the two points of view become in fact identical. Indeed, the fluctuations are in this limit completely uncoupled between themselves and from the macroscopic evolution and may therefore be treated indifferently, as internal events or as consequences of an external noise source.

REFERENCES

[1] G. Nicolis and G. Auchmuty, Proc. Nat. Acad.Sci. (USA) 71, (1974).
[2] G. Nicolis, in Proc. Symp. in Appli.Math. Vol., 27, Amer. Math. Soc. (1974).
[3] G. Auchmuty and G. Nicolis, submitted to Bulletin Math. Biology (1974).
[4] M. Herschkowitz-Kaufman, submitted to Bulletin Math. Biology (1974).
[5] L. Cesari, "Asymptotic Expansions and Stability Problems in Ordinary Differential Equations", Springer Verlag, Berlin (1963).
[6] P. Glansdorff and I. Prigogine, "Thermodynamics of Structure, Stability and Fluctuations", Wiley-Interscience, New York (1971).
[7] G. Nicolis and J. Portnow, in Chem.Rev. 73, 365 (1973), give a review of oscillating reactions including the Belousov- Zhabotinski reaction. A more detailed discussion of the latter is found in a review paper by R. Noyes and R. Field, Ann. Rev.Phys. Chem. (in the press).
[8] A.T. Winfree, Science 181, 937 (1973).
[9] A. Goldbeter, Proc. Nat. Acad. Sci. (USA) 70, 3255 (1973).
[10] D. Sattinger, "Topics in Stability and Bifurcation Theory", Lecture Notes in Mathematics, Vol. 309, Springer Verlag, Berlin (1973).
[11] P. Hanusse, C.R.Acad. Sci. (Paris), C274, 1245 (1972).
[12] J. Tyson and J. Light, J. Chem. Phys. (in the press).
[13] R. Thom, "Stabilité Structurelle et Morphogénèse", W.A. Benjamin, Reading, Mass. (1972).
[14] A.T. Barucha-Reid, "Elements of the theory of Markov Processes and their applications", Mc Graw-Hill, New York (1960).
[15] D. Mc Quarrie, Suppl. Rev. Series in Appl. Prob., Methuen, London (1967).
[16] G. Nicolis and A. Babloyantz, J. Chem. Phys. 51, 2632 (1969).
[17] G. Nicolis and I. Prigogine, Proc. Nat. Acad. Sci. (USA) 68, 2102 (1971).
[18] G. Nicolis, J. Stat. Phys. 6, 195 (1972).
[19] F. Schlögl, Lecture notes in this volume and references therein.
[20] D. Walls et al., preprint, Univ. of Waikato (1974).
[21] L.D. Landau and E.M. Lifschitz, "Statistical Physics", Pergamon Press, Oxford (1958).
[22] G. Nicolis, M. Malek-Mansour, K. Kitahara and A. Van Nypelseer, Phys. Letters, in the press.
[23] M. Malek-Mansour, G. Nicolis, A. Van Nypelseer and K. Kitahara, submitted for publication in Physica.
[24] I. Prigogine, G. Nicolis, R. Herman and T. Lam, submitted for publication in Cooperative Phenomena.
[25] Y. Kuramoto, Progr. Theor. Phys., 52, (1974).
[26] J. Frenkel, "Kinetic Theory of Liquids", Dover Publ., New York (1955).
[27] N.G. Van Kampen, Adv. Chem. Phys., 15, 65 (1969).

Cooperative Phenomena, H. Haken, ed.

NONLINEAR TRANSPORT IN PHONON SYSTEMS

M. Wagner
University of Stuttgart, Stuttgart, Germany

1. Introduction and Motivation

In the present time physics seems to undergo a phase-transition, which has some similarity to the one which is incorporated in the personality of Kepler. Right in the middle of his scientific life we have the change from a purely geometric view of physics to a specific kind of dynamical view. In Kepler's famous book "Harmonia mundi" this is illustrated by his cognition that "harmonia" is not only possible in a purely geometrical sense but also in a dynamical one. The geometrical harmony led him to explain the planetary orbits as lying on the surfaces of those spheres given respectively by the smallest and largest radius of the Platonian regular polyhedra, built successively into each other. Although, by chance, this description worked rather well, Kepler later came to the correct description as incorporated in his famous laws. It is noteworthy,however, that these laws were conceived by him as being harmonical, although this harmony was of a new kind.

After Kepler in the course of centuries the fundamental dynamical equations in all physical disciplines have been established (Newton, Maxwell, Boltzmann, Schrödinger). Nevertheless, except for very specific examples, there has been the general trend to seek only for those solutions of the equations, which belonged either to extremely simple systems, or which were manifestations of an utmost simplification of a complicated problem. To give examples, there was never a great effort in Classical Mechanics, to look for the more sophisticated dynamical behaviour of more than 3 bodies. And although already the original Boltzmann equation of Statistical Mechanics was nonlinear, there has not been a great willingness to look at this nonlinearity and to seek "unusual" kinds of solutions.

In our generation there now seems to be again a new kind of phase transition in physical thinking. Very similar to the situation of Kepler, it is closely attached to harmony and symmetry, but again in a new sense. Though, the stage of development is not yet advanced enough to define its nature in a simple way.

One important feature of the new situation is the fact that the "splendid isolation" of physicists, which has been so pleasing through many years, no longer can be maintained. It has turned out that there are close connections between the dynamical phenomena in engineering, biology, chemistry, physics and even sociology. Haken [1] has given the name "Synergetics" to this interdisciplinary field, which encircles dynamical phenomena of similar nature. At this meeting a representative selection of these problems, and some approaches to their solutions are presented. The motivation for my interest in nonlinear transport, and in particular phonon transport, is twofold. On the one hand it seems important to know what kind of nonlinear kinetic structures can be established on the basis of microscopic physical principles. Naturally, at the present stage, rather crude approximations have to be made, if one seeks for simple forms. In any case physical principles restrict the manifold of nonlinear structural forms. Then the question arises,what kind of "transport phase transition" may arise. In particular there is the fascinating problem, whether microscopic control systems (or"microscopic machines") can be established by mere transport or even may organize themselves in space, if there is transport.

On the other hand there is the big difficulty, to justify the crude approximations which are indispensable to arrive at simple non-linear structures. Therefore it has seemed important to us to study a series of simple transport systems, which can be handled exactly.

Since the pure transport approach is not handled in other lectures, it has seemed reasonable to me, not to start at a too an advanced level. Therefore, the greater part of my lecture is devoted to the basic concepts of phonon transport theory. These in the end will lead in a straightforward way to nonlinear kinetic structures. The simple example of a harmonic linear chain coupled to a local exciton is discussed in some detail. It is shown, how at a critical phonon flux a "transport phase transition" may arise.

2. The Peierls-Boltzmann Equation

Since the fundamental work of Peierls [2] there has been a great development in phonon transport theory. In particular, during the last decade, highly sophisticated theoretical formulations have been given [3,4]. Yet, their application is restricted to the linear regime. Our task at this lecture will be, to develop a more crude theory, which, however, offers itself more easily to construct nonlinear structures.

The most direct approach to formulate a phonon transport theory is via the definition of a phonon density operator

$$\hat{n}(\underset{\sim}{k}\lambda, \underset{\sim}{r}, t) = \sum_{\underset{\sim}{k}'\lambda} \Big\{ a^{+}(\underset{\sim}{k}\lambda, t)\, a(\underset{\sim}{k}'\lambda', t) \exp[-i(\underset{\sim}{k} - \underset{\sim}{k}')\cdot \underset{\sim}{r}] \cdot \chi(\underset{\sim}{k}\lambda, \underset{\sim}{k}'\lambda') + C.C. \Big\} \tag{1}$$

such that the energy density of the mode $\underset{\sim}{k}\lambda$ is given by $\hat{\mathcal{E}}(\underset{\sim}{k}\lambda, \underset{\sim}{r}, t) = \omega(\underset{\sim}{k}\lambda)\, \hat{n}(\underset{\sim}{k}\lambda, \underset{\sim}{r}, t)$ $(\hbar = 1)$. Now, on physical grounds, the definition of $\hat{n}(\underset{\sim}{k}, \underset{\sim}{r}, t)$ [5] is not uniquely given. Let there be two particles, one of them within and the other outside a given region. Then there is some arbitrariness in the statement, which part of their interaction energy has to be counted to belong to the region inside. If exactly one half is taken to belong to the inner region, $\chi(k, k')$ will be a Wigner fuction [4]. But, as Hardy [6] has shown also a Gauss function is a good choice,

$$\chi(\underset{\sim}{k}\lambda, \underset{\sim}{k}'\lambda') = \frac{1}{2V}\, \delta_{\lambda\lambda'}\, e^{-\alpha|\underset{\sim}{k} - \underset{\sim}{k}'|^2} = \chi(k, k')\, \delta_{\lambda\lambda'} \tag{2}$$

where V is the total volume. Yet, there are also many other good choices. In any case, $\chi(k, k')$ restricts the deviation of $\underset{\sim}{k}'$ from $\underset{\sim}{k}$ in the wave packet. Our choice (2) has the advantage of being rather flexible (parameter α).

Let us assume now that the total Hamiltonian be given by

$$\hat{H} = \hat{H}_p + \hat{H}_s + \hat{\mathcal{U}} \tag{3}$$

where $\hat{H}_p$ is the diagonal, harmonic part of the lattice vibrations, $\hat{H}_s$ a diagonalized Hamiltonian for some singular additional degrees of freedom and $\hat{U}$ the representative for all interactions within the phonon system and between phonons and singular motions. Then we have

$$[\hat{n}(\underset{\sim}{k}\lambda, \underset{\sim}{r}, t), \hat{H}_p] = \Big\{ -\sum_{k'} (\omega(\underset{\sim}{k}\lambda) - \omega(\underset{\sim}{k}'\lambda))\, a^{+}(\underset{\sim}{k}\lambda, t)\, a(\underset{\sim}{k}'\lambda, t) \exp[-i(\underset{\sim}{k} - \underset{\sim}{k}')\cdot \underset{\sim}{r}]\, \chi(k, k') + C.C. \Big\} \tag{4}$$

and because of restriction (2) we are allowed to make a series expansion,

$$\omega(\underset{\sim}{k}'\lambda)-\omega(\underset{\sim}{k}\lambda)=[\nabla_{\underset{\sim}{k}}\omega(\underset{\sim}{k}\lambda)]\cdot(\underset{\sim}{k}'-\underset{\sim}{k})+O(\alpha^{-1}V^{-1}\partial^2_{\underset{\sim}{k}}\omega(\underset{\sim}{k}\lambda)) \tag{5}$$

Inserting this into (4) and employing the identity

$$i(\underset{\sim}{k}'-\underset{\sim}{k})\exp[i(\underset{\sim}{k}'-\underset{\sim}{k})\cdot\underset{\sim}{r}] = \nabla_{\underset{\sim}{r}}\exp[i(\underset{\sim}{k}'-\underset{\sim}{k})\cdot\underset{\sim}{r}] \tag{6}$$

we have

$$[\hat{n}(k,\underset{\sim}{r},t),\hat{H}_p] = -i\underset{\sim}{v}(k)\nabla_{\underset{\sim}{r}}\hat{n}(k,\underset{\sim}{r},t)+O(\alpha^{-1}V^{-1}\partial^2_{\underset{\sim}{k}}\omega(k)) \tag{7}$$

where

$$\underset{\sim}{v}(\underset{\sim}{k}\lambda) = \nabla_{\underset{\sim}{k}}\omega(\underset{\sim}{k}\lambda) \tag{8}$$

is the phonon group-velocity. If there would be no interaction $\hat{U}$, the Heisenberg equation of motion, in view of eq. (7), could be written in the form

$$[\frac{\partial}{\partial t}+\underset{\sim}{v}(k)\nabla_{\underset{\sim}{r}}]\hat{n}(k,\underset{\sim}{r},t) = O(\alpha^{-1}V^{-1}\partial^2_{\underset{\sim}{k}}\omega(k)) \qquad (\text{for } \hat{U}=0) \tag{9}$$

This pertains to the harmonic, translationally invariant crystal. The r. h. s. of eq. (9) disappears, provided α is chosen large enough, or, if there is no dispersion (i.e. $\nabla\nabla = 0$), (Debye approximation)). In this case eq. (9) has the structure of a continuity equation. Hence, the density (1) indeed pertains to quasi-particles, which in the purely harmonic translat. invariant crystal can move with velocity $\underset{\sim}{v}(k)$ and satisfy a continuity equation. Moreover, density (1) has the integral property

$$\int_V d^3\underset{\sim}{r}\ \hat{n}(k,\underset{\sim}{r},t) = \hat{N}(k,t) = a^+(k,t)a(k,t) \tag{10}$$

which again is consistent with the interpretation of a density, since $\hat{N}(k,t)$ is the total number of phonons of species k.

In the general case $U \neq 0$ the Heisenberg equation for $\hat{n}(k,\underset{\sim}{r},t)$ reads

$$[\frac{\partial}{\partial t}+\underset{\sim}{v}(k)\cdot\nabla_{\underset{\sim}{r}}]\hat{n}(k,\underset{\sim}{r},t) = \frac{1}{i}[\hat{n}(k,\underset{\sim}{r},t),\hat{U}] + O(\alpha^{-1}V^{-1}\partial^2_{\underset{\sim}{k}}\omega(k)) \tag{11}$$

where the transcription (7) has been used. Eq. (11) is the most general "Boltzmann" equation for phonons. It has to be supplemented by the kinetic equations for the singular excitation degrees,

$$\frac{\partial}{\partial t}\hat{N}_s = \frac{1}{i}[\hat{N}_s,\hat{U}] \tag{12}$$

where $N_s = a_s^+a_s$. Eqs. (10) and (12) are coupled to each other. Naturally, the r. h. s. of (11) ("collision term"), in principal also incorporates higher forms of phonon densities, e. g. two phonon-densities, etc. For these further kinetic equations arise and in succession a Bogoliubov hierarchy can be established. Naturally, this will be avoided here. Instead, the collision term will be simplified in such a way that already the first equation of the hierarchy is disentangled from the rest.

3. Simplification of the Collision Term

If the commutator (4) would be averaged over the whole space, it would disappear. This is a consequence of the wave-packet description (1). But, as seen from expr. (7), this commutator plays the role of the phonon source term,

$$\frac{1}{i}[\hat{n}(k,\underset{\sim}{r},t), \hat{H}_p] \approx -\mathrm{div}_{\underset{\sim}{r}} \hat{\underset{\sim}{j}}(k,\underset{\sim}{r},t) \tag{13}$$

where

$$\underset{\sim}{j}(k,\underset{\sim}{r},t) = \underset{\sim}{v}(k)\,\hat{n}(k,\underset{\sim}{r},t) \tag{14}$$

Until now, however, we have tacitly assumed that our system is "closed" and described by the "internal" Hamiltonian (3). Yet, we want to have an "open" transport system. Hence we have to allow for regions, where phonons are created or annihilated by external stimulii. To this end we would have to introduce an additional Hamiltonian H_{ext} for the coupling to the external world, and this would yield additional terms on the r. h. s. of eq. (11). Formally, these terms could be written as

$$\mathrm{div}_{\underset{\sim}{r}}^{(ext)}\, j(k,\underset{\sim}{r},t)$$

E. g. such terms would arise at the surfaces, if we impose a thermal gradient onto the system. Though, since it is of great advantage, to avoid the explicit introduction of H_{ext}, we henceforth exclude all external source and sink regions from the spatial averaging procedure. But then the spatial mean over expr. (13) will no longer disappear. However, as a consequence, this term then mathematically no longer may be viewed as a homogeneous one, but as an inhomogeneous one, which represents the external energy stimulus. We may thus look at it as the "driving force" which produces the phonon flux, or simply as the "transport", which is given by external sources,

$$\frac{1}{i}[\hat{n}(k,\underset{\sim}{r},t), \hat{H}_p]_{Av.\underset{\sim}{r}} = -\frac{1}{V}\hat{\Phi}(k,t) \tag{15}$$

On the other hand, there is no complication for the spatial mean of the collision term. By use of the form of (1) we have

$$[\hat{n}(k,\underset{\sim}{r},t), \hat{U}]_{Av.\underset{\sim}{r}} = \frac{1}{V}[\hat{N}(k,t), \hat{U}] \tag{16}$$

where $\hat{N}(k,t) = a^{+}(k,t)\,a(k,t)$. But, thinking of the Heisenberg eq. of motion, the r. h. s. of eq. (16) just represents the time variation of $N(k,t)$ as initiated by the internal interaction $\hat{U}$,

$$\frac{\partial}{\partial t}\hat{N}(k,t)/_{coll.} = \frac{1}{i}[\hat{N}(k,t), \hat{U}] \tag{17}$$

Let us introduce now the density matrix $\hat{\rho}(t)$ of the system. If we assume that its nondiagonal part $\hat{\rho}_{nd} = 0$ for $t = 0$, its diagonal elements $\rho_\mu(t)$ will satisfy the Zwanzig [7] master equation

$$\frac{\partial}{\partial t}\rho_\mu(t) = \sum_\nu \int_0^t d\tau[\rho_\nu(t-\tau)W_{\nu\mu}(\tau) - \rho_\mu(t-\tau)W_{\mu\nu}(\tau)] \tag{18}$$

where $|\mu\rangle$ are the eigenstates of $\hat{H}_p + \hat{H}_s$, and $W_{\mu\nu}$ is defined later. By means of eq. (18) the collision term may be rewritten as

$$\frac{\partial}{\partial t} N(k,t)/_{coll.} = \frac{\partial}{\partial t} \sum_{\mu} \varrho_{\mu}(t) N^{\mu}(k)$$

$$= \sum_{\mu} N^{\mu}(k) \sum_{\nu} \int_0^t d\tau \, [\varrho_{\nu}(t-\tau) W_{\nu\mu}(\tau) - \varrho_{\mu}(t-\tau) W_{\mu\nu}(\tau)] \tag{19}$$

where $N(k,t) = \langle \hat{N}(k,t) \rangle$ is the expectation value of $\hat{N}(k,t)$. Now, the first term of the last expression can be rearranged in the following way

$$\sum_{\mu\nu} N^{\mu} \varrho_{\nu} W_{\nu\mu} = \sum_{\mu,\nu}^{(N^{\mu}=N^{\nu})} N^{\nu} \varrho_{\nu} W_{\nu\mu} + \sum_{\mu,\nu}^{(N^{\mu}=N^{\nu}+1)} (N^{\nu}+1) \varrho_{\nu} W_{\nu\mu}$$

$$+ \sum_{\mu\nu}^{(N^{\mu}=N^{\nu}-1)} (N^{\nu}-1) \varrho_{\nu} W_{\nu\mu} + \sum_{\mu\nu}^{(N^{\mu}=N^{\nu}+2)} (N^{\nu}+2) \varrho_{\nu} W_{\nu\mu} + \ldots \tag{20}$$

$$= \sum_{\mu\nu} N^{\nu} \varrho_{\nu} W_{\nu\mu} + \sum_{m=0}^{+\infty} m \left[\sum_{\mu\nu}^{(N^{\mu}=N^{\nu}+m)} \varrho_{\nu} W_{\nu\mu} - \sum_{\mu\nu}^{(N^{\mu}=N^{\nu}-m)} \varrho_{\nu} W_{\nu\mu} \right]$$

The first term of (20) just cancels the very last term of (19). Therefore, inserting eq. (20) into (19) and collecting all results of this section, we are left with the following phonon transport equations

$$\frac{\partial}{\partial t} N(k,t) + \Phi(k,t) = \sum_{m=0}^{\infty} m \int_0^t d\tau \sum_{\mu} \varrho_{\mu}(t-\tau) \left\{ \sum_{\nu}^{(N^{\nu}(k)=N^{\mu}(k)+m)} W_{\mu\nu}(\tau) - \sum_{\nu}^{(N^{\mu}(k)=N^{\mu}(k)-m)} W_{\mu\nu}(\tau) \right\} \tag{21a}$$

and in a completely analogous manner the kinetic equation for the singular degrees of freedom can be written as

$$\frac{\partial}{\partial t} N(s) = \sum_{m=0}^{\infty} m \int_0^t d\tau \sum_{\mu} \varrho_{\mu}(t-\tau) \left\{ \sum_{\nu}^{(N^{\nu}(s)=N^{\mu}(s)+m)} W_{\mu\nu}(\tau) - \sum_{\nu}^{(N^{\nu}(s)=N^{\mu}(s)-m)} W_{\mu\nu}(\tau) \right\} \tag{21b}$$

4. Statistical Assumptions and Higher Golden Rules

The kernel of the Zwanzig master equation (18) is given as [7]

$$W_{\mu\nu}(\tau) = -[\hat{L}_u \, e^{-i\tau(\hat{L}_o + (1-\hat{D})\hat{L}_u} \, \hat{L}_u]_{\mu\mu\nu\nu}$$

$$= -\sum_{ij} \left\{ U_{\mu i} [G_{i\mu j\nu} U_{j\nu} - G_{i\mu\nu j} U_{\nu j}] + c.c. \right\} \tag{22}$$

where

$$\hat{\hat{G}}(\tau) = \exp[-i\tau(\hat{\hat{L}}_o + (1-\hat{\hat{D}})\hat{\hat{L}}_u] \tag{23}$$

and $\hat{\hat{L}}_o$, $\hat{\hat{L}}_u$ and $\hat{\hat{L}} = \hat{\hat{L}}_o + \hat{\hat{L}}_u$ are the Liouville "tetradic" operators belonging to $\hat{H}_o$, $\hat{U}$ and $\hat{H}$ respectively, e. g.

$$(\hat{\hat{L}}_u)_{i\alpha j\beta} = U_{ij}\,\delta_{\alpha\beta} - U_{\beta\alpha}\,\delta_{ij} \tag{24}$$

whereas D is the diagonalization projector, i.e.

$$(\hat{\hat{D}}\,\hat{\hat{A}})_{i\alpha j\beta} = \delta_{i\alpha}\,A_{i\alpha j\beta} \tag{25}$$

Utilizing the Goldberger-Adams theorem [8], we have

$$G_{i\alpha j\beta}(\tau) = \sum_{n=0}^{\infty} G^{(n)}_{i\alpha j\beta}(\tau) \tag{26}$$

where

$$G^{(n)}_{i\alpha j\beta}(\tau) = (-i)^n e^{-i(E_i - E_\alpha)\tau} \int_0^{\tau} d\tau_1 \dots \int_0^{\tau_{n-1}} d\tau_n \cdot \left\{ [(1-\hat{D})\,\hat{L}_u(\tau_1)] \dots [(1-\hat{D})\,\hat{L}_u(\tau_n)] \right\}_{i\alpha j\beta} \tag{27}$$

and $\hat{\hat{A}}(\tau)$ denotes the interaction representation,

$$\hat{\hat{A}}(\tau) = \exp(i\hat{\hat{L}}_0\tau)\,\hat{\hat{A}}\,\exp(-i\hat{\hat{L}}_0\tau) \tag{28}$$

Our statistical postulates will now be the usual ones, as discussed somewhat carefully by Jancel [9]. As a first postulate we have already introduced $\rho_{nd}(0)=0$. The second is a kind of Markofficity assumption, which we specify in the form

$$\rho_\mu(t-\tau)\,W_{\mu\nu}(\tau) \approx \rho_\mu(t)\,W_{\mu\nu}(\tau) \tag{29}$$

Then the integrands of eqs. (21a,b) are simply $W_{\mu\nu}(\tau)$. The 3rd postulate is a kind of irreversibility assumption, which we mathematically represent in the equation

$$\int_0^t d\tau \, \cos(E_i - E_j)\tau \approx \pi\,\delta(E_i - E_j) \tag{30}$$

With the background of this postulates we are able to write down the integrals of the form

$$\int_0^t W^{(n)}_{\mu\nu}(\tau)\,d\tau \tag{31}$$

where $W_{\mu\nu}^{(n)}$ is supposed to pertain to $\hat{\hat{G}}^{(n)}$ in the expansion (26). In particular we immediately arrive at

$$\int_0^t W^{(0)}_{\mu\nu}(\tau)\,d\tau = 2\pi\,U_{\mu\nu}\,U_{\nu\mu}\,\delta(E_\mu - E_\nu) \tag{32}$$

which is the wellknown "Fermi golden rule" and corresponds to the Pauli master equation. We note that there is direct energy conservation for the transition $\mu \rightarrow \nu$. Hence, in such an approximation those transition processes, in which energy is not directly conserved, would be excluded from the kinetic equations. But if there is a finite flux of phonons, there may well be a cumulation of energy in such a way that excitations arise by a succession of energy non-conserving intermediate processes. These are described by "higher golden rules". We will write down only simplified versions of these rules. To be more specific, we assume that a higher golden rule become effective only, if the next lower one can not be satisfied. So, if we have

$$U_{ij}\,\delta\,(E_i - E_j) \approx 0 \tag{33}$$

the integrands $W^{(2)}_{\mu\nu}(\tau)$ become effective, and so forth. We then end up with the results

$$\int_0^t W^{(2)}_{\mu\nu}(\tau)\,d\tau = 2\pi \sum_{\alpha\beta} U_{\mu\alpha} U_{\alpha\nu} U_{\nu\beta} U_{\beta\mu} \frac{\delta(E_\mu - E_\nu)}{(E_\beta - E_\mu)(E_\alpha - E_\mu)} \tag{34}$$

$$\int_0^t W^{(4)}_{\mu\nu}(\tau)\,d\tau = 2\pi \sum_{\substack{\alpha\beta \\ \alpha_1\beta_1}} U_{\mu\alpha} U_{\alpha\alpha_1} U_{\alpha_1\nu} U_{\nu\beta_1} U_{\beta_1\beta} U_{\beta\mu} \frac{\delta(E_\mu - E_\nu)}{(E_\mu - E_\alpha)(E_\mu - E_{\alpha_1})(E_\nu - E_{\beta_1})(E_\nu - E_\beta)} \tag{35}$$

All integrals, where n is an odd number, disappear. In exprs. (34,35) we note that no "direct"energy conservation is required, but that there are intermediate states which do not conserve energy.

5. Factorization and Nonlinear Structure

It is advantageous to illustrate the further procedure at a specific example. Let us assume that a predominant part of the nondiagonal Hamiltonian $\hat{U}$ be given either by

$$\hat{U} = (a_s + a_s^+) \sum_k \lambda(k)(b(k) + b^+(-k)), \quad \lambda(-k) = -\lambda(k) \tag{36a}$$

or by

$$\hat{U} = (a_s + a_s^+) \sum_{k_1 k_2 k_3} \lambda(k_1 k_2 k_3)(b(k_1) + b^+(k_1))(b(k_2) + b^+(-k_2))(b(k_3) + b^+(-k_3)) \tag{36b}$$

where a_s^+ represents a single local excitation, whereas the b's are the phonon operators. Let us assume further that a 3-phonon process is necessary for the excitation of the local degree of freedom,

$$\omega_s = \omega(k_1) + \omega(k_2) + \omega(k_3) \tag{37}$$

Then in case (36a) the higher order golden rule (35) would be indispensable, whereas in case (36b) already the Fermi golden rule would suffice. For the collision term in case (36a) we would have then

$$\frac{\partial}{\partial t} N(k,t)/_{coll.} = 2\pi \sum_{\mu, k', k''} A(kk'k'')\rho_\mu(t)\,\delta(\omega_s - (\omega(k) + \omega(k') + \omega(k''))$$
$$[(N^\mu(s))^3 (N^\mu(k)+1)(N^\mu(k')+1)(N^\mu(k'')+1) - (N^\mu(s)+1)^3 N^\mu(k) N^\mu(k') N^\mu(k'')] \tag{38a}$$

where

$$A(kk'k'') = \lambda(k)^2 \lambda(k')^2 \lambda(k'')^2 \left[\frac{1}{\omega(s)+\omega(k)}\left(\frac{1}{\omega(s)-\omega(k')} + \frac{1}{\omega(s)-\omega(k'')}\right) + cycl\right]^2 \tag{39}$$

In case (36b) we would have

$$\frac{\partial}{\partial t} N(k,t)/_{coll.} = 2\pi \sum_{\mu,k'k''} \lambda(kk'k'')^2 \varrho_\mu(t)\, \delta(\omega_s - (\omega(k_1) + \omega(k_2) + \omega(k_3)))$$

$$\cdot\left[(N^\mu(s))^3 (N^\mu(k)+1)(N^\mu(k')+1)(N^\mu(k'')+1) - (N^\mu(s)+1)^3 N^\mu(k) N^\mu(k') N^\mu(k'')\right] \tag{38b}$$

The close similarity to expr. (38a) is evident. Whence, in both cases the further procedure is identical. It is also immediately evident that thermal equilibrium,

$$\bar{N}(s) = [\exp(\beta\omega_s) - 1]^{-1} \qquad \beta = (k_B T)^{-1}$$
$$\overline{N(k)} = [\exp(\beta\omega(k)) - 1]^{-1} \tag{40}$$

is a stationary solution. (For this solution the collision terms (38a,b) disappear). It is clear that after the summation over μ has been performed, the forms (38a,b) will not be of a product form, namely (e.g.for (38a))

$$\frac{\partial}{\partial t} N(k,t)/_{coll.} = 2\pi \sum_{k'k''} A(kk'k'')\, \delta(\omega(s) - (\omega(k) + \omega(k') + \omega(k'')))$$

$$\left[\overline{(N(s))^3 (N(k)+1)(N(k')+1)(N(k'')+1)} - \overline{(N(s)+1)^3 N(k) N(k') N(k'')}\right] \tag{41}$$

From this we see that at this stage we only can proceed, if we intuitively postulate a factorization prescription. This prescription naturally has to be such that the thermal equilibrium remains one of the stationary solutions (necessary condition). In the chosen example the most straightforward prescription would be

$$\frac{\partial}{\partial t} N(k,t)/_{coll.} = 2\pi \sum_{k'k''} A(kk'k'')\, \delta(\omega(s) - (\omega(k) + \omega(k') + \omega(k'')))$$

$$[\overline{(N(s))^3} \cdot (\overline{N(k)}+1)(\overline{N(k')}+1)(\overline{N(k'')}+1)$$
$$- (\overline{N(S)+1})^3\ \overline{N(k)} \cdot \overline{N(k')} \cdot \overline{N(k'')}] \tag{42}$$

Yet, even if this is adopted, the factorization is not yet accomplished in the local kinetic variable $\overline{N(s)}$. To achieve this, we finally introduce an "eigen-temperature" T(s) for the local excitation,

$$\overline{N(s)} = [\exp(\beta(s)\,\omega(s)) - 1]^{-1}, \quad \beta(s) = (k_B T(s))^{-1} \tag{43}$$

which yields

$$\overline{N(s)^2} = \overline{N(s)} + 2(\overline{N(s)})^2$$
$$\overline{N(s)^3} = \overline{N(s)} + 6(\overline{N(s)})^2 + 6(\overline{N(s)})^3 \tag{44}$$

This completes the establishment of the kinetic equations. In case (36a) they read [10] (vid. (21a,b)):

$$\frac{\partial}{\partial t} N(k,t) + \Phi(k,t) = 2\pi \sum_{k'k''} A(kk'k'')\, \delta(\omega(s) - (\omega(k) + \omega(k') + \omega(k'')))$$

$$[(6N(s)^3 + 6N(s)^2 + N(s))(N(k)+1)(N(k')+1)(N(k'')+1) - (6N(s)^3 +$$
$$+ 12N(s)^2 + 7N(s) + 1)\, N(k) N(k') N(k'')] \tag{45a}$$

$$\frac{\partial}{\partial t} N(s) = 2\pi \sum_{kk'k''} A(kk'k'')\, \delta(\omega(s) - (\omega(k) + \omega(k') + \omega(k'')))$$

$$[(6N(s)^3 + 12N(s)^2 + 7N(s) + 1)\, N(k)\, N(k')\, N(k'') - \tag{45b}$$

$$- (6N(s)^3 + 6N(s)^2 + N(s))\,(N(k)+1)\,(N(k')+1)\,(N(k'')+1)]$$

This is now the final system of coupled nonlinear equations which has to be solved.

6. Dissipation and Transport Phase Transition

We now switch to a more scetchy way of description. Let us investigate the possibility of stationary solutions. The condition

$$\frac{\partial}{\partial t} N(s) = 0 \tag{46}$$

from (45b) yields [11]

$$N(s)^3 + (1 - \sigma)\, N(s)^2 + (\tfrac{1}{6} - \sigma)\, N(s) - \tfrac{1}{6}\sigma = 0 \tag{47}$$

with the abbreviation

$$\sigma = b/a = \sigma\,(N(k),\, N(k'),\, N(k'')) \tag{48}$$

where

$$b = \sum_{kk'k''} A(kk'k'')\, \delta(\omega(s) - (\omega(k) + \omega(k') + \omega(k''))\, N(k) N(k') N(k'')$$

$$a = \sum_{kk'k''} A(kk'k'')\, \delta(\omega(s) - (\omega(k) + \omega(k') + \omega(k''))) \tag{49}$$

$$\times [1 + N(k) + N(k') + N(k'') + N(k)N(k') + N(k')N(k'') + N(k'')N(k)] \tag{50}$$

In view of expr. (39), i.e. $A(kk'k'')>0$, we note that $b>0$, $a>0$ and hence $\sigma>0$. For the discussion of the cubic eq. (47) we then may follow the way of Cardan, which is found in any standard textbook [11]. If we further apply Descartes' rule of signs it can be shown [11] that there never can be more than a single positive real root of eq. (47).

However, before continuing our discussion, we should face the fact that the kinetic equations (45a,b) are rather unphysical, since they allow only for a single kind of collision processes, i.e. those specified by the interaction (36a). If we do not want to loose physical reality, we are bound to account somehow for all other collision processes which may be present. We choose the most simple way, which is to hump together all forgotten collision processes into a single additional relaxation time. This, for instance, would mean that the r. h. s. of (45b) would have to be supplemented by an additional term of the form $-(\tau(s))^{-1}(N(s) - N(s)_T)$. By this means the factor $(\frac{1}{6} - \sigma)$ in the cubic eq. (47) would be replaced by $(\sigma_0 - \sigma)$, where

$$\sigma_0 = \frac{1}{6}\left(1 + \frac{1}{a\,\tau(s)}\right) \tag{51}$$

Now, if $\sigma_0 > 1$, Descartes' rule of signs would allow 3 real positive solutions in the region $1 < \sigma > \sigma_0$. Whether all 3 exist and, if they do, whether one of the 2 "non-thermodynamic" ones becomes physically effective, requires a lengthy mathematical discussion. After their existence has been established, it has to be ascertained that the non-thermodynamic branches can be stable. In this case there would be one or two "transport-phase-transitions" from the thermodynamic to other stationary (or oscillatory) branches.

It is thus clear that if we have found stationary solutions, we have to start a detailed mathematical discussion of stability. In general, a discussion of this kind is cumbersome. In principle, however, there is no difficulty, since there are well-developed mathematical methods, be means of which the calculation can be performed [12]. We do not go into any details of the calculation at this place. The results are given in Figs. 1 and 2. In Fig. 1 the local excitation n(s) above thermal equilibrium is shown for different values of the effective additional relaxation time $\tau(s)$. In Fig. 2 the result for the stability discussion of the solution is given. The sign +(-) indicates that in its region any solution tends to an increasing (decreasing) value. Φ is the average phonon flux. We mention that for the stability calculation Ljapunov's first method has been used. By direct computation it has been found that there are extended stability regions on both sides of the stationary solutions. We realize that for increasing flux Φ there is a critical value Φ_{cu} where the solution "jumps" from the "thermodynamic" branch to a "non-thermodynamic" one above. On the other hand, if the flux continuously decreases from a value above Φ_{cu}, there again is a critical value Φ_{cl}, where the non-thermodynamic solution falls down to the thermodynamic one. Features of particular interest are the "phase-transition" behaviour and the "hysteresis" in the transition region. The detailed calculation and the stability discussion has been performed by F. Kaiser [13].

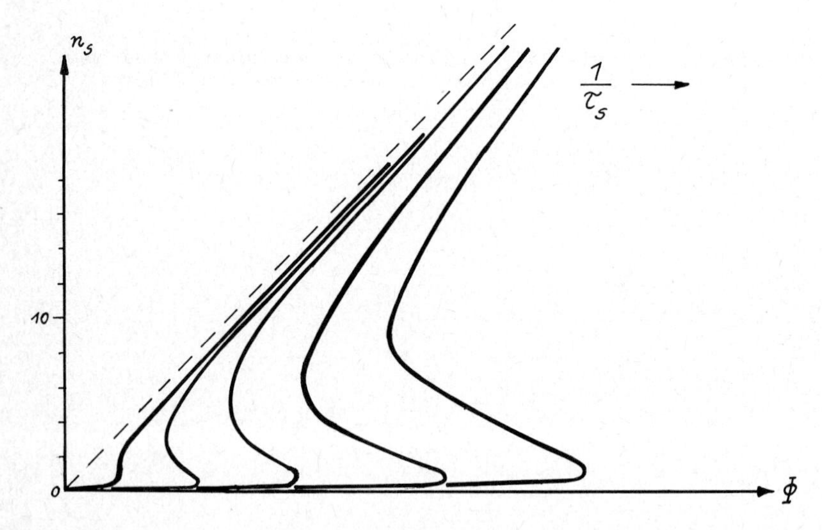

Fig. 1: Excitation of a local mode coupled to a one-dimensional phonon system. n_s is the occupation deviation from thermal equilibrium. Φ is the average phonon flux and τ_s the relaxation time for additional scattering processes required for stability. For each curve τ_s is a fixed parameter.

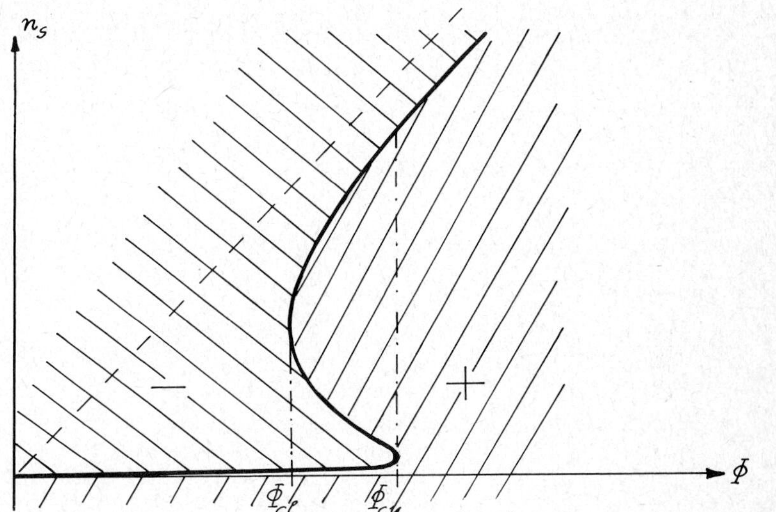

Fig. 2: Transport phase-transition for a local mode coupled to a one-dimensional phonon system. The + and -sign regions indicate that there a non-stationary solution would tend to higher or lower values respectively. For increasing flux the critical value is Φ_{cu}, for decreasing it is Φ_{cl} . The transition exhibits a hysteresis.

To give an interpretation of the preceding result it is useful to conceive the local excitation as a cumulative phenomenon. In our particular case 3 phonons combine in succession to create the high-energetic local quantum. An important feature is the stabilization of the local energy storage by means of the phonon-flux. In this context we again refer to the paper [11] of the author and to another one [14] in which a possible significance of such a storage for biological systems is indicated. It is interesting to note that the stabilization requires the introduction of additional relaxation paths, e. g. a coupling to local heat baths. But in physically realistic systems such additional paths are always present and indeed can never be avoided. So, our result seems to make physical sense. Also the order of magnitude for the flux near the critical region does not seem to be exorbitantly high for realistic values of the relevant chain parameters.

A series of extremly complicated questions arise, if one tries to discuss the involved approximations. Though, local excitations via energy flux actually are very familiar in physics. E.g. any classical harmonic lattice can excite "local modes" coupled to it and of a frequency above the phonon-bands, provided the external force has just the high frequency ("direct resonance"). Considered as a transport system, also in this case low-frequency modes would cumulatively create the local mode. Here again it is quite illustrative to note that the local excitation is only stable, if there is an additional relaxation way. To circumvent the problem of approximations as well as to establish experience for their examination it would be highly desirable to study exactly solvable models. In these cases one would not use transport equations, but other methods. One of the most promising is the canonical transformation method. To give an impression of the application of this formalism, let us consider a linear, harmonic chain, which at two sites, $\nu = \pm r$, is coupled to equi-energetic localized excitons. The latter be described by the operators a_i, a_i^+ (i = 1,2) and their excitation be $\omega(s)$. Let us further take the coupling to be of the bilinear form

$$\hat{U} = \sum_{i,k} \lambda_i(k)\,(a_i + a_i^+)\,(b(k) + b^+(-k)) \tag{52}$$

where

$$\lambda_1(k) = \lambda_2(k) = K\;\omega(k)^{-1/2}\; e^{-ikra}\; \sin^2 \frac{ka}{2} \tag{53}$$

and $\omega(k)$ are the phonon frequencies. We now perform the exponential transformation $\tilde{H} = \exp(-S)\,H\,\exp(S)$, which diagonalizes the Hamiltonian up to order K^2, if we choose

$$\hat{S} = \sum_{i,k} \Big[\frac{\lambda_i(k)}{\omega(s)+\omega(k)} (a_i b(k) - a_i^+ b^+(k)) + \frac{\lambda_i(k)}{\omega(s)-\omega(k)} (a_i b^+(-k) - a_i^+ b(k)) \Big]$$

Since by this the dynamical problem is solved, the phonon kinetics in the chain is now easily calculated for any initial condition. E. g. if we start with an exciton on the left at time t = 0, i.e. $\Psi(0) = a_1^+|0\rangle$, we have

$$|\langle a_2^+ | \Psi(t)\rangle|^2 = \sin^2 \tfrac{1}{2}\,\Delta\omega t \tag{54}$$

where

$$\Delta\omega = 2K^2 \sum_k \frac{\sin^4 \frac{1}{2} ka \cos 2kra}{\omega^2(k) - \omega^2(k)} \tag{55}$$

and the phonon excitation is given by

$$\langle b^+(k)|\Psi(t)\rangle = K\,\omega(k)^{-1/2}\,(\omega(s)-\omega(k))^{-1} \sin^2 \tfrac{1}{2} ka \times \Big\{ e^{i\omega(s)t} \Big[e^{ikra} \cos \tfrac{1}{2}\Delta\omega t + i e^{-ikra} \sin \tfrac{1}{2}\Delta\omega t \Big] - e^{i\omega(k)t + ikra} \Big\} \tag{56}$$

We can learn from these results that at time $\tau = (\Delta\omega)^{-1} \cdot \pi$ the exciton to the right, a_2^+, will be fully excited. The energy oscillates between the exciton locations, but it can only be transmitted via phonons. Nevertheless, as can be seen from projection (56), the phonon excitation and the phonon flux remains very small. But the phase-correlation of the phonon modes is completely fixed. In more general terms we may say that even a small phonon flux is able to create a high singular excitation, if there is a large phase-correlation between the phonon modes. The more this correlation is destroyed, the greather the flux has to be chosen to create a local excitation. In the given example the energy-transfer mechanism can be viewed as a non-adiabatic resonance phenomenon between the effective splitting $\Delta\omega$ of the excitonic modes and the phonon excitations. If there are less correlated but stronger fluxes a kind of Fano-mechanism may take over, in which a multi-phonon occupation ("continuum") is contrasted with the singular (local) modes. Both mechanisms can be considered as cumulative excitation processes.

Recently Sigmund and the author [15] have derived a prescription to write down exponential transformations for many different kinds of interactions. Thus, a direct investigation of phonon-transport in the above manner, i.e. without relying on transport equations, seems rather promising for many simple systems. In this way a direct critical study of the assumptions incorporated in transport formalisms seems possible. Two problems seem to be of the highest urgency. One is that of the validity of a factorization procedure. The other pertains to the large manifold of fluxes. As we can see from the preceding considerations, the same total flux can be reached in very different ways, depending on the nature of the phase-correlations between the transport modes. So, a systematic study of these correlations with respect to the flux seems highly desirable.

References

[1] H. Haken, Synergetics, ed. H. Haken, Teubner, Stuttgart, 1973
[2] R. Peierls, Ann. Phys. 3, 1055 (1929)
[3] P. C. Kwok, P. C. Martin, Phys. Rev. 142, 495 (1966)
[4] R. Klein, R. K. Wehner, Phys. Kond. Mat. 10, 1 (1969)
[5] Henceforth, $\underset{\sim}{k}\lambda$ will be abbreviated by k whenever it is convenient to do so.
[6] R. J. Hardy, Phys.Rev. 132, 168 (1963)
[7] R. Zwanzig, Physica 30, 1109 (1964)
[8] M. L. Goldberger, E. N. Adams, J. Chem. Phys. 20, 240 (1952)
[9] R. Jancel, Foundation of Classical and Quantum Statistical Mechanics, Oxford, Pergamon 1969
[10] For clarity in eqs. (40, 41, 42, 43, 44) a "bar" has been introduced to denote the expectation value. It is discarded again in the following.
[11] M. Wagner, Cooperative Phenomena, ed. H. Haken and M. Wagner, Springer Heidelberg, 1973
[12] See e. g. the work of Andronov, Vitt, Khaikin
[13] F. Kaiser, Ph. D. Thesis, University of Stuttgart
[14] M. Wagner, Synergetics, ed. H. Haken, Teubner Stuttgart, 1973
[15] E. Sigmund and M. Wagner, Z. Phys. 268, 245 (1974)

Cooperative Phenomena, H. Haken, ed.

HIERARCHICAL SYSTEMS IN VISUAL PERCEPTION

Bela Julesz
Bell Laboratories
Murray Hill, N.J., U.S.A.

Although the human visual system is perhaps the most complex structure in the known universe, this does not mean that all perceptual tasks require the full use of its vast processing capabilities. There are some fundamental processes in visual perception that use relatively simple structures, and here I review some efforts by my co-workers and me to clarify the organization of these structures using only psychological means. This review will be restricted to three processes: a)figure-ground perception, b)texture discrimination, and c)global stereopsis, that exhibit increasingly complex structure, yet are much simpler than form recognition. In these psychological studies the mathematical methods or reasoning used are familiar to the physicist, which was an important consideration in selecting them rather than other work treating with phenomena that are more enigmatic but not amenable to mathematical theorizing.

PERCEPTION VERSUS COGNITION

In this review the subject matter will be restricted to perceptual phenomena, and the higher symbolic processes (cognition) will be ignored. In general this distinction is not easily made, and I hope a few demonstrations will clarify this point. In Fig. 1, one of the line drawings is singly connected while the other is not.

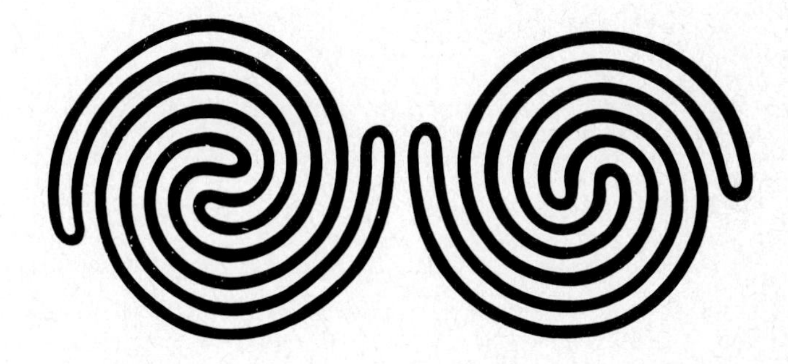

Fig. 1 Demonstration of our inability to perceive visually connected line drawings. (Minsky and Papert, 1968)

This fact cannot be spontaneously perceived. One has to trace the lines point by point with his fingers (or perhaps by slow scanning eye movements) and remember the point from where he started in order to convince himself that he was able to return to this starting point. Any visual task which cannot be performed spontaneously, without effort or deliberation, will be disregarded as a perceptual task. By the way, that connectivity in Fig. 1 cannot be perceived by our visual system is not surprising in light of the study by

Minsky and Papert [1] who showed that one would need an automaton (perceptron) composed of many n-th order linear decision networks to solve this task when there are n picture elements (receptors) of the retina (input array). Since n for the human fovea (the center of the retina with the best resolution) is about 10^4, and from neurophysiological considerations we know that a processor cannot be of (10^4)th order complexity, we can understand why connectivity is a stimulus property beyond the extraction capabilities of the visual system.

The second example elucidates even better what perception is or is not. In Fig. 2 the left top array consists of randomly portrayed black and white cells. The right top array appears also random.

Fig. 2 Demonstration that the 4 identically repeated quadrants in the upper right array appear as random as the random upper left array. When the 4 identical quadrants are mirrored across the axes, the 2-fold symmetry in the lower right array can be perceived. (Julesz, 1971)

Only by scrutinizing the stimulus feature by feature do we realize that all 4 quadrants are identical. On the other hand, if we take the same four quadrants but mirror them across a horizontal and vertical axis, we derive at the bottom left an array in which the two-fold symmetry is effortlessly perceivable. In this example the redundancy of the four times repeated quadrants and of the mirrored quadrants is the same, however, symmetry is perceived - repetition is not. The scrutinizing for the latter requires cognitive processes beyond the subject matter of this review.

FIGURE-GROUND PERCEPTION

The simplest visual task is to find a moving object in a still background. If there are only one or two such moving objects in the visual field at a given moment, then the size and velocity of these objects is adequate to determine some important properties about these objects. Indeed, the frog's visual system can cope with such a task as shown by Lettvin et al. [2]. There are neural units in the frog's retina and optic tectum that fire only when an object of a certain size enters a certain region of the retina (receptive field). Some of the units fire only for small objects, others for extended objects. Whether these types of units or feature extractors are "bug" and "snake" detectors, or whether some combination of such simple receptors constitutes a complex unit that might elicit attack or escapement behavior, goes beyond the scope of such brief review. Here it suffices that the frog's visual system is blind to still targets and detects only moving ones. The human peripheral visual system (outside the fovea) is also insensitive to still targets, and has very poor spatial resolution, yet is very sensitive to moving or flickering targets even of small sizes. Our peripheral vision can detect and locate such moving targets without being able to resolve or recognize what they are.

The task of detecting a target when both target and background are stationary is much more difficult. In general, it cannot be solved without relying on the enigmatic processes of semantics. Without having familiarity with objects in our environment, one could not organize the random patches in Fig. 3 into a familiar form.

Fig. 3 Disorder-order transition of random patches into a dalmatian dog. (Photograph by R. C. James; Carraher and Thurston, 1966)

There are situations that are much simpler than the disorder-order transition of Fig. 3. For instance, the figure-ground reversal of a modified Rubin [3] display (see Fig. 4) is perceived either as two faces or as a goblet, and one suspects that this switching between 2 perceptual states does not require semantic processings. Indeed, when Fig. 4 is viewed upside down one can still perceive figure-ground reversals. It appears as if one had two independent channels, one for white areas and one for black areas, and the visual system would shift attention to only one of them at a given time. After some inspection time the attended channel would be neurally adapted to (fatigued) and would release lateral inhibition of the unattended channel (ground), which in turn could become dominant. In my book (Julesz [4]) an operator theory for figure-ground perception has been formulated. The essence of this theory is as follows: The E visual environment is separated into two parts, the figure (F) and the ground (G):

$$E = F + G = F + (E-F)$$

The act of perception is to assign an operator P that acts on the stimulus:

$$PE = PF + P(E-F) = PF + PG$$

Now every operator (Q) has at least one eigenfunction (R) and eigenvalue (λ), such that $QR = \lambda R$. We assume that the perceptual operator P has two eigenfunctions F and G with the corresponding eigenvalues λ_1 and λ_2 respectively, thus

$$PE = \lambda_1 F + \lambda_2 G$$

The peculiarity of figure-ground perception is that the perceptual process assigns the eigenvalues $\lambda_1 = 1$ and $\lambda_2 = 0$. In other words F is attended to, is vivid, and can be memorized, while (E-F) is ignored, pale, and unmemorizable. There are some instances, as in Fig. 4, when F and (E-F) can alternately become the figure; thus either λ_1 or λ_2 becomes 1, but at a given instant

$$\lambda_1 + \lambda_2 = 1$$

Thus the roles of figure and ground reverse.

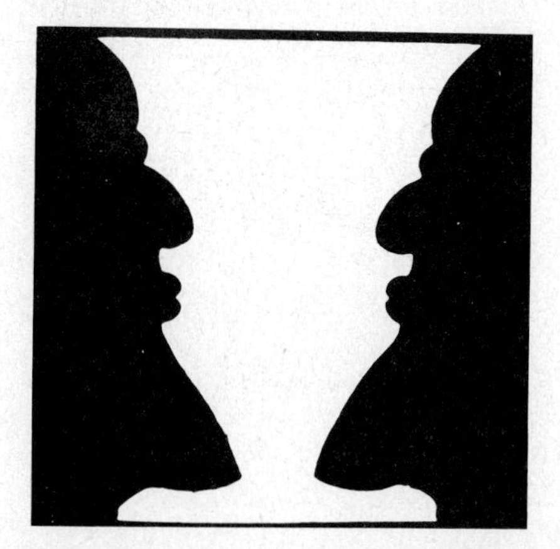

Fig. 4 Figure-ground reversal of a modified Rubin (1915) display.

The Gestaltist school, and particularly Rubin, studied the perceptual factors underlying figure perception. In general, some of these factors are rather complex and difficult to define in precise terms. However, one can restrict the stimulus to random textures with various area fractions between black and white as shown in Fig. 5 reprinted from a study by Frisch and Julesz [5]. For this simplified case one can ask whether subjects regard the random textures as a black tabletop (ground) with white specks (figures) on it, or a white tabletop with black specks on it as the area fraction varies. Figure 5 shows three white to black area fractions ϕ with 10%, 50%, and 90%, both for two tessellations (a Cartesian grid and a triangular grid), respectively. Here the regular tessellations of the plane have their squares or triangles colored black or white according to a binomial distribution.

Fig. 5 Random textures with 10%, 50%, and 90% area fraction of Cartesian and triangular tessellation. (Frisch and Julesz, 1968)

These textures are described by statistical geometry, particularly by a generalization of the classical Buffon needle problem (Frisch and Stillinger [6]. This method is based on the random (Poisson distribution) throwing of polygons of n vertices $\{n\}=(\underline{r}_1, r_2, \ldots, \underline{r}_n)$, $n \geq 1$ (n=1 a point, n=2 a line, n=3 a triangle, etc.) and determining the probability $\gamma_n = \gamma\{n\})$, these n vertices all lie in the white region of the texture. The statistical geometry is contained, so to say, in the hierarchy $\gamma_1, \gamma_2, \gamma_3, \ldots,$ and specific geometric properties can be expressed naturally by certain functionals of the γ_n's (Novikoff [7]; Frisch and Stillinger [6].

Now it can be shown (Frisch and Julesz [5]) that the response function of subjects (when viewing ambiguous random textures and perceiving figure-ground reversals) is a functional of γ_n and all γ_n, except γ_1 depend on the tessellation used. For instance, a simple parameter that can be derived from γ_2 is the perimeter per unit area (lla). It can be shown that the triangular lattice has a (lla) which is $\sqrt{3}$ times larger than the value for a square lattice and this is $\sqrt{3}$ times larger than the value for the hexagonal lattice. It is also reasonable to assume that the psychological response would change accordingly, since a change of $\sqrt{3}$ is not negligible. However, when Frisch and Julesz determined the psychological response for figure-ground reversals, they found no differences for the various tessellations used. This means that figure-ground reversal depends only on γ_1, which in turn is ϕ the white to black area fraction.

Thus, for figure-ground perception, we have the simplest case where the system is throwing just point detectors (n = 1) on the retinal representation of random textures. That such detectors really exist was shown by the neurophysiologists (Kuffler [8]; Hubel and Wiesel [9a,9b] in various stages of the central nervous system of the cat and the monkey. They found in the optic nerve and lateral geniculate nucleus neural units that fired when a small disk with a certain retinal position and diameter was stimulated (receptive field). If this retinal stimulation surpassed an optional disk size the firing did not increase but actually decreased (owing to an inhibitory annulus shaped region around this disk). In the cortex of the cat and the monkey neural units were found whose retinal receptive fields were elongated ellipses with certain width and orientation at some retinal position. Some of these units responded optimally for stimulus features with increasing complexities. Thus the hierarchy of n-gon throwing can be regarded as a simplified model of the hierarchy of feature extractors that exist in the nervous system. The insight that figure-ground perception of random textures depends only on the simplest such extractors with small disk shaped receptive fields is rather unexpected. It shows that for some fundamental perceptual processes only the simplest analyzers are used, and the more complex feature extractors do not participate.

TEXTURE DISCRIMINATION

The next complexity takes us to texture discrimination. In 1962, I studied texture discrimination by generating side by side two random textures with different statistics and asked whether they could be discriminated without effort or deliberation as a spontaneous process. Figure 6 shows a Markov-texture, where the two

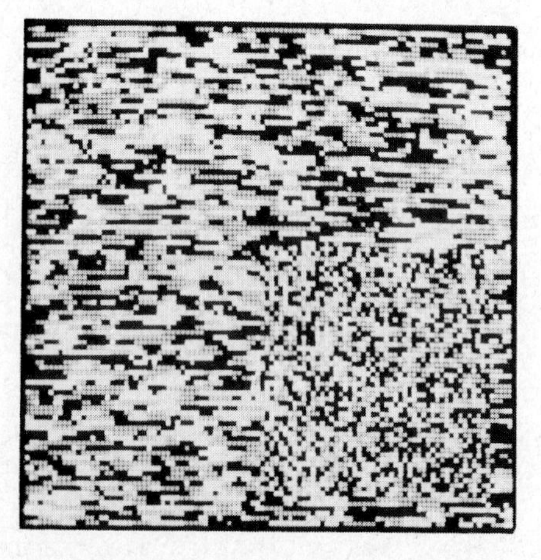

Fig. 6 Markov textures side by side with identical first-order but different second-order statistics. Texture discrimination is easy. (Julesz, 1962)

textures have identical first-order statistics (first order probability distribution) but different second-order statistics (second-order joint probability distribution), and discrimination is easy; whereas Fig. 7 shows a texture of seven letter words where one-half is English, the other half nonsense - a fact that can be determined by scrutiny - yet one cannot organize the two halves into two separate entities.

PUNCHED METHODS SCIENCE COLUMNS NIATREC YLKCIUQ DEHCNUP SDOHTEM

PRECISE SUBJECT MERCURY GOVERNS ECNEICS YFICEPS ESICERP TCEJBUS

OXIDIZE COLUMNS CERTAIN QUICKLY SDOHTEM SDROCER EZIDIXO SNMULOC

CERTAIN RECORDS EXAMPLE SCIENCE STCIPED HSILGNE NIATREC SDROCER

GOVERNS MERCURY SPECIFY PRECISE TCEJBUS DEHCNUP SNREVOG YRUCREM

SPECIFY METHODS COLUMNS MERCURY ELPMAXE YLKCIUQ YFICEPS SDOHTEM

EXAMPLE CERTAIN DEPICTS ENGLISH ECNEICS ESICERP ELPMAXE NIATREC

PUNCHED QUICKLY METHODS EXAMPLE YFICEPS YRUCREM DEHCNUP YLKCIUQ

OXIDIZE ENGLISH SUBJECT RECORDS ELPMAXE SNREVOG EZIDIXO HSILGNE

PRECISE MERCURY PUNCHED CERTAIN SNMULOC TCEJBUS ESICERP YRUCREM

EXAMPLE SUBJECT OXIDIZE GOVERNS HSILGNE SDROCER ELPMAXE TCEJBUS

PUNCHED METHODS ENGLISH COLUMNS NIATREC ESICERP DEHCNUP SDOHTEM

SCIENCE DEPICTS SPECIFY PRECISE EZIDIXO YLKCIUQ ECNEICS STCIPED

ENGLISH CERTAIN RECORDS SCIENCE STCIPED ELPMAXE HSILGNE NIATREC

GOVERNS PRECISE QUICKLY METHODS YFICEPS YRUCREM SNREVOG ESICERP

Fig. 7 English and nonsense word textures side by side. No texture discrimination. (Julesz, 1962)

In 1962, I raised several questions for some of which the answers came very slowly. First of all, I wanted to know whether texture discrimination could be described by the statistical properties of the textures alone, or whether one has to introduce some additional factors such as the idiosyncrasies of the various feature extractors. If under certain conditions one could describe texture discrimination by statistical methods, I wanted to know the answers to two more questions: 2)Would it be possible to generate textures side by side such that their k-th order joint probability distributions are identical but their k+1)-th order joint probability distributions differ; 3)If such stochastic processes can be generated and portrayed as textures, what is the smallest (k+1), for which no texture discrimination can be obtained by human subjects?

The first question seems to be hopelessly difficult. If the visual system contains a hierarchy of specific feature extractors, then the statistical descriptors have to match these features. It seems rather unlikely that all of these ad hoc features such as clusters of certain shapes, sizes, orientations, etc. could be described by the usual statistical parameters. On the other hand, many of the simple features manifest themselves through the first- and second-order statistics. So I became interested in generating textures with identical first- and second-order statistics, but different third- and higher-order statistics. That such Markov processes exist was proven by Rosenblatt and Slepian [10] provided the stochastic variables aad at least three (luminance) values. So I turned to the third question and generated such Markov textures side by side and found that, with identical first- and second-order statistics but different third-order statistics, these texture pairs were not discriminable.

These findings did not seem conclusive enough for several reasons. First, Markov processes are inherently one dimensional, and in order to generate two dimensional textures one has to introduce some arbitrary one-dimensional scanning procedure. Obviously, the same Markov process when plotted in successive rows from left to right yields a very different texture from the same process being plotted in a dense spiral starting from the center to the periphery. Thus the various scanning pathways influence at least as much the appearance of the textures as the constraints imposed by the Markov process. Second, the influence of third-order statistical constraints on simple feature formation is not known, which in turn can be detected by the simple feature extractors.

One could have tried methods of statistical geometry where the randomly thrown n-gons might be more closely related to visual features than the parameters derived from n^{th}-order joint probability distributions. Unfortunately, the mathematical problems of generating textures that have identical n-gon statistics, but have different (n+1)-gon statistics are not yet solved.

However, in the last year through the efforts of two mathematicians at Bell Laboratories, E. N. Gilbert and L. A. Shepp, several methods were devised that enable us to generate non-Markov textures in two dimensions with identical first- and second-order statistics but different third- and higher-order statistics. All these methods use texture pairs composed of identical micropatterns - either regularly spaced or thrown at random (with or without random rotation) - except that the micropattern b in Texture B is some transformation of the micropattern a in Texture A. For details see Julesz et al. [11]. There only two examples will be shown. Figure 8 shows two textures derived by repeating micropattern a and b on a regular spacing, where b is derived from a by a 90 degree

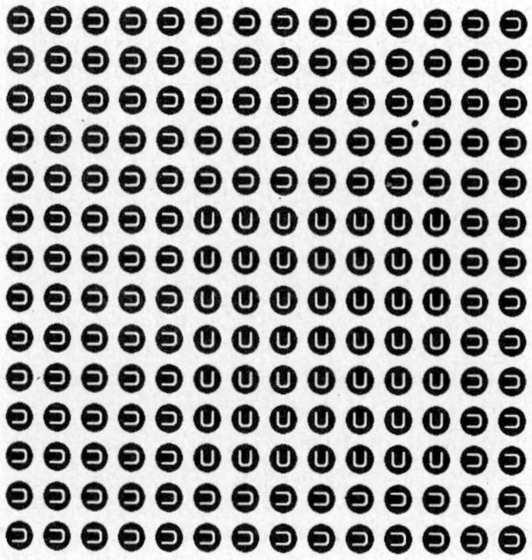

Fig. 8 Texture discrimination between textures having identical first-order but different second-order statistics. (Julesz, et al. 1974)

rotation. Obviously, the second-order statistics for Textures A and B must be different, since there are line segments (dipoles) whose end-points fall on the white U-shaped area in Texture A that have no corresponding dipoles (with the same orientation) in Texture B, or if they do have then they occur with different statistics.

Figure 8 yields strong texture discrimination. However, if any micropattern b is derived from micropattern a by 180 degree rotation then the corresponding textures A and B have identical second- (and first-) order statistics. This can easily be verified by generating only texture A which is viewed by two observers, one in regular viewing position, the other standing on his head. Since the two observers see the same dipoles - only their end-points interchange - thus the two-point probabilities of coverage (dipole statistics0 [and one-point probabilities of coverage (point statistics)] are identical for the two observers, thus they are identical for textures A and B. This, however, does not work for three- or more points. Figure 9 shows such a case, and no texture discrimination can be obtained, although by scrutinizing the micropatterns the u and n shaped micropatterns appear very different.

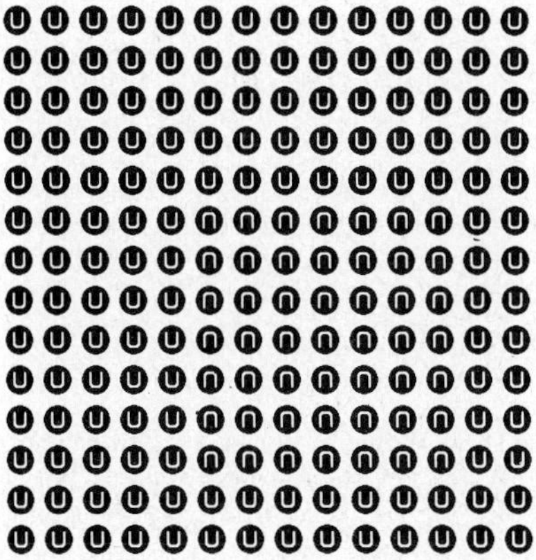

Fig. 9 No texture discrimination between textures having identical first- and second-order statistics. (Julesz, et al. 1974)

It can also be shown that if textures are derived from throwing micropatterns in all possible orientations, then Texture B which is derived from Texture A by taking for micropattern b the mirror image of micropattern a, has the same first- and second-order statistics as Texture A but different third- and higher order statistics. The verification is similar to the previous case, except the second observer views Texture A from behind as if the plane on which micropatterns a were thrown were transparent. Obviously the point and dipole statistics must be the same for both observers, which is not the case for three- or more points.

Figure 10 shows such an example, where texture A is composed of the letter R thrown in all possible orientations, whereas Texture B consists of the mirror image of R, the Russian letter "ya" in all possible orientation. As the inspection of Fig. 10 demonstrates texture discrimination is impossible, although with scrutiny one can determine whether each letter is an R or its mirror image.

There are many such demonstrations given in this recent study (Julesz, et al.[11]) where the micropatterns that constitute the

Fig. 10 Demonstration that a texture composed of micro-patterns that are mirrored in the other texture cannot be discriminated. First- and second-order statistics are identical. (Julesz, et al. 1974)

texture pairs look very different, but texture discrimination is very difficult or impossible. In all these cases the second-order statistics were kept identical. However, there are some counter-examples. For instance, it can be shown that a black ring with a tiny black hole in it has the same short-term γ_2 statistics as a black ring with a tiny black hole outside of the ring. Since the visual system is probably computing correlations in a short range (note that in Figs. 9 and 10 no discrimination occured though the micropatterns were regularly placed, thus the second-order statistics were identical only in a short range), one would hope that texture pairs composed of such dual micropatterns do not yield texture discrimination. However, as Fig. 11 shows, texture discrimination can be experienced.

This counter-example just indicates that one cannot say in general that no texture discrimination is possible when the second-order statistics are identical. However, one has to carefully inspect the few counter-examples. It seems that for these cases the single feature extractors were able to distinguish between the constituent micropatterns. Indeed, a simple unit with a disk-shaped receptive field, having an annulus-shaped inhibitory region around the disk, can easily discriminate between a hole inside or outside a ring. On the other hand, if the micropatterns cannot be discriminated by the simple units, then no other unit complex, hypercomplex, etc. is employed in visual texture discrimination. What is even more, the global texture discrimination stage *can only compare the output of two simple feature extractors at any given time*. Thus figure-ground perception of random arrays requires only the statistical average of simple feature extractors while texture discrimination utilizes the weighted statistical average of comparators that compute the output of only 2 simple feature extractors. Both processes are rather simple and do not exhibit all the properties of cooperative phenomena. The third example, stereopsis (stereoscopic depth perception) of random-dot stereograms (Julesz [12]) is still adequately simple, yet exhibits such cooperative phenomena.

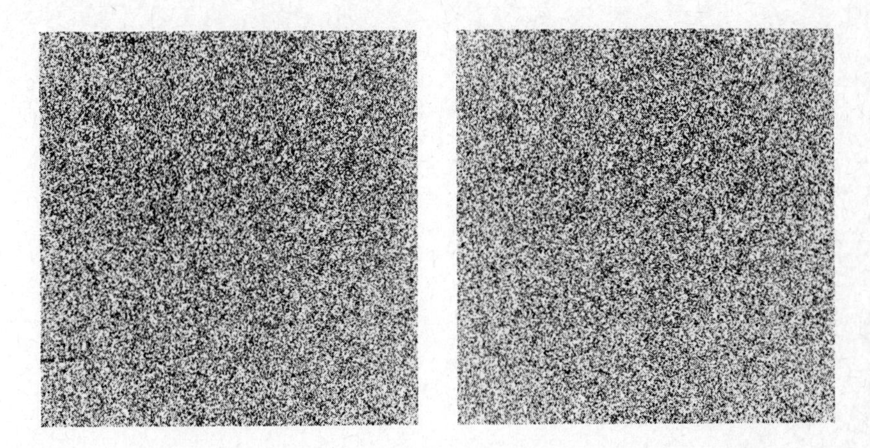

Fig. 11 Random-dot stereogram, which when viewed stereoscopically portrays a surface in vivid depth. (Julesz, 1960)

GLOBAL STEREOPSIS

When Fig. 11 is viewed stereoscopically, with the help of a prism, or by crossing (or everting) one's eyes, after some seconds the random dots in the two eyes' views suddenly coalesce into a unified view and a surface jumps out in vivid depth from the surround. Those who cannot fuse stereo images without aid, but have access to my book (Julesz [4]) can take advantage of the red-green anaglyphs and special viewer that permits the easy fusion of many computer-generated random-dot stereograms.

Figure 12 shows an example of multiple stable states (Julesz and Johnson [13]). Here the same random-dot stereogram does portray two different surfaces, a tent-shaped surface behind or a flat plane in front. One can perceive either one organization or the other but not both. Finally, the hysteresis of stereoscopic fusion was shown by the technique of binocularly stabilized retinal images (Fender and Julesz [14]). With the aid of close-fitting contact lenses with a mirror mounted on them, we presented retinally stabilized images for both the left and right eye. As the eyes moved or converged, the images stayed in a fixed position on the retinas. Thus, if the random-dot stereograms were not in registration, no fusion could be obtained. If the experimenter physically shifted the stereo pair on the retinas when the images came within 6 minutes of arc alignment (Panum's fusional area), the images suddenly coalesced into a fused percept. After this fusion, however, the two images could be slowly pulled apart in the horizontal direction by as much as 120 min of arc, at which point they suddenly jumped apart. After this break, the images had to be brought back within 6 min of arc registration in order to be fused again. The

Fig. 12 Multiple stable states in stereopsis. After stereoscopic fusion the percept is either a wedge behind or a flat plane in front.

hysteresis loop for a random-dot stereogram, such as Fig. 11 is shown in Fig. 13. If the target is just one vertical line in the

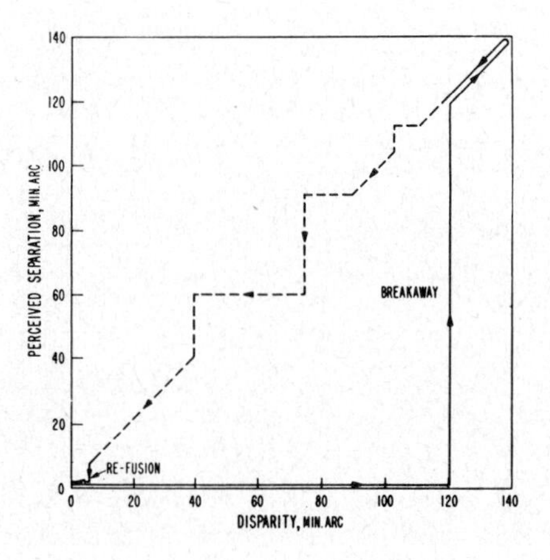

Fig. 13 Hysteresis phenomenon of a random-dot stereogram during binocular retinal stabilization. (Fender and Julesz, 1967)

two eyes' view, the hysteresis loop is greatly reduced, as shown in Fig. 14. The dependence of hysteresis on the complexity (number of dots) of the stimulus is characteristic of cooperative phenomena in general.

These experiments indicate also that the horizontal shifts required to bring corresponding points of the left and right retinal projections into registration are not the result of convergence movement of the eyes alone. While large binocular disparities are reduced by convergence-divergence movements of the eyes, a few minutes of arc disparity (Panum's fusional area) is cortically registered. The existence of cortical shift was already known by Dove [15].

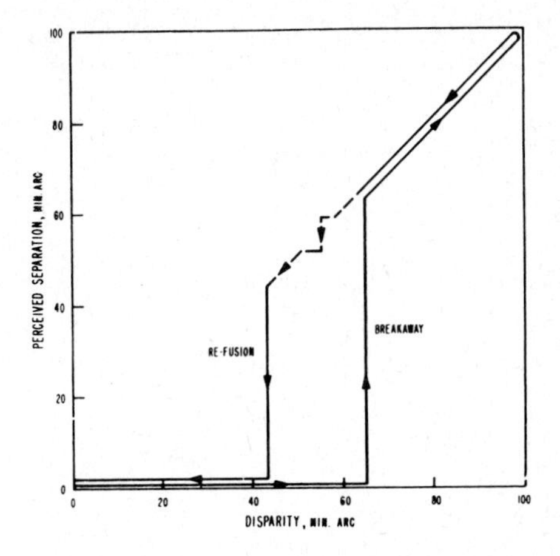

Fig. 14 Hysteresis phenomenon of a vertical line target during binocular retinal stabilization. (Fender and Julesz, 1967)

After this brief review of stereopsis we can formulate a model. This model was described in great detail elsewhere (Julesz [4]). It uses only simple elements such as magnets and springs; however, exhibits all the cooperative phenomena described and its workings can be understood without mathematical background. This is in contrast to Ising-models, which in spite of their apparent simplicity require sophisticated mathematical treatment. Of course, this model of stereopsis is conceptually simpler than the Ising-models, nevertheless it gives an easy grasp of order-disorder transitions, multiple stable states and fusion.

Imagine that little magnetic dipoles are mounted in ball-joint bearings and can rotate in any direction (see Fig. 15). Local stereopsis (sensing the depth of a dot) corresponds to the amount of horizontal rotation of a given dipole. These magnetic dipoles are the cortical representations of the stimuli cast on the retinas. If a retinal element of the left eye receives a white dot, its corresponding dipole is turned so that its north pole faces us, while for a black dot the south pole is turned toward us. The black and white random-dot array on the left and right retinas therefore corresponds to a left and right array of dipoles with similarly ordered north and south poles.

Thus the ordering of the dipoles is forced on the system and is not the "order parameter" we want to establish. The order parameter is the number of locked dipoles between the left and right dipole arrays that correspond to binocular fusion. Before these dipole arrays interact, an important nonlinear operation has to be performed. This operation is the "secret" of the success of the model: the already polarity-ordered dipole arrays are suddenly coupled to their adjacent neighbors by mechanical springs. The two ordered and coupled dipole arrays should be imagined in close proximity sliding over each other, as shown in Fig. 15.

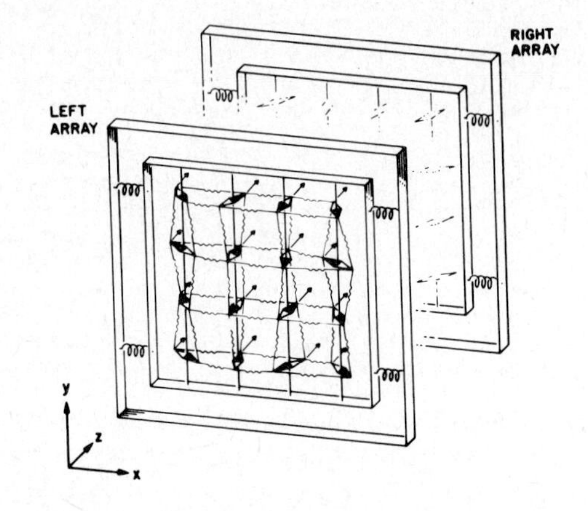

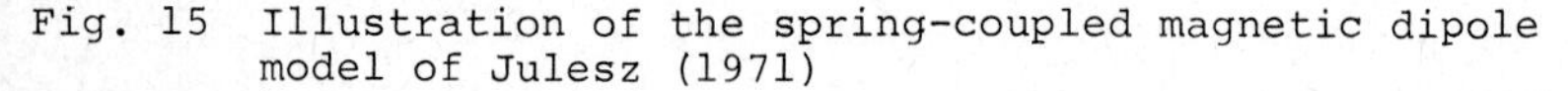

Fig. 15 Illustration of the spring-coupled magnetic dipole model of Julesz (1971)

If the patterns on the retina are identical and the two dipole arrays overlap, then the south and north poles of the corresponding dipoles interlock. Identical arrays of inter-locked dipoles in one dimension are shown in Fig. 16A.

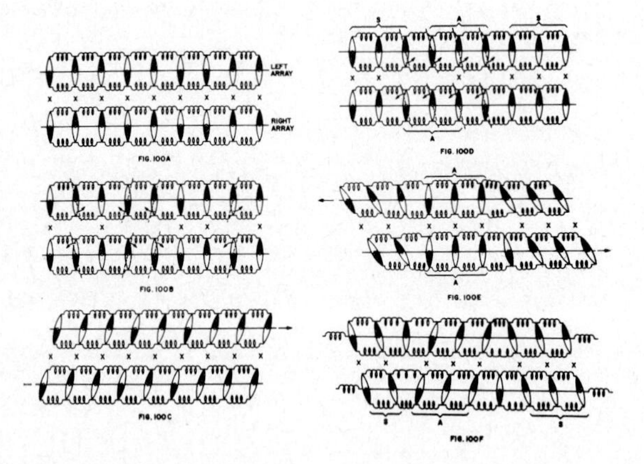

Fig. 16 The spring-coupled dipole model in one dimension under various conditions described in text. (Julesz, 1971)

Let us try to explain the fusion of a random-dot stereogram such as Fig. 11. As you will remember, before the arrays are shifted in alignment, by chance 50 percent of the dipoles become interlocked. Because of spring coupling, the other 50 percent are forced to face the corresponding dipoles with identical polarities in the other array. This is shown in Fig. 16B, where 50 percent of the dipoles attract each other, and 50 percent repel each other, making the global attraction force between the arrays zero. Because there is no attraction, the arrays can easily be slid into any other position. However, in a totally interlocked array, the pulling of the arrays causes the dipoles to turn together, as shown in Figure 16C.

In order to fuse the arrays of Fig. 11, we first align, say, the background areas. This shift, when it is larger than Panum's fusional limit, is achieved by the convergence-divergence movements of the eye; within Panum's limit the shifts are neurally executed. In the model the arrays are physically moved with respect to each other in the horizontal direction. Thus, when the two arrays are exactly aligned, the corresponding dipoles in the background area (with zero binocular disparity) will interlock. Once interlocked they exert a large force of attraction on each other.

The center area, however, which has a diamond shape (with nonzero disparity), contains dipoles that partly attract and partly repel each other (Fig. 16D). When, as the second step, we shift the center area into alignment, the magnets of the center area will interlock; however - and this is the crucial aspect of the model - the already interlocked dipoles of the diamond turn while remaining interlocked (Fig. 16E). If, finally, we assume that the two dipole arrays are suspended in a spring-loaded frame (as shown in Fig. 16F), after the two areas are interlocked they will be in an equilibrium position.

We note again that sensing the degree of rotation in the horizontal direction corresponds to local stereopsis. The dipoles, of course, can also turn in the vertical direction, but this rotation is not sensed. The limit of horizontal rotation that still gives rise to stereopsis is Panum's fusional limit.

A COOPERATIVE MODEL OF STEREOPSIS

We are now in a position to understand most of the phenomena of stereopsis. For instance, if we expand one array by 10 percent, the dipoles in this array will turn on their ball joints and try to face their corresponding dipoles. The model also explains the hysteresis effect of Fender and Julesz [14]. After the dipole array becomes interlocked it exerts a great force that causes the springs (Fig. 16F) to give way (at least for 120 min arc displacement) - but only if the pulling is slow. When the pulling of the stimuli proceeds at a fast rate, the inertia of the springs will cause the arrays to move away from each other, and after a critical distance is reached the dipoles become unlocked. With proper masses attributed to the dipoles, spring constants, and frictions, this model can be regarded as an analog computer.

Although space does not permit me to go into all the details of this model (for which, see Julesz [4]), let me note how it can clarify the problem of perceptual learning of fusion. When the stereograms are simple, as in Fig. 11, there are only two steps required for fusion: first, aligning the background and then trying a nasal or temporal shift in order to find the other depth plane. If by chance the center area is fused first, a quick cortical shift will fuse the background as well. The only ambiguity is due to the dichotomy of performing a nasal or temporal shift.

However, for stereograms that contain complex surfaces (having many depth planes)this dichotomous decision has to be made for each depth plane. A wrong decision causes a false localization, and often one must backtrack several decisions and start from scratch. Nevertheless, if this decision tree of nasal-temporal forks is learned, refusion becomes greatly facilitated.

The model also explains the problems of ambiguities. If a solution is reached--i.e. all dipoles become interlocked--it is impossible to obtain another global solution until the interlocked state is destroyed by drastic means. The only sure way to find another solution is to unlock the local interlocked dipoles, which can be done by using electromagnets with short time constants rather than permanent magnets.

After the local elements become unlocked, the global fusional process has another chance to search for a different global solution. This explains why in the central nervous system adaptation to stimuli occurs at very early stages! Were adaptation a highly central phenomenon we would never be able to "shake" a global solution once it had been obtained. On a higher level, we would never be able to enjoy a pun or a joke based on ambiguous meanings.

In spite of its powers, this model has some serious limitations. First of all, it is not a neurophysiological model, and it remains to be seen how the local binocular disparity-sensitive cortical units interact through complex facilitatory and inhibitory connections. Obviously, a magnetic dipole corresponds to a hypercomplex stereo unit that integrates the output of many local binocular disparity-sensitive cortical units that are tuned to many disparity values but belong to one retinal position. On the other hand, when the springs force many adjacent dipoles to turn together at a constant angle, as if they were a hyperdipole, the corresponding hypercomplex neural unit can be regarded as integrating the output of many local binocular disparity-sensitive cortical units that are tuned to one disparity value but belong to many retinal positions. While the columnar organization of the cortex might suggest the existence of hypercomplex stereo units that evaluate the output of local disparity-sensitive units within a column, it should be stressed that no such global units have yet been found, and a neurophysiological model of stereopsis can be based only on mere speculation.

Second, there is recent evidence that the input is first analyzed by spatial frequency filters, and it is the output of these filters of similar frequencies which is binocularly combined. It is easy to incorporate such filters into the model by postulating that the dipoles correspond both to retinal elements of given coordinates and given spatial frequencies. Finally, in 1964, I verified experimentally that stereopsis has to be a parallel process and cannot be a serial one as is the model described here (see Julesz [4] for details).

While it is relatively easy to incorporate other features into the model to cope with these limitations, I think the model as described should suffice to illustrate my initial claim that this model of stereopsis is conceptually much simpler than Ising models. In the Ising models the long-term order is obtained as a result of the workings of the model. In the model of stereopsis the long-term order is imposed on the system from outside by the patterns cast on the retinas. The order parameter corresponds to the extent that these patterns become binocularly matched by the system.

The model can be further generalized from two to n dimensions. The two arrays that are the cortical representations of the two retinal stimulus patterns could correspond to any two correlated events. For instance, one array could be the n-dimensional feature-space of objects stored in memory while the other array could be the n-dimensional feature space of objects presented as input stimuli to an organism. The model then could function as a feature-recognition automaton. Once a random search in n dimensions resulted in an interlocked array, the automaton could search in other directions, since the interlocked arrays will remain so as long as the search (shift) stays within some limit. It seems plausible that related features should be stored not too distant from each other in memory.

One nice aspect of the stereopsis model is the simplicity of the search. The cortical shift is a single, scalar quantity within Panum's area (which for foveal targets is ±6 min arc). Stereo acuity--the finest perceivable disparity change--is a few seconds of arc. Thus, there are about ±100 binocular disparity units that can easily be searched through in a random fashion. This simplicity permits the study of how perceptual learning may affect random shifting--one of the problems I am working on at present.

I have deliberately chosen a generalization of the model in the direction of higher mental processes. Here some of the cooperative features of the model are useful assets but not adequate for a workable process. Most likely, in a high-dimensional space, random search would last too long to be tolerated in practical situations. It seems that an important criterion of intelligent behavior (whether in living organisms or machines) is the built-in heuristics that substitute for random search some more efficient method.

At present these enigmatic, goal-oriented heuristic algorithms are beyond our grasp. In chess, for instance, grandmasters reportedly do not examine a single move further ahead than lower-ranked masters nor do they examine more moves per position (Hearst [16]). The grandmaster is somehow able to "see" the core of the problem immediately and preselect the best moves for further scrutiny. In our terminology, he knows more or less the right direction and amount of the shift around which the corresponding patterns on the board interlock with patterns from memory.

How this preconscious, automatic selection process operates is still a mystery and prevents progress in the field of artificial intelligence (Dreyfus [17]),although recently Simon and Chase [17] have proposed that grandmasters during their training build up a chunk-memory in excess of 30,000 "chunks" of familiar chess configurations, such as a "short-castled Black King" or a "Knight-pinned-on Queen-by Bishop," This "vocabulary" of familiar chess configurations is comparable to the word-recognition vocabularies of persons able to read English and might explain the "intelligence" of the preselection process in an unexpectedly blunt way.

In summary, we have seen how cooperativeness can lead to some remarkable feats, such as the slow fusion of Fig. 11. Similarly, cooperativeness, coupled with random search, suffices to obtain stable states in tolerable time limits in ferromagnetism or in the conformation of protein molecules. In systems of higher dimensions, however, random search is inadequate. Somehow semantic information permits intelligent organisms to abandon random search for a heuristic search. The essence of this intelligence is still a great mystery (even if in many cases the intelligence might be based on recognizing a subpattern within a vocabulary of memorized chunks). Yet, until now, many of the cooperative phenomena were mistaken for intelligence. By trying to separate the cooperative structures of a system from heuristic search procedures, I think we can come closer to the enigmatic problems of semantics and intelligence.

[1] M. Minsky and S. Papert, Perceptrons, Cambridge, Mass.: MIT Press (1968).

[2] J. Y. Lettvin, H. R. Maturana, W. H. Pitts, and W. S. McCulloch, Proc. Inst. Radio Engrs. 47 (1959) 1940

[3] E. Rubin, Visuell wahrgenommene Figuren. Copenhagen: Gyldenalske Boghandel. Reprinted as "Figure and ground" in Readings in Perception, ed. D. C. Beardslee and M. Wertheimer. Princeton, N.J.: Van Nostrand 1958, p.194.

[4] B. Julesz, Foundations of Cyclopean Perception, Chicago, Ill.: University of Chicago Press (1971).

[5] H. L. Frisch and B. Julesz, Perception & Psychophys. 1 (1966) 389.

[6] H. L. Frisch and F. H. Stillinger, J. Chem. Phys. 38 (1963) 2200.

[7] A. E. J. Novikoff, Integral geometry as a tool in pattern perception. In Principles of Self Organization, ed. H. von Foerster and G. W. Fopf. New York: Pergamon Press, p.11.

[8] S. W. Kuffler, J. Neurophysiol. 16 (1953) 37-68.

[9a] D. H. Hubel and T. N. Wiesel, J. Physiol. 160 (1962) 106.

[9b] D. H. Hubel and T. N. Wiesel, J. Physiol. 195 (1968) 215.

[10] M. Rosenblatt and D. Slepian, J. Soc. Indust. Appl. Math 10 (1962) 537.

[11] B. Julesz, H. L. Frisch, E. N. Gilbert, and L. A. Shepp, Inability of humans to discriminate between visual textures that agree in second-order statistics - revisited. Perception, to be published, 1974.

[12] B. Julesz, Bell System Tech. J. 39 (1960) 1125-62.

[13] B. Julesz and S. C. Johnson, Bell System Tech. J. 49 (1968) 2075-83.

[14] D.H. Fender and B. Julesz, J. Opt. Soc. Am. 57 (1967) 819.

[15] H.W. Dove, Ann. Phys. series 2 110, (1841) 494.

[16] E. Hearst, Psychol. Today 1, (1967) 35.

[17] H. L. Dreyfus, What Computers Can't Do, New York: Harper & Row (1972).

[18] H. A. Simon and W. G. Chase, Am. Sci. 6 (1973) 394.

[19] R. Carraher and J. Thurston, Optical Illusions and the Visual Arts, New York, Van Nostrand Reinhold (1966).

Cooperative Phenomena, H. Haken, ed.

MATHEMATICAL MODELS OF NEURAL TISSUE

Hugh R. Wilson
University of Chicago
Chicago, Illinois, U.S.A.

INTRODUCTION

Much of contemporary brain research is devoted to the study of the neural basis of such functions as sensory information processing, pattern recognition, learning, and memory. Any one of these functions certainly involves the spatio-temporally organized activity of from 10^3 to 10^6 neurons. This fact presents major methodological and conceptual difficulties for the neurophysiologist, whose techniques are largely limited to recordings of activity from one neuron at a time. It is precisely this difficulty which the theoretician may hope to overcome. Using mathematics as a tool, it is possible to develop models embodying the properties of single neurones as well as anatomical features of different regions of the brain. Appropriate aspects of the dynamics of such models may then be compared both with single cell responses and with perceptual data on pattern recognition, etc. It is to be hoped that this approach will lead to a deeper understanding of the phenomena involved and to an increased sophistication of modeling techniques.

The bulk of this paper will be devoted to the discussion of a mathematical model for the types of neural tissue sheets characteristic of the cortex of the brain. The model is certainly oversimplified (although it may be a fairly accurate representation of some of the most primitive cortical regions), yet it incorporates a number of general features of cortical organization. Following this, an extension of the model to deal with temporal variations of parameter values (a crude form of learning) will be briefly discussed. This latter work deals with certain aspects of visual perception and is a first step toward the development of a more sophisticated neural tissue model tailored to the properties of the visual system.

CORTICAL TISSUE MODEL

Virtually all sensory systems in the brain may be thought of as a series of thin sheets of neural tissue such that the output of each sheet provides the input to the next level of the system. Each of these sheets may have an extensive surface area, but they are usually limited in thickness to 2-3mm. For example, the first three levels of the visual system in mammals are the retina, the lateral geniculate nucleus, and the visual cortex (area 17). Each of these structures projects to the next one, and importantly, these projections are neighborhood preserving. Thus, retinal loci which are proximate will be represented by regions in both the lateral geniculate nucleus and the cortex which are also close. The model to be developed here reflects some of the general organizational principles of the higher levels of such neural systems -- the levels represented in the visual system by the lateral geniculate nucleus and the cortex. As a detailed derivation of the model is available elsewhere (Wilson and Cowan, [1], [2]), only the basic assumptions and the significance of the various terms in the mathematical formulation will be discussed here.

A fundamental assumption of the modeling philosophy developed here is that cortical and thalamic neural tissue may be represented functionally as series of two dimensional sheets. There are at least two lines of evidence supporting this assumption. In a classic series of studies, Hubel and Wiesel [3],[4] found that all cells lying perpendicularly under a small element of surface area of the visual cortex received input from the same small region of visual space and that all of these cells were selectively sensitive to the same features of the visual stimulus. Each of these "cortical columns" was found to extend throughout the depth of the cortex. The assumption that the cortex is functionally two dimensional, therefore, amounts to no more than the assumption that the cortical column is the fundamental processing and output unit of the cortex.

A second line of evidence in support of the two-dimensional hypothesis derives from the anatomical structure of the cortex (Colonnier [5]; Szentagothai [6]; Chow and Leiman [7]). Very briefly, there are two basic cell types in the cortex: pyramidal cells and stellate cells. The cell bodies of the pyramidal cells are distributed in depth throughout the cortex. Each of these cells, however, sends a process, the apical dendrite, all the way to the cortical surface. As it is along this process that pyramidal cells receive much of their input from other cells, it follows that these cells are ideally suited to integrate all activity occurring within a cylinder running perpendicularly up to the surface. Furthermore, the axon or output line of each pyramidal cell sends an array of side branches or collaterals coursing obliquely upward to the cortical surface. In virtue of this anatomical arrangement, it would appear that the pyramidal cells within a column of cortical tissue are very strongly coupled in both their input and output relations. In the language of physics, it may be said that the anatomically determined correlation length in the radial dimension of the cortex is equal to the thickness of the tissue. It is therefore natural to deal with this densely interconnected aggregate of cells as a unit and to collapse our description of the cortex to two dimensions.

Implicit within this description of the cortex as composed of radially organized columns is the notion of redundancy of cell function. If the cells within the cortex are radially interconnected as suggested, then it follows both that there must be a redundancy of cell function and that it must be the total activity of the column rather than the independent activity of each single neuron which is of functional significance. This proposed redundancy is itself functionally significant in two ways. First, it provides the brain with an immunity to the random daily death of some thousand neurons. In the second place, it allows for the reliable and repeatable processing of information despite random fluctuations in single cell excitability. This latter aspect of columnar redundancy is readily recognized by the neurophysiologist, who finds it necessary to average between 50 and 100 trials in order to obtain reproducable data from a single cell. For obvious reasons of survival an animal would not have time for such temporal averaging, so spatial averaging has evolved instead, suggesting a kind of ergodic principle operative in the nervous system.

These redundantly organized columns are certainly not without structure, however, for within each are to be found a variety of cell types. A given cell type may be simultaneously classified in three ways. (1) The cells may be either exclusively excitatory or else exclusively inhibitory in their action upon other cells. As the terms imply, excitatoy cells stimulate all cells to which they send impulses; while inhibitory cells tend to prevent the firing of the cells they contact. (2) Cells of a given type are either output cells, with axons that leave the cortical sheet and project to other parts of the brain; or else they are interneurones, with axons which ramify relatively locally within the cortical sheet of origin. (3) Finally, a particular cell type is specified by the strength and spatial arrangement of its interconnections with the other cell types present in the tissue. Cell types which differ according to this scheme of classification will often differ in the value of such parameters as time constants, average thresholds, and so forth, but these differences do not affect the classification.

To do complete justice to the physiological complexity of the cortical column in some of the more highly evolved areas of the brain, ([3],[4]) it will eventually be necessary to evolve a model comprising a dozen or more cell types. However, at this stage in the development of neural modeling it would be extremely difficult to intelligently specify the relevant connections among so many cell types and even more difficult to solve such a model once formulated. Accordingly, this paper will be restricted to the simplest significant case: a neural tissue composed of two cell types. These two cell types are excitatory output cells (E-cells) and inhibitory interneurones (I-cells) and may be thought of as model pyramidal cells and stellate cells respectively.

All possible interactions will be allowed between these two cell types, namely: E-E, E-I, I-E, and I-I. These interactions will be specified by four connectivity functions designated $\beta_{ei}(|x-x'|)$, etc. For simplicity these functions have been taken to be functions of the distance between cells only and not of the absolute position of either cell. There is some anatomical evidence suggesting that this is a reasonable first approximation ([8]). Also, in accord with the assumption of the functional two-dimensionality of the cortex, the connectivity functions are assumed to be functions of distance along the cortical surface only.

The only remaining background necessary to the development of the model is the mention of a few relevant properties of individual nerve cells, an excellent elementary treatment of which may be found in Katz [9]. When a synapse onto a dendrite of a cell becomes active, it generates a potential change in the dendrite known as a postsynaptic potential or PSP. Excitatory postsynaptic potentials (EPSP's) depolarize the membrane, while inhibitory postsynaptic potentials (IPSP's) result in hyperpolarization. If the net PSP, which results from a combination of these opposing influences, exceeds a threshold value, one or more impulses are generated and transmitted along the axon to other cells. Immediately after the firing of an impulse the cell becomes refractory for a brief period: it is temporarily unable to generate further impulses. These three single cell properties, namely: combination of PSP's, existence of thresholds, and the refractory period, are the only ones to be considered here. For simplicity it will be assumed that EPSP's are additive and IPSP's are subtractive in their effect upon the net PSP.

Let us turn now to the mathematical formulation of a cortical tissue model. In view of the assumption that it is the total activity of all cells of a given type within a cortical column which is the functionally significant variable, let $E(x,t)$ be defined as the proportion of excitatory cells becoming active in the column located at x per unit time. $I(x,t)$ is similarly defined for the inhibitory cells. These variables will be referred to as excitatory and inhibitory activities respectively.

As the cells within a column are densely interconnected, it may be assumed that the same mean postsynaptic potential affects all of them. On the additivity assumption the postsynaptic potential generated in E-cells at point x and time t will be:

$$PSP_e(x,t) = \int_{-\infty}^{\infty}\beta_{ee}(|x-x'|)E(x',t)dx' - \int_{-\infty}^{\infty}\beta_{ie}(|x-x'|)I(x',t)dx' + P(x,t) \qquad (1)$$

or:

$$PSP_e(x,t) = \beta_{ee}\otimes E - \beta_{ie}\otimes I + P(x,t)$$

where $\otimes$ denotes spatial convolution.
The first convolution integral in this equation is the sum of EPSP's generated by other E-cells weighted by the strength of their connections to the cells at point x, while the second convolution integral similarly represents the summation of IPSP's. The function $P(x,t)$ represents any external input or stimulus to the tissue. A similar expression represents the PSP generated in I-cells.

The proportion of cells which are excited above their thresholds will naturally be a function of the net PSP. Let us assume that there is a distribution of thresholds among the E-cells. Then if this distribution is designated by $D_e(\theta)$, it is obvious that the proportion of E-cells that are above threshold,

$S_e(PSP_e)$, will be given by:

$$S_e(PSP_e) = \int_0^{PSP_e} D_e(\theta)d\theta, \tag{2}$$

where the threshold is assumed to be measured in the same units as the PSP. It is evident from the properties of distribution functions that the function S(PSP) will be a monotonically increasing function bounded by zero and unity. If D(θ) is a unimodal distribution, then S(PSP) will be a sigmoid function such as that shown in Figure 1. Since multimodal distributions can always be decomposed into a sum of unimodal distributions, S(PSP) will be taken to be sigmoidal in this paper.

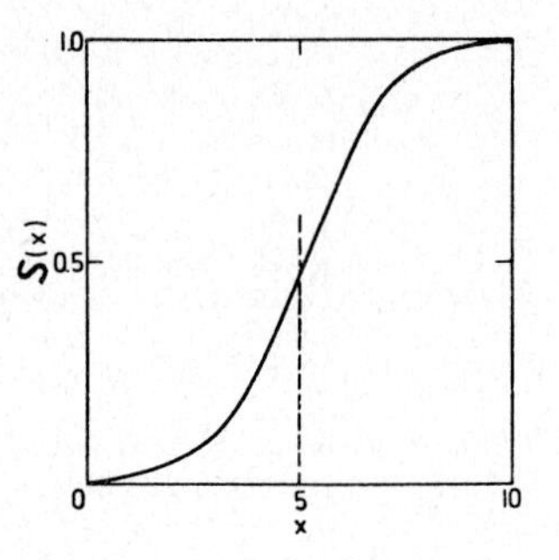

Figure 1: A typical sigmoid function. The particular function shown here is the logistic curve: $S(x) = [1 + \exp(-a(X-\theta))]^{-1}$ with a = 1 and θ = 5.

If there were no refractory period, then S(PSP) would give the proportion of cells which would fire in response to the given PSP. However, it is only those cells which have not recently fired which are available to fire at any given time. This proportion of sensitive E-cells is approximately [1-rE(x,t)], where r is the duration of the refractory period. Therefore, the proportion of E-cells which will actually become active is just:

$$[1 - E(x,t)]\ S_e(PSP_e)$$

Entirely analogous considerations apply to I-cells.

Combining all of these considerations leads directly to the following pair of equations which govern the neural tissue dynamics:

$$\tau\frac{\partial E}{\partial t} = -E + [1-rE]S_e[\beta_{es}\otimes E - \beta_{ie}\otimes I + P(x,t)] \tag{3}$$

$$\tau\frac{\partial I}{\partial t} = -I + [1-rI]S_i[\beta_{ei}\otimes E - \beta_{ii}\otimes I] \tag{4}$$

The time derivatives and exponential decay terms in Equations (3) and (4) are due to the finite time course of postsynaptic potentials. The dynamics are thus governed by a pair of coupled, nonlinear integro-differential equations in which spatial interactions are represented by convolution integration.

Before examining the properties of solutions of these equations, it is necessary to specify analytic forms for S(PSP) and the connectivity functions. The particular forms employed in this study were the following:

$$S(PSP) = \frac{1}{1 + \exp[-a(PSP-\theta)]} \tag{5}$$

$$\beta(|x-x'|) = b \exp(-|x-x'|/\sigma) \tag{6}$$

Differing values of the parameters a, θ, b, and σ were assigned to the different occurrences of these functions in Equations (3) and (4). No particular significance should be assigned to these particular analytic forms aside from the fact that (5) is sigmoid and (6) is a monotonically decaying distance function. Both computer calculations and some analytic results [1] on a simpler form of Equations (3) and (4) suggest that the qualitative properties of solutions are unaffected by the particular mathematical forms chosen for S(PSP) and the connectivity functions. It may be mentioned, however, that Equation (6) is consistent with anatomical data by Sholl [8].

PROPERTIES OF SOLUTIONS

Equations (3) and (4) are too complex to be soluble by analytic methods, and therefore results have been obtained by computer simulation. However, some analytic insight may be gained by considering only the temporal variation of the solutions in the absence of spatial interactions. The relevant equations are:

$$\tau\frac{dE}{dt} = -E + (1-E)S_e[aE-bI+P] \tag{7}$$

$$\tau\frac{dI}{dt} = -I + (1-I)S_i[cE-dI] \, , \tag{8}$$

where spatial convolution is no longer present. These equations may be thought of as representing the dynamics of a single cortical column within which the neurones are sufficiently close together and densely interconnected so that local spatial variations of the response are not significant.

A detailed analysis of the properties of Equations (7) and (8) has been carried out elsewhere [1] using the techniques of the qualitative theory of nonlinear differential equations. The major results of the analysis are that multiple asymptotically stable steady states are possible for one class of parameter values, while a second range of values results in the existence of asymptotically stable limit cycle oscillations. In the former case, simple and multiple hysteresis phenomena arise in response to appropriate variations of P in Equation (7). All of these results have been shown to be dependent upon the existence of both positive and negative feedback in Equations (7) and (8). As will be seen, these dynamical phenomena are observed with the addition of spatial structure in the computer solutions to Equations (3) and (4).

The properties of Equations (3) and (4) have been studied for a variety of different sets of parameter values. For physiological reasons several constraints were placed on the parameters [2], the most important of these being that the space constant of the E-I connections must be larger (longer range) than that of the E-E connections, i.e., $\sigma_{ei} > \sigma_{ee}$. This means that the inhibitory or negative feedback has a greater spatial spread than the positive feedback of E-cells onto themselves. As will be seen, the effect of this constraint is to localize the responses to patterned stimulation, thus giving rise to self-organizing and cooperative spatial activity. Finally, only the responses of the model to one-dimensional stimuli have been investigated in detail. This has permitted the economy of examining the responses of Equations (3) and (4) in only one spatial dimension.

As with Equations (7) and (8), it has been found that the properties of Equations (3) and (4) fall into several classes depending upon the parameter values used.

In particular, three distinct classes of solutions have been found: spatially inhomogeneous stable steady states, oscillations, and active transients. As the parameters which characterize these different modes are essentially anatomical in nature (connectivities, thresholds, etc.), it is apparent that the three dynamical modes represent three distinct anatomical specializations of the tissue model. As will be indicated, the dynamics of these differing anatomies appear to be well correlated with physiological properties of differing regions of the brain.

The spatially inhomogeneous stable steady state mode is characterized by self-maintained activity in the tissue following the cessation of stimulation. Figure 2 shows examples of two such states triggered by different spatial stimuli.

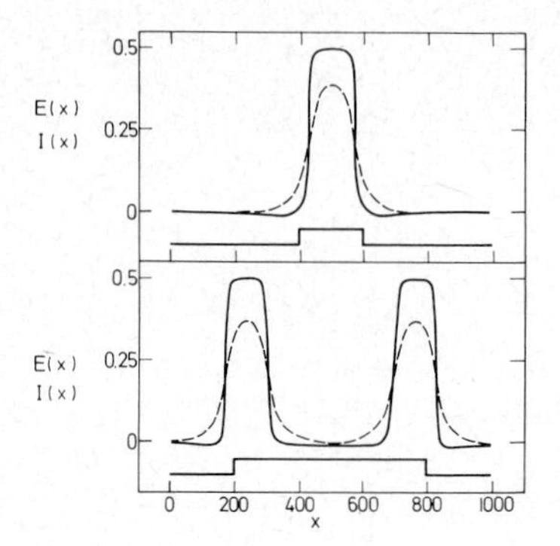

Figure 2: Spatially inhomogeneous stable steady states of neural activity generated in response to two different stimuli. E(x) is represented by a solid line and I(x) by a dashed line. The rectangular inset below each graph shows the spatial configuration of the stimulus used to excite the tissue to each of the steady states. Once established by brief stimulation, the neural activity is self-maintained.

In both cases the stimulus, P(x,t) in Equation (3), was briefly presented and then removed. As the stimulus was above threshold in these cases, the neural response continued to grow even after the cessation of the stimulus until the spatially inhomogeneous stable steady states shown in the diagram were attained. The spatial configuration of each stimulus was a rectangular pulse, narrow in the first case and much broader in the second, as indicated in the diagram. It is apparent that following narrow stimuli, the neural response is a single peak centered at the location of the stimulus. Following the wider stimulus, however, two peaks of activity are generated marking the locations of the edges of the initial stimulus. In both cases the activity is self-maintained due to the short range positive feedback, while it remains localized as a result of the longer range negative or inhibitory feedback.

The possible significance of this type of activity in the neural tissue model follows from the observation that the activity, once generated, is actively self maintained, and that the spatial configuration of the activity retains information about the size and shape of the stimulus. In fact, preliminary studies of the response of the model to two dimensional stimuli indicate that the steady state neural activity encodes information about the contours of the original stimulus. It may be mentioned here that contour enhancement or accentuation is by now a well known phenomenon in vision [10] and has been suggested to be operative also in the auditory system [11]. Accordingly, it may be hypothesized that spatially inhomogeneous stable steady states of neural activity represent a form of active, short term memory capability of cortical tissue.

The notion of an active, short term memory is perhaps best illustrated by the physiological experiments of Fuster and Alexander [12]. These workers trained monkeys to respond in one of several different ways depending upon what visual cue they were given. In order to receive a reward, however, the monkey had to remember the briefly presented cue until a second signal indicated that it was time to respond. Recordings from prefrontal cortex revealed that many neurons fired at rates up to ten times the resting rate during the entire period between cue presentation and response, often as long as one minute. This is exactly the type of result to be expected on the basis of the neural tissue model and suggests a tentative identification of the spatially inhomogeneous stable steady state mode with the properties of prefrontal cortex.

A second range of parameter values in Equations (3) and (4) leads to the generation of spatially localized limit cycle type oscillations in response to maintained stimulation. The main characteristics of this type of response are shown in Figure 3. In response to a maintained, time independent stimulus with a

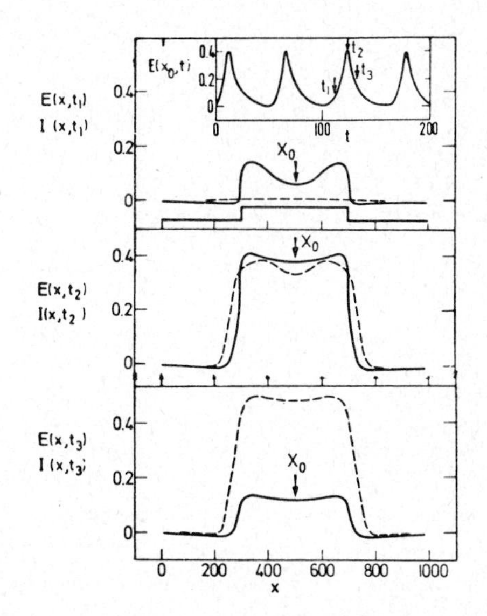

Figure 3: Three phases in a spatially localized limit cycle oscillation in response to a maintained stimulus with the rectangular profile shown at the bottom of the first diagram. Solid line is E(x); dashed line is I(x). Inset at the top shows temporal details of the oscillation at the point marked x_0 in the other diagrams. The times t_1, t_2, and t_3 show the points on the cycle at which the spatial profiles are plotted.

rectangular spatial profile, the neural activity oscillates with a highly non-linear waveform. Due to long range inhibition, the oscillation remains localized in the region stimulated, and in addition the neural responses at the edges of the stimulus are enhanced. As the stimulus used in these studies has only two characteristics -- width and intensity, it was of interest to see whether the frequency of the neural oscillation was dependent upon either of them. The results are shown in Table 1.

Table 1: DEPENDENCE OF SPATIALLY LOCALISED LIMIT CYCLE FREQUENCY UPON STIMULUS INTENSITY AND STIMULUS WIDTH

Stimulus Intensity:	Stimulus Width:	Frequency of Limit Cycle Response: (sec^{-1})
2.5	80	14
2.5	600	14
5	80	18
5	400	18
5	600	18
10	400	22
10	600	23

It is apparent that the frequency is independent of stimulus width but increases monotonically with increasing stimulus intensity. Evidence for the encoding of stimulus intensity into the frequency of neural oscillations has been obtained in recordings from thalamic nuclei by Poggio and Viernstein [13]. This suggests that the oscillatory mode of the neural tissue model may be related to the properties of various thalamic nuclei.

In order to further investigate the relationship of the oscillations of the model to those of the thalamus, a further experiment was tried. Purpura [14] has observed that under appropriate conditions neurones in certain parts of the thalamus will respond only to every second stimulus from a train of pulses. As is shown in Figure 4, the neural tissue model also exhibits this frequency

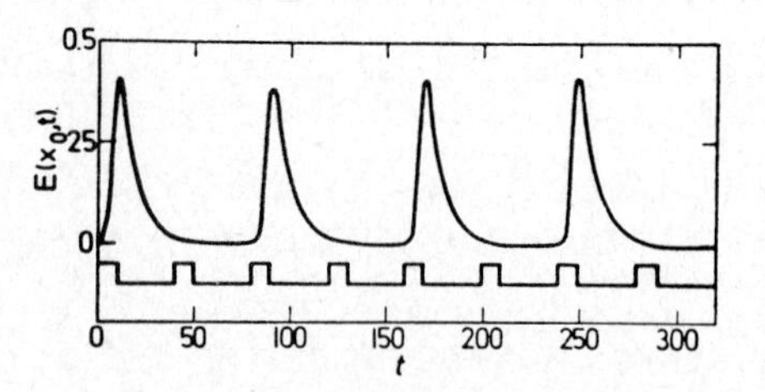

Figure 4: Frequency demultiplication in the neural response to a stimulus pulse train. Each stimulus pulse had a narrow, rectangular spatial profile. $E(x_0,t)$ is the response at the center of the region stimulated.

demultiplication. As the model parameters used in obtaining this result were identical to those used in Figure 3 and Table 1, further support is lent to the association of the oscillatory response mode of the model with the properties of thalamic nuclei.

The final type of dynamic response exhibited by the model has been termed the active transient. In the active transient mode subthreshold stimuli are suppressed, while superthreshold stimuli are briefly amplified and then die out. This amplification is a result of local positive feedback and may continue to develop long after the stimulus has been removed. Following this amplification, however, the neural activity always decays. This temporary amplification effect as well as several other elementary properties [2] suggested that the active transient mode might have characteristics similar to those of sensory neocortex, and particularly of visual cortex. Accordingly, an attempt was made to reproduce the features of several experiments in visual perception using active transient dynamics, three of which will be mentioned below. It is to be emphasized that the model parameters were not changed during these experiments, and that the differing responses are the result of differing modes of stimulation only.

In a series of experiments on the perception of gratings composed of alternating light and dark vertical bars, Campbell and Robson [15] made the following discovery. If a grating whose bars had a square wave intensity profile was slowly increased in contrast the bars first appeared blurred and indistinguishable from the bars of a grating with a sinusoidal intensity profile. It was only at higher contrasts that the sharp edges of the square wave grating appeared. This phenomenon, namely the existence of separate thresholds for the perception of bars and the perception of edges, was observed only for an intermediate range of bar widths (spatial frequencies). At high frequencies (narrow bars) distinct edges never appeared, while at very low frequencies (broad bars) sharp edges were visible whenever the bars themselves were visible.

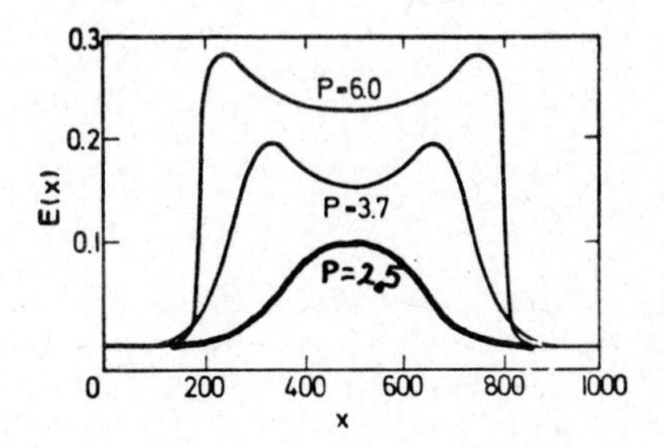

Figure 5: Maximum active transient mode response to a rectangular bar as a function of intensity. In all cases the bar width was the same. Edge enhancement is not present in the response to the bar with P = 2.5.

Figure 5 shows the response of the active transient mode to a single rectangular bar briefly presented at each of three different intensities. It is apparent that at the intensity P = 2.5 the neural response does not show edge enhancement, while at the higher intensities edge enhancement is clearly visible. This result reflects the shifting balance between positive feedback, which favors the development of a single peak in the neural activity, and recurrent lateral inhibition, which favors edge enhancement.

Under the assumption that edge enhancement is necessary for the perception of rectangularity, this result is in qualitative agreement with the findings of Campbell and Robson [15]. Although not shown here, it is also true that the response of the model to very narrow bars never displays edge enhancement, while the response to sufficiently wide bars shows edge enhancement as soon as the stimulus is above threshold. While Campbell and Robson [15] have developed a different interpretation of their results, it is clear that the model reproduces the qualitative features of the phenomenon.

A second well-known visual phenomenon is that of metacontrast or backward masking [16]. In a typical experiment of this type a bar (the target) is flashed briefly at some location in the visual field, and is followed a short time later by the flashing of a pair of bars (the mask) on either side of the position at which the target had been presented. If the interval between the target and mask is not too long the subject perceives only the pair of masking bars. Thus, the neural response to the target must be suppressed by the response to the mask even though the mask is presented later and does not spatially overlap the target area. The effectiveness of this masking decreases with increasing time between target and mask and with increasing spatial separation between them.

A simulation of such a metacontrast experiment is shown in Figure 6. The spatial arrangement of the stimulation is shown in the inset at the upper right, where

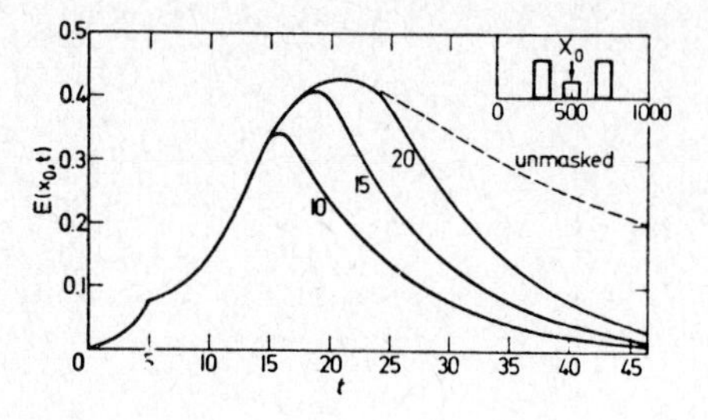

Figure 6: Metacontrast effects in the active transient mode. $E(x_0,t)$ is the excitatory response at the point of target presentation, as indicated in the inset. Figures on the curves are the times at which masking bars are presented at the flanking locations shown in the inset. The dashed line indicates the unmasked result.

the target is flanked by the pair of more intense masking bars. The graph shows the time course of the peak of the neural response to the target, in the absence of a mask, and when the mask follows by 10, 15, and 20 time units. As the target stimulus was only on for 5 units of time, there was no temporal overlap between stimulus and mask. Although not shown, the model shows decreased masking when the mask is at a greater distance from the target. The evident success of the model in reproducing the basic elements of metacontrast is again due to two features: positive feedback, which generates a delayed active transient response to the target, and long range inhibitory feedback which functions to suppress responses at neighboring locations.

The final visual phenomenon to be discussed here is the spatial hysteresis phenomenon discovered by Fender and Julesz [17],[18] during studies of binocular vision. In these experiments a special optical device was used so that the experimenter could precisely control the positions of patterns falling on the two retinas independently of any eye movements made by the subject. At the beginning of the experiment patterns were projected onto corresponding positions on the left and right retinas. Under these conditions the subject easily fused the two images and perceived a single image in depth. The two patterns were then slowly pulled away from corresponding positions on the two retinas until the separation became so great that the subject could no longer fuse them and instead saw a double image. Following this breakaway the patterns were again brought back towards corresponding points until the subject was able to refuse them and again perceived a single image in depth. The striking result was that the separation at which breakaway occurred was much greater than the separation necessary for refusion: the visual system exhibited binocular hysteresis in the spatial domain!

As there is ample evidence that the site of this binocular hysteresis phenomenon is the visual cortex [18], an attempt was made to simulate the phenomenon using the active transient mode of the cortical tissue model. The two retinal inputs were simulated by a pair of sharply peaked Gaussians. These stimuli were first superimposed and then pulled slowly apart at a rate that allowed the neural response to remain essentially in equilibrium. Three stages of the neural response to this stimulus are shown in Figure 7. Starting from superposition, the Gaussians were pulled apart until their peaks were at the locations marked by the

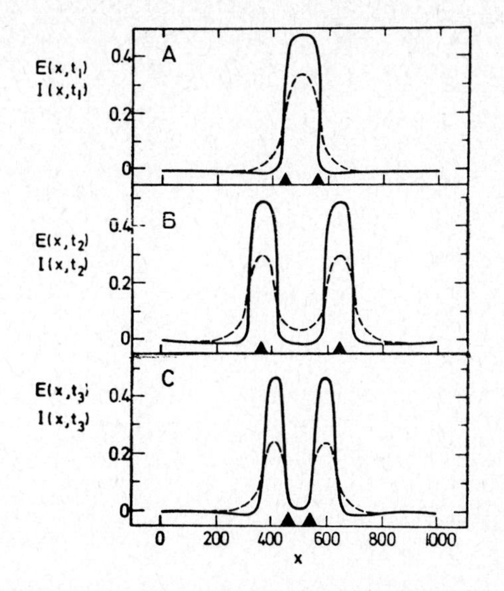

Figure 7: Neural activity at three stages of the Fender-Julesz hysteresis loop. $E(x,t)$ is plotted as a solid line and $I(x,t)$ as a dashed line. A, B, and C indicate the points on the hysteresis loop in Fig. 8 at which the activities are plotted. Positions of the maxima of the Gaussian stimuli are indicated under each graph by the solid triangles. Note that although the stimuli are further apart in A than in C, the response in A is a single peak, whereas the response in C exhibits twin peaks.

dark triangles in A, yet the neural response still showed a single peak. As the Gaussians were pulled further apart a point was reached at which the single peak of the neural response broke apart into a pair of peaks as shown in B. When the Gaussians were brought back together, however, it was possible to move them much closer together while still retaining two peaks in the neural response, as in C. If the single peak and twin peak responses are taken to represent a fused image and double vision, respectively, then the model traces out the hysteresis loop in Figure 8. This is directly analogous to the results of Fender and Julesz [17],[18].

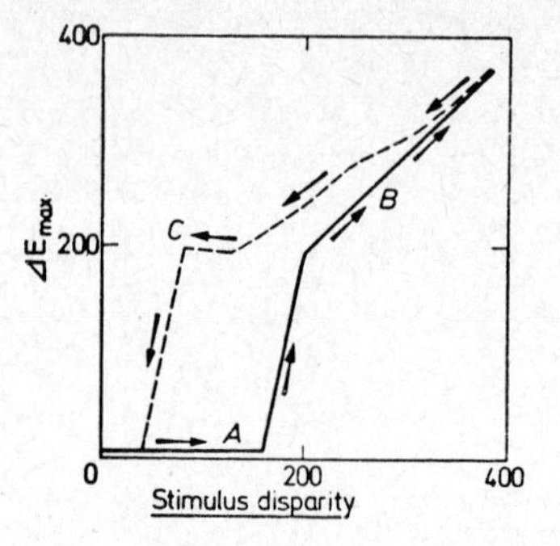

Figure 8: Simulation of the Fender-Julesz experiment on hysteresis in binocular vision. Stimulus disparity is the distance between maxima of the two displaced Gaussian stimuli, while ΔE_{max} is the separation between maxima in the neural response. Arrows show the hysteresis loop generated by first pulling the stimuli apart and then bringing them back into register.

ADAPTATION TO SPATIAL PATTERNS

The variety of responses of the cortical tissue model and particularly the success of the active transient mode in accounting for a variety of visual phenomena is impressive considering the conceptual simplicity of Equations (3) and (4). However, the model is obviously far too simple to deal with the wealth of complexity of mammalian cortical and thalamic tissue. If more sophisticated models are to evolve, it will be necessary to specialize them to deal with the anatomy and physiology of particular parts of the brain. As the visual system has been more thoroughly studied, both physiologically and perceptually, than any other part of the brain; it is a natural system to select for more detailed modeling. In this section a mathematically elementary approach to the understanding of the class of visual phenomena known as adaptation will be developed. Once the general principles are understood, the necessary modifications may readily be incorporated into Equations (3) and (4). Further extensions of the basic model to deal with visual phenomena are briefly mentioned in the discussion.

It has long been known that prolonged viewing of a visual stimulus causes a subject to become less sensitive to that stimulus, a process known as adaptation. Recently, a new adaptation effect has been discovered [19],[20]. If a spatial grating with vertical bars having a sinusoidal luminance profile is viewed for about two minutes, the threshold for perception of gratings of the same or similar spatial frequency is increased by a factor of at least two. This adaptation effect has four major characteristics: the threshold elevation is greatest at the adapting frequency and is limited to frequencies within about one octave of the adapting frequency; the magnitude of the effect is greater at higher adapting frequencies; threshold elevation curves become narrower at higher adapting frequencies; and the effect is dependent on the relative orientations of the adapting and test gratings.

The discoverers of this adaptation effect originally suggested that it was due to the fatiguing of neurones which were tuned to the spatial frequency of the adapting grating. This interpretation, however, is open to a variety of criticisms, and recent experiments cast doubt on its validity (See [21] for a discussion). An alternative hypothesis which suggests itself is that threshold elevation following grating adaptation may be the result of a temporary modification of the strengths of inhibitory synapses in the visual system. This may be stated as a specific hypothesis:

> The effectiveness of inhibitory synapses may be modified as a function of the correlation between recent presynaptic and postsynaptic activity. A positive correlation increases the strength of the synapse, while a negative correlation decreases the synaptic strength.

The assumption that synaptic modification is a function of correlated activity is crucial for the model. It was first suggested as a basis for memory by Hebb in 1949 [22], and has since been used by a number of other authors (See [21]). The new idea in the present context is that this mechanism might be used by the brain to filter out environmental constancies as well as to retain relevant associations.

In order to use the synaptic modification hypothesis to model grating adaptation, it will be necessary to incorporate the modifiable synapses into a neural tissue model. As there is evidence that the adaptation effect is cortical [20], it would be natural to utilize the model expressed by Equations (3) and (4). For convenience, however, a very simple linear network model has been used in order to obtain analytic results. Details of the calculations may be found elsewhere [21], and only the basic equations and results are reported here.

Letting $\Delta W(x,x',t)$ represent the change in weight of a synapse between an inhibitory cell at x' and an excitatory cell at x, the correlation hypothesis may be written as:

$$\Delta W(x,x',t) = \int_0^t h(t-t')\, E(x,t')\, I(x',t')\, dt' \qquad (9)$$

In this equation $h(t-t')$ gives the time course of adaptation.

The simplest possible network within which to incorporate the modifiable syanpses of Equation (9) is one involving simple feedforward excitation and inhibition with spread functions $\beta_e(x)$ and $\beta_i(x)$ respectively. Due to the use of feed-forward interactions and the linearity requirement, the network equations assume a form which is highly simplified compared to Equations (3) and (4):

$$\tau\frac{\partial E}{\partial t} = -E + \beta_e \otimes S(x,t) - \beta_i \otimes I - \int_{-\infty}^{\infty}\beta_i(x-x')\, \Delta W(x,x',t) S(x',t)dx' \qquad (10)$$

$$\tau\frac{\partial I}{\partial t} = -I + S(x,t) \qquad (11)$$

Here $S(x,t)$ is the stimulus to the network, and $\otimes$ again denotes convolution. The influence of changes in inhibitory synaptic weights is represented by the final integral in Equation (10). As $\Delta W(x,x',t)$ is the change in weight per synapse between cells located at x and x', the integrand is the product of this factor and $\beta_i(x-x')$, which gives the number of synapses between these cells that are available for modification. This change in the network connectivity is then multiplied by the stimulus, $S(x',t)$, and integrated to give the net effect of the synaptic modification on the network response.

As adaptation is a process with a time constant on the order of a minute whereas the time constants for the neural response are on the order of 100 msec., it will only be necessary to examine Equations (10) and (11) under steady state conditions. This assumes also that any temporal variation of the stimulus is sufficiently slow. Under these conditions Equations (10) and (11) reduce to the single equation:

$$E(x,t) = \beta_e \otimes S(x,t) - \beta_i \otimes S(x,t) - \int_{-\infty}^{\infty}\beta_i(x-x')\, \Delta W(x,x',t)S(x't)dx' \qquad (12)$$

Equations (9) and (12) complete the formulation of the adaptive neural network model. To simulate adaptation experiments, the response of the network is obtained for $\Delta W(x,x',t)$ equal to zero (the unadapted state). Equations (9) and (12) are then solved for an adapting stimulus, $S_a(x,t)$, to obtain $\Delta W(x,x',t)$. Once this has been done the response of the adapted network to any desired stimulus may readily be calculated. In order to obtain solutions to these equations it is obviously necessary to specify β_e, β_i, and the nature of the time dependence of Equation (9). Based on the data of Campbell, Carpenter, and Levinson [23], the connectivity functions were chosen to be:

$$\beta_e(x-x') = \frac{a}{a^2 + (x-x')^2}; \qquad \beta_i(x-x') = \frac{b}{b^2 + (x-x')^2} \tag{13}$$

As the present purpose is only to look at the final or steady state solutions of the adapting network, it is not necessary to specify $h(t-t')$ beyond assuming that it is a monotonically decreasing function of its argument, i.e., the effects of adaptation should be temporary.

To model adaptation to spatial gratings Equations (9) and (12) were solved for a stimulus of the form $\cos(2\pi\omega_a x)$. During adaptation the subject is required to move his eyes in such a way as to average over all phases of the grating [20]. Under these experimental conditions the temporal integration in Equation (9) may be replaced by integration over all phases of the stimulus, and Equations (9) and (12) may be solved analytically [21]. A comparison of the response of the unadapted model with the response following adaptation to a grating of spatial frequency 12 cycles/degree is shown in Figure 9. To conform to standard practice both axes of the graph are plotted on logarithmic coordinates.

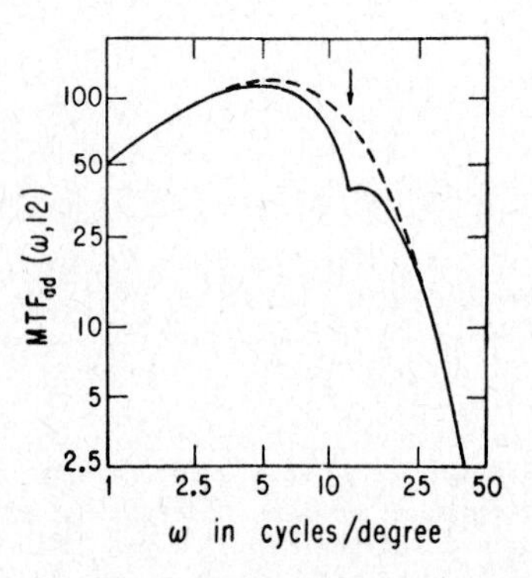

Figure 9: Effect of adaptation to 12 cycles/degree sinusoidal grating (arrow). The solid curve shows the response of the adapted model to gratings of differing spatial frequencies. The dashed line is the response of the unadapted model for comparison. Note that the adaptation effect is greatest at 12 cycles/degree.

The threshold elevation curves reported in the literature are the ratios of the curves (rather, the logarithm of the ratio) obtained before and after adaptation. A family of such curves obtained from the model for adaptation at 2, 4, 8, and 16 cycles/degree is plotted in Figure 10. Also plotted are some data obtained by Graham [24]. In agreement with the experimental data, the model curves peak at the adapting frequency and are both narrower and taller at higher frequencies. Furthermore, the fit of the data to the curve shapes is acceptable. The model

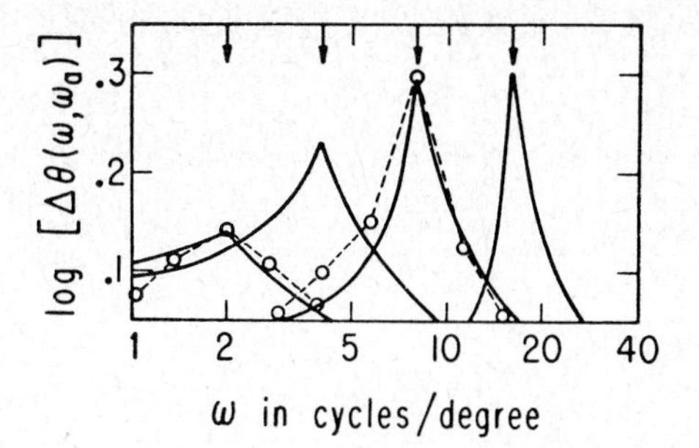

Figure 10: Threshold elevation curves for adaptation at 2, 4, 8, and 16 cycles/ degree (arrows). Each curve is the ratio of the unadapted to the adapted grating response as shown in Figure 9. Open circles are data for adaptation at 3 and 7.5 cycles/degree obtained by Graham [24]. The data have been translated and normalized for comparison with the 2 and 8 cycle/degree model curves.

results plotted in Figure 10 actually involve a further hypothesis, namely that synaptic weights change in such a way that the total inhibitory synaptic strength is conserved. This hypothesis, however, only affects adaptation curves at low frequencies, and the interested reader is referred to the original paper for details [21]. Finally, the model also correctly predicts the results of adaptation to square waves, single bars, and tilted gratings [21]. Synaptic modification thus provides an effective means for filtering out constant and redundant aspects of the visual environment, thereby emphasizing novelty.

DISCUSSION

The basic neural tissue model embodied in Equations (3) and (4) is sufficiently complex to exhibit fundamental aspects of neural cooperativity: hysteresis, asymptotically stable spatially inhomogeneous steady states, spatial limit cycle type oscillations, etc. The basis of this activity is the balance between the competing forces of excitatory and inhibitory feedback in a system with threshold properties. Excitatory feedback is responsible for the self-generating and self-maintaining aspects of the responses to superthreshold stimulation, while long range inhibitory feedback results in the localization of such responses and the consequent maintenance of spatial structure in the neural activity. This suggests an important difference between the nature of cooperative phenomena in physics, such as phase transitions, and those in neural systems. In physical systems it would appear that while there is typically an energy barrier or threshold to phase transitions and also a positive feedback generated by interparticle interactions, there is nothing analogous to inhibitory feedback. In consequence, there is seldom macroscopic spatial structure present in either of the two states between which the transition occurs.

A further point of interest here is that Gierer and Meinhardt [25] have shown that many aspects of pattern formation in hydra may be successfully modeled by a pair of coupled nonlinear diffusion equations in which one chemical substance provides short range positive feedback and the other provides longer range (in virtue of its diffusion constant) negative feedback. This raises the exciting possibility that the balance between spatially distributed excitatory and inhibitory feedback may be a fundamental principle of dynamical organization in biological systems and that some of the processes underlying both pattern formation and pattern recognition may be fundamentally similar.

The model presented in this paper is, of course, the simplest one possible which still demonstrates functionally significant neuronal cooperativity. As demonstrated in the previous section, however, the basic approach is amenable to expansion to encompass more detailed phenomena such as grating adaptation. The next step in the evolution of such models is a natural one: a consideration of the dynamics of a neural tissue containing several different populations of excitatory cells. This will be necessary, for example, in order to deal with the phenomena of stereopsis and depth perception in detail. Based on data gathered by Richards [26] it is reasonable to hypothesize the existence of three populations of E-cells: one optimally sensitive to the depth at which the two eyes are converged, one most sensitive to depths less than this, and one most sensitive to greater depths. Such a model, involving three E-cell populations and perhaps just one I-cell population, may be naturally formulated using four equations of the form of Equations (3) and (4). The key to the success of such a model, however, lies in the specification of the 16 different connectivity functions. Fortunately, experimental studies of the conditions under which hysteresis occurs provide strong constraints on the possible connectivity patterns so that the problem is actually quite tractable. There are also indications that tissue models comprising four spatially distributed cell types may provide a natural framework for the study of motion perception and other visual phenomena. Such models, therefore, may be expected to epitomize the next stage in the evolution of neural tissue modeling.

REFERENCES

[1] H. R. Wilson and J. D. Cowan, Biophys. J. 12 (1972) 1.

[2] H. R. Wilson and J. D. Cowan, Kybernetik, 13 (1973) 55.

[3] D. H. Hubel and T. N. Wiesel, J. Physiol. 160 (1962) 106.

[4] D. H. Hubel and T. N. Wiesel, J. Physiol. 195 (1968) 215.

[5] M. L. Colonnier, Article in Brain and Conscious Experience, J. C. Eccles, ed., Springer: New York (1965).

[6] J. Szentagothai, Article in Handbook of Sensory Physiology, vol. VII/3B R. Jung, ed., Springer: New York (1973).

[7] K. L. Chow and A. L. Leiman, The structural and Functional Organization of the Neocortex. Neurosciences Research Program Bulletin 8 (1970) 153.

[8] D. A. Sholl, The Organization of the Cerebral Cortex. Methuen: London (1956).

[9] B. Katz, Nerve, Muscle, and Synapse. McGraw Hill: New York (1966).

[10] F. Ratliff, Mach Bands. Holden-Day: London (1965).

[11] I. C. Whitfield, The Auditory Pathway. Williams and Wilkins: Baltimore (1967)

[12] J. M. Fuster and G. E. Alexander. Science 173 (1971) 652.

[13] G. F. Poggio and L. J. Viernstein, J. Neurophysiol. 27 (1964) 517.

[14] D. R. Purpura, Article in The Neurosciences: Second Study Program. F. O. Schmitt, ed., Rockefeller University Press: New York (1970).

[15] F. W. Campbell and J. G. Robson, J. Physiol. 197 (1968) 551.

[16] D. Kahneman, Psychol. Bull. 70 (1968) 404.

[17] D. Fender and B. Julesz, J. opt. Soc. Am. 57 (1967) 819.

[18] B. Julesz, Foundations of Cyclopean Perception. University of Chicago Press: Chicago (1971).

[19] A.Pantle and R. Sekuler, Science, 162 (1968) 1146.

[20] C. Blakemore and F. W. Campbell, J. Physiol. 203 (1969) 237.

[21] H. R. Wilson, "A Synaptic Model for Spatial Frequency Adaptation", Submitted to J. Theor. Biol.

[22] D. O. Hebb, The Organization of Behavior. John Wiley: New York (1949).

[23] F. W. Campbell, R. Carpenter, and J. Z. Levinson, J. Physiol. 204 (1969) 283.

[24] N. Graham, Vision Res. 12 (1972) 53.

[25] A. Gierer and H. Meinhardt, Kybernetik, 12 (1972) 30.

[26] W. Richards, J. opt. Soc. Am. 61 (1971) 410.

Cooperative Phenomena, H. Haken, ed.

COLLECTIVE PHENOMENA IN BIOLOGICAL SYSTEMS

by H. Fröhlich
Department of Electrical Engineering, University of Salford, Salford M5 4WT

1. Macroscopic systems which exhibit an organised dynamic behaviour - such as a machine, a laser, a biological system - are frequently governed by a few collective degrees of freedom which dominate the rest. Such behaviour even extends to superfluids and superconductors where it finds its expression in the existence of macroscopic wavefunctions which impose a long range coherence. It should be noticed that quantum mechanical phases are closely connected with velocity fields even in a very typical 'classical' fluid. For consider a fluid composed of one type of particles described by a wave operator $\psi(X)$ and let[1)]

$$\Omega_1(x', x'') = \langle \psi^+(x'') \psi(x') \rangle = \sigma(x', x'') e^{i\chi(x', x'')} \tag{1}$$

be its first reduced density matrix where $\langle \; .... \; \rangle$ indicates the appropriate quantum mechanical average. Then

for $x' = x'' = x$, one has $\chi(x, x) = 0$, and $\sigma(x, x)$

represents the macroscopic density. Definition of the hydrodynamic velocity field $\underset{\sim}{v}(x)$, necessarily involves Planck's constant $\hbar$ and the particle mass m by

$$\underset{\sim}{v}(x) = \lim_{x' = x''} \frac{\hbar}{2m} (\partial' - \partial'') \chi(x', x'') \, . \tag{2}$$

Also if the phase term is factorised, i.e. if

$$\chi(x', x'') = S(x') - S(x'') \tag{3}$$

then

$$\underset{\sim}{v}(x) = \frac{\hbar}{m} \operatorname{grad} S(x) \quad , \text{ i.e. } \operatorname{curl} \underset{\sim}{v} = 0 \tag{4}$$

and the material is superfluid.

It should be realized that like the density $\sigma(x, x)$, the velocity field $\underset{\sim}{v}(x)$ is a macroscopic quantity but its definition involves the phase $\chi(x', x'')$ which has the same range 2π as has a microscopic phase - unlike the amplitude $\sigma(x', x'')$ which by

$$\int \sigma(x, x) \, d^3x = N \tag{5}$$

is normalized to the total number of particles and thus in a macroscopic case is very large in contrast to the microscopic case.

The powerful influence of phases has led to the hypothesis that coherent vibrations are at least partly responsible for the organised behaviour of biological systems.[2) 3)] In recent years this idea has been applied to large molecular systems and attempts are at present in progress to verify this hypothesis.

In other frequency regions coherent vibrations under the names of entrainment or phase locking have for many years been invoked to describe biological rythms.[4)] As a working hypothesis this idea has been successful but to my knowledge the actual physical vibrations have in no case yet been demonstrated.

Clearly coherent vibrations provide an order of motion which can be superimposed on an otherwise disordered system. Another nondynamic state of order may also be imposed on an otherwise disordered system namely excitation of a strong electric polarisation similar to that which exists in ferroelectric materials.[5)] Both types of excitation will be assumed to be excited by supply of energy(pumping) arising from metabolic processes, and stabilized by non linear couplings.

Model calculations in both cases exhibit the importance of non linear features. Thus if a system of polar modes - non linearly coupled to a heat bath - is excited by supply of energy, then above a certain rate of supply it exhibits features similar to that of Bose condensation i.e. very strong excitation of a single mode. In this state then the supplied energy is not thermalized but available in non thermal form.

Excitation of strong electric polarisation would lead at short range to strong forces on charged particles (ions or electrons) and might thus assist in lowering activation energies of certain processes.

At long range the presence of ions usually leads to screening of static electric charges so that the polarisation would not be noticeable. In conjunction with the vibrations, however, it could lead to very strong forces provided they are in a frequency range in which they are not strongly absorbed.

It would be possible without great difficulty to develop to a considerable extent theoretical models exhibiting the above mentioned features. I feel, however that the first task should be the discussion of circumstances in which they would exhibit themselves most clearly and the establishment of the experimental conditions required to prove their existence.

2. For this purpose it will be useful to estimate some relevant orders of magnitude. Frequencies of vibrations of very large sections of giant molecules or membranes are largely governed by size and 10^{-6} cm is a frequently recurring

magnitude. The interaction forces usually may be expected to give rise to elastic properties corresponding to sound velocities of order $10^5 - 10^6$ cm/sec. We then expect frequencies in the range $10^{11} - 10^{12}\ s^{-1}$.

Typical electric fields should be those present inside a membrane which exhibits a potential difference of 0.1 volt over a distance of 10^{-6} cm causing a field of order 10^5V/cm. Similarly the electric field due to a singly charged ion at a distance of 10^{-6}cm is of the same order, but at a distance of 3×10^{-7}cm (size of a small protein) it is ten times larger. Furthermore the dipole moment of the small protein myoglobin has been measured with high accuracy [6] to $\mu \simeq 200D = 2 \times 10^{-16}$ electrostatic c.g.s. units. Assuming the molecule to be homogeneously polarised and to be spherical with a radius 3×10^{-7}cm would yield an electric field

$$E = 4\pi\mu / \text{volume} \simeq 6 \times 10^6\ V/cm \qquad (6)$$

In such a field the interaction energy with an amino acid with dipole moment $\mu_0 \simeq 15D$ is (T represents room temperature).

$$\mu_0 E \simeq 6\,kT \qquad (7)$$

Investigations of the existence of low frequency vibrations in the above mentioned range requires use of Laser Raman effect. Experiments by Peticolas on crystals of small proteins have indeed exhibited vibrations [7] [8] with frequencies close to $10^{12}\ s^{-1}$. They disappear on denaturisation and are thus connected with the conformation of these materials. The Rayleigh scattering of water which is always present in the crystals unfortunately causes a very strong broadening of the primary line so that detailed interpretation of the experiments will have to await the development of more refined experimental techniques. In solutions the predominance of water is so much stronger that so far it has been impossible to pick up the Raman bands. For the same reason the excitation of the vibrations in biological systems by their activity cannot be expected to be discovered by the available experimental techniques, and this might be the reason for the failure of a recent attempt.

Experiments to establish the existence of the proposed excited meta stable ferroelectric state will require application of very strong fields which transform the polarised meta stable state into the stable one. Some evidence for the existence of such a state arises from investigations on nerve membranes which suggest a drastic change in the permeability for Na^+ ions of proteins dissolved in the membrane with or without electric field. This suggests the use of fields of the order of 10^5V/cm. Clearly experimental difficulties will again arise through the always present water. Some indirect evidence for the existence of strongly polarised states in some biological materials does exist though the interpretation of these experiments is not yet completed.

3. The action of enzymes is a field in which the postulated collective excitations might find application.[9] For while analysis of enzyme structure has led in many cases to the identification of an active site it has been said that[10]

".... the essential mystery of their enormous catalytic power remains."

The attraction of the substrate to the enzyme, and the subsequent chemical reactions lead to the liberation of (free) energy which in the first place might be used to excite both the coherent vibrations of the enzyme and the meta stable polarised (ferroelectric) state. Energy stored in this way will then not be thermalised but will be available at a later stage. Existence of electric fields connected with the polarisation might be instrumental in decreasing activation energies which determine rates of reactions. Excitation of the coherent vibrations apart from storing energy might be of importance for selective long range interaction with other molecules.

It is of great interest that from a detailed study of many enzymatic reactions D. E. Green[11] independently has reached conclusions which perfectly fit into the above framework. In hi words "The protein, as a unit, may be visualized as a vast storehouse of energy in three interconvertible forms: electrical, mechanical, and chemical". While this is very encouraging it does not absolve us from finding direct physical evidence for the existence of the postulated excitations. A method that could promise success with regard to the excitation of the strongly polarised state would consist in following enzymatic reactions dielectrically. Excitation and relaxation of strong polarisation will lead to electrical pulses which it should be possible to observe. Furthermore while the polarised state is excited the complex will have a strongly increased dipole moment which will lead to an increase in dielectric constant given by

$$\Delta \varepsilon = f(\varepsilon) \frac{4\pi \overline{m^2} n}{kT} \tag{8}$$

Here $\overline{m^2}$ is the mean square of the (excited) dipole moment, n the number of excited enzyme molecules per unit volume. $f(\varepsilon)$ is a factor expressing the influence of the medium; its knowledge is required only when the stage for detailed quantitative analysis has been reached. In preliminary experiments it can be considered to be of order unity. It will be noticed that an easily measurable $\Delta\varepsilon = 0.01$ would require $n = 10^{14}$ for a dipole moment of 1000D.

4. Interesting collective phenomena have been detected in the activity of haemoglobin[12] on the replacement of hydrogen ^{1}H by its isotope ^{2}H and in Agar.[13] Again it would be interesting to detect by direct measurement the relevant vibrations.

Attention should also be drawn to the striking biological effects (like mutations) of pulses of micro wave radiation found by

Heller.[14] This frequency region is by a factor 10^4 lower than that found in small proteins. Possibly, however, DNA might show resonances in the micro wave region. Investigations in this direction might, therefore, yield important results.

REFERENCES

1) H. Fröhlich, Rivista del Nuovo Cimento 3, 490 (1973)
2) H. Fröhlich, Intern. J. Quantum Chem. 2, 641, (1968)
3) H. Fröhlich, in Synergetics, Stuttgart 1972, p.641
4) cf. reference 22, 23, 24 in E. M. Dewan, IEEE Trans. Automat. Contr. AC17, 655 (1972)
5) H. Fröhlich, J. Collect. Phenomena 1, 101 (1973)
6) P. Schlecht, Biopolymers, 8, 757 (1969)
7) K. G. Brown, S. C. Erfurth, E. W. Small and W. L. Peticolas, Proc. Nat. Acad. Sci. USA 69, 1467 (1972)
8) L. Genzel, F. Keilmann, M. W. Makinen, T. P. Martin, G. Winterling, personal communication.
9) H. Fröhlich, Nature, 228, 1093 (1970)
10) D. E. Koshland, and K. E. Neet, Ann. Rev. Biochem. 37, 359, 380 (1968)
11) D. E. Green, New York Acad. Sci. 274, in print, 1974) and personal communication.
12) A. Cupane, M. U. Palma and E. Vitrano, J. Mol. Biol. 81, in print,(1974)
13) G. Aiello, M. S. Micciancio-Giammarinaro, M. B. Palma-Vittorelli, M. U. Palma in Cooperative Phenomena (ed. Haken & Wagner) Springer 1973, 395
14) H. H. Heller, Symposium on Biological Effects of Microwave Radiation, Richmond, Virginia, Sept. 1969, p. 116

Cooperative Phenomena, H. Haken, ed.

DYNAMICS OF INTERACTING SOCIAL GROUPS

Wolfgang Weidlich

Institut für Theoretische Physik der Universität Stuttgart, Germany

I. Introduction and General Outline

The structural analogy between some quantitative aspects in the development of the society and cooperative phenomena in physics is not self evident, as the fundamental laws governing so different fields of science as physics and sociology are completely different. Therefore we first have to ask for the unifying level of consideration for both sciences. This level is the statistical behaviour of systems with a large number of interacting units. The probabilistic laws governing the statistical dynamics of such systems lead to many common features independently of the nature and the items of the "microscopic" interaction between the elementary units. In particular we may define macroscopic collective variables in systems composed of many subunits. The *"macroscopic observables"* characterize the system on a global level. The macroscopic observables may often be considered as an *order parameter* in the following sense: The interacting subunits of the system contribute to the formation and sustentation of a collective variable (orderparameter) which vice versa orders and organizes the interactions between the individual subunits. Systems which thus are selfstabilizing by *feedback-processes* may be called *dynamical selfconsistent systems.*

A short survey over the anorganic as well as the organic world shows how universal this concept is. Starting from very high temperatures and proceeding to high, moderate and low temperatures we find in the anorganic world a sequence of condensation-processes leading to the formation of more and more complex subsystems. The strong interaction *within* the bound condensed systems may approximately be treated as a selfconsistent field (=orderparameter) organizing the structure of the bound system (SCFA). On the other hand there remain (usually weaker) restinteractions *between* the bound units establishing the properties of the macroscopic system consisting of many units. The restinteraction again may approximately be treated as a selfconsistend field = mean field (SCFA). The following table gives a coarse account of these relations:

Macroscopic System	Hadronic Matter	Plasma	Gas	Condensed Matter
Units	Elementary Particles	Nuclei Electrons	Atoms, Molecules	Liquid Solid
Interaction *within* Units	Strong interactions	Strong nuclear interactions	Electromagn. interaction	Electromagn. van der Waals
SCFA		Shell-model	Hartree-Fock	Band-theory Crystal-field
Interaction *between* Units	Strong interactions	Electromagn. interaction	van der Waals force	Cohesion, Elasticity Gravitation
SCFA		**Mean-**	**Field-**	**Theories**

In the organic world we have by far more complex relations between a hierarchy of multistable levels. However, also here the concept of "orderparameters" and of self-consistent dynamical regulative feedback-processes proves to be very useful.
The interaction between the system of multistable levels typically has the following structure: The fluctuations at the lower ("microscopic") level trigger the dynamics - in particular also branching- and diversification-processes (e.g. mutation) - of the higher level which is in turn organized and controlled by the orderparameters on this level. On the other hand, the lower level is partially influenced by feedback-processes from the higher level tending to sustain the organization of the higher level. In particular we have such an interrelation between the level of the individual members of a society and the situation of civilization and culture:
The members of the society contribute by their ideas, activities and decisions to the different aspects of the civilization and the social athmosphere, which in turn influences and regulates the attitude and behaviour of the individual.
In the following table we give a coarse account of the dynamic interrelations between the levels of the organic world:

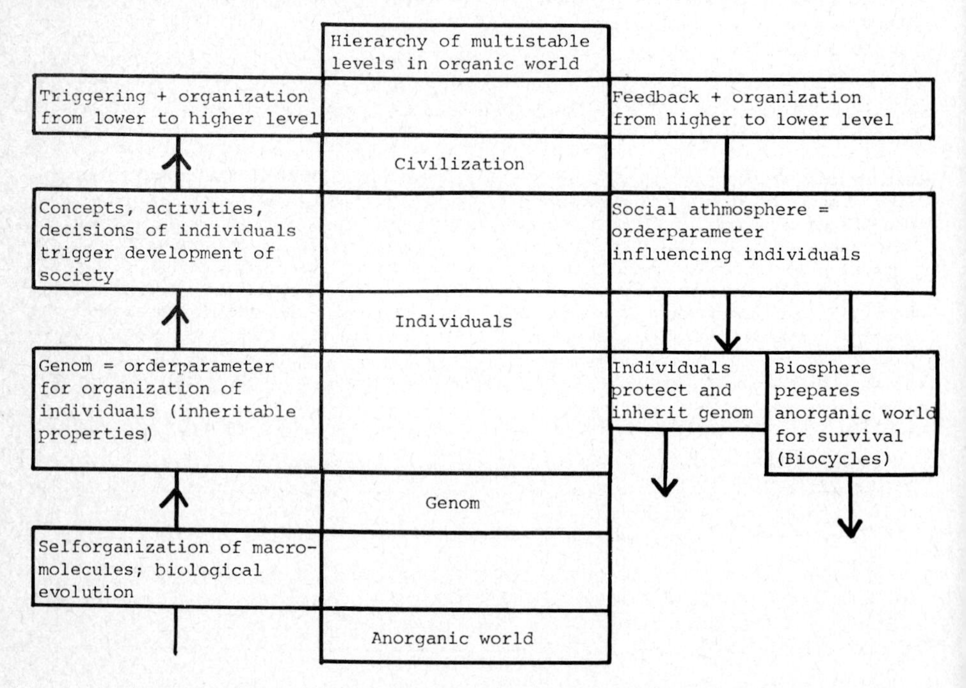

In the sense of the foregoing general consideration we now try to treat the dynamics of society as a selfconsistent macroscopic system, where the interacting individuals by their attitudes and activities contribute to the development of the "situation" which in turn influences probabilistically attitudes and actions of the individuals.

In section II we treat the "situation" as a quasi constant environment determining in a constant way the structure of the behaviour-probabilities of the individuals. On the other hand in this section we fully take into account the statistical aspects of the development of the society. This amounts to setting up a Fokker-Planck equation for the probability distribution over group configurations. In a simple example we shall solve this equation and show its applicability to the description of polarisation processes, in social groups. Further we will see the structural analogies between phase transitions in physics and society.

In section III we want to take into account the mutual coupling between a time dependent "situation" and the "attitudes" of the members of social groups. Of course, we have to make idealized but plausible assumptions about the situation and the coupled equations of motion. For simplicity we will neglect in this section the statistical aspects and restrict our consideration to the equations for meanvalues. The aim of this extended treatment is to gain a first crude semi quantitative insight into the problem, under which circumstances developments in society tend towards a stationary state, or, alternatively, tend to unstable revolutionary processes.

II. Fokker-Planck Equation of Interacting Social Groups

The first step in the statistical treatment of society is the transition from the "microscopic point of view" which takes into account the behaviour and attitude of the single individual to a "macroscopic level". To this aim we assume that we may divide society into Z homogeneous groups, where the members of one group have the same social, economic and educational background. Further we assume that the continuous variety of possible opinions and ways of behaviour may be "coarsegrained" into a number of Ω disjunctive "attitudes". We denote with $n_{j,\alpha}$ the number of members of group j (j = 1...Z) with attitude α (α = 1... Ω). $n_j = \sum_{\alpha=1}^{\Omega} n_{j,\alpha}$ is the number of members of group j. According to the coarse grained level of consideration we have to go over to a probabilistic description of the possible change of attitudes of the individual members of society. In this sense we introduce as a fundamental quantity the transition probability $p^{(i)}_{\alpha\to\beta}(\{n_{j,\gamma}\})$ for a member of group i to change from attitude α to attitude β. This probability may depend on the group configuration in the society characterized by the variables $\{n_{j,\gamma}\}$. It may further depend on the constant or variable constituents of the objective situation in which the society lives. In this section we will, however, consider the "situation" as a given time independent constant determining the parameters in the functions $p^{(i)}_{\alpha\to\beta}(\{n_{j,\gamma}\})$ The explicit form of these functions should be found by a preceding sociological analysis taking into account the mentality of the groups and the situation. We will restrict ourselfes to plausible assumptions about $p^{(i)}_{\alpha\to\beta}$.

Let us now introduce the probability-distribution function $f(\{n_{j,\gamma}\},t)$ describing the probability to find at time t in the society a group configuration $\{n_{j,\gamma}\}$. The equation of motion for $f(\{n_{j,\gamma}\},t)$

$$\frac{\partial f(\{n_{j,\gamma}\},t)}{\partial t} = \sum_{\alpha\neq\beta}^{\Omega} \sum_{i=1}^{Z} \Big[n'_{i,\alpha} \cdot p^{(i)}_{\alpha\to\beta}(\{n_{j,\gamma}\}'_{i,\alpha,\beta}) f(\{n_{j,\gamma}\}'_{i,\alpha,\beta},t) \tag{2.1}$$

$$- n_{i,\alpha} \cdot p^{(i)}_{\alpha\to\beta}(\{n_{j,\gamma}\}) f(\{n_{j,\gamma}\},t) \Big]$$

with $\{n_{j,\gamma}\}'_{i,\alpha,\beta}$ arising from $\{n_{j,\gamma}\}$

by the substitution $n_{i,\alpha} \to n'_{i,\alpha} = n_{i,\alpha} + 1$; $n_{i\beta} \to n'_{i\beta} = n_{i\beta} - 1$

is then easily derived from a probability balance-consideration: The first term on the right hand side of (2.1) is the *increase* per time unit of the probability of configuration $\{n_{j,\gamma}\}$ arising from a change of neighbouring configurations $\{n_{j,\gamma}\}'_{i,\alpha,\beta}$ into $\{n_{j,\gamma}\}$ by transitions of people to other attitudes, and the second term describes the *decrease* per time unit of the probability of configuration $\{n_{j,\gamma}\}$ by change into neighbouring configurations arising from transititions of people to other attitudes. The next step is the introduction of quasi continuous variables

$$x_{j,\gamma} = n_{j,\gamma}/n_j \;;\; \sum_{\alpha=1}^{\Omega} x_{j,\alpha} = 1 \;;\; 0 \leq x_{j,\gamma} \leq 1 \;;\; \Delta x_{j,\gamma} = \frac{1}{n_j} \tag{2.2}$$

where we assume, that $n_{j,\gamma} \gg 1$. Then we may expand the right hand side of (2.1) up to second order in $\Delta x_{j,\gamma} = 1/n_j$. This approximation yields a Fokker-Planck equation for $f(\{x_{j,\gamma}\} t)$ of the following form:

$$\frac{\partial f(\{x_{j,\alpha}\},t)}{\partial t} = \sum_{\alpha \neq \beta}^{\Omega} \sum_{i=1}^{Z} \Big[\Big(\frac{\partial}{\partial x_{i,\alpha}} - \frac{\partial}{\partial x_{i,\beta}} \Big) + \tag{2.3}$$

$$+ \frac{1}{2n_i} \Big(\frac{\partial}{\partial x_{i,\alpha}} - \frac{\partial}{\partial x_{i,\beta}} \Big)^2 \Big] \Big(x_{i,\alpha} \, p^{(i)}_{\alpha \to \beta}(\{x_{j,\gamma}\}) f(\{x_{j,\gamma}\},t) \Big)$$

The meanvalues $x_{k,\varepsilon}$ (t) of the "macroscopic observables" $x_{k,\varepsilon}$ are defined by

$$\bar{x}_{k,\varepsilon}(t) = \int f(\{x_{j,\gamma}\},t)\, x_{k,\varepsilon}\, d\tau \; ; \quad d\tau = \prod_{\gamma,i} dx_{i,\gamma} \tag{2.4}$$

We obtain their equation of motion, if we insert eq. (2.3) into the expression for $d\bar{x}_{k,\varepsilon}/dt$ and perform some partial integrations, where we assume that $f(\{x_{j,\gamma}\},t)$ vanishes at the boundary of the domain of integration. The result is

$$\frac{d\bar{x}_{k,\varepsilon}}{dt} = \sum_{\substack{\alpha=1 \\ (\alpha \neq \varepsilon)}}^{\Omega} \int \Big(x_{k,\alpha}\, p^{(k)}_{\alpha \to \varepsilon}(\{x_{j,\gamma}\}) - x_{k,\varepsilon}\, p^{(k)}_{\varepsilon \to \alpha}(\{x_{j,\gamma}\}) \Big) f(\{x_{j,\gamma}\},t)\, d\tau \tag{2.5}$$

If the probability distribution $f(\{x_{j,\gamma}\},t)$ is a sharply peaked function, we have only small fluctuations of the $x_{j,\gamma}$ around their meanvalues. In this case we may approximately calculate the right hand side of eq. (2.5) and obtain the following closed set of ordinary differential equations for the meanvalues

$$\frac{dx_{k,\varepsilon}(t)}{dt} = \sum_{\alpha \neq \varepsilon}^{\Omega} \Big(\bar{x}_{k,\alpha}\, p^{(k)}_{\alpha \to \varepsilon}(\{\bar{x}_{j,\gamma}\}) - \bar{x}_{k,\varepsilon}\, p^{(k)}_{\varepsilon \to \alpha}(\{\bar{x}_{j,\gamma}\}) \Big) \tag{2.6}$$

Now we want to apply the general Fokker-Planck equation (2.3) to the simple case of one group with a cooperative internal interaction between its members. We assume, that only two attitudes are possible. Then the Fokker-Planck equation (2.3) for the case $Z = 1$, $\Omega = 2$ takes the form

$$\frac{\partial f(x,t)}{\partial t} = - \frac{\partial j(x,t)}{\partial x} = \tag{2.7}$$

$$= - \frac{\partial}{\partial x} \big[K_1(x) f(x,t) \big] + \frac{1}{2} \frac{\partial^2}{\partial x^2} \big[K_2(x) f(x,t) \big]$$

where we have introduced the variables

$$x = \frac{1}{2}(x_2 - x_1) \quad ; \quad (x_1 + x_2) = 1 \tag{2.8}$$

and the coefficients

$$K_1(x) = \Big\{ \Big(\frac{1}{2} - x\Big) p_{1\to 2}(x) - \Big(\frac{1}{2} + x\Big) p_{2\to 1}(x) \Big\} \quad ; \tag{2.9}$$

$$K_2(x) = \frac{1}{n} \Big\{ \Big(\frac{1}{2} - x\Big) p_{1\to 2}(x) + \Big(\frac{1}{2} + x\Big) p_{2\to 1}(x) \Big\} .$$

Because of (2.7), the quantity

$$j\ (x,t) \equiv K_1\ (x) f\ (x,t) - \frac{1}{2}\frac{\partial}{\partial x}\left[K_2\ (x) f\ (x,t)\right] \tag{2.10}$$

may be interpreted as a probability current, which therefore has to fulfill the boundary condition

$$j\ (\pm 1/2\ ;\ t) = 0 \tag{2.11}$$

For the stationary solution f_{st} (x) we have $\partial j/\partial x = 0$ which gives, together with (2.11), j_{st} (x) = 0. The first order differential equation (2.10) for f_{st} (x) can easily be solved with the result

$$f_{st}\ (x) = c\ K_2^{-1}\ (x)\ \exp\left\{2\int^{x}\frac{K_1\ (y)}{K_2\ (y)}\ dy\right\} \tag{2.12}$$

The time dependent solutions of the Fokker-Planck equation (2.7) may also be found, at least numerically. It can be shown that the Fokker-Planck operator has only negative eigenvalues and the eigenvalue 0 corresponds to the unique stationary solution. Further it is easy to solve the meanvalue-equation which in our case has the form

$$\frac{d\bar{x}(t)}{dt} = K_1\ (\bar{x}) \tag{2.13}$$

and the solution

$$(t-t_o) = \int_{\bar{x}_0}^{\bar{x}}\frac{dy}{K_1\ (y)} \tag{2.14}$$

If we want to have explicit results we have to make explicit assumptions about the transition probabilities $p_{1\to 2}$ (x) ; $p_{2\to 1}$ (x). For instance we may put

$$p_{2\to 1}\ (x) = \gamma\exp\ (-k\ x-h)\quad ;\ p_{1\to 2}\ (x) = \gamma\ \exp\ (k\ x+h) \tag{2.15}$$

This ansatz describes a mutual interaction between the members of the group such that the decision behaviour of the individual is influenced by the attitudes of all members. k and h are open parameters which because of their meaning in (2.15) may be called k = "adaptation parameter"; h = "preference parameter". The form (2.15) is analogous to the flipping probability of spins σ_i = ± 1 at site i in the Ising model of ferromagnetism which is given by

$$p\ (\sigma_i \to -\sigma_i) = \gamma\ \exp\left\{-\frac{H\sigma_i + \sum_j J_{ij}\sigma_j}{kT}\right\}\ ;\quad \sigma_i = \pm 1 \tag{2.16}$$

where H = external magnetic field, J_{ij} = spin-spin interaction constant; T = temperature.

The explicit solution f_{st} (x;k,h) which is found with (2.12) by using (2.15) and (2.9) now changes its structure fundamentally at a value k_o of k defined by

$$\left.\frac{\partial\ K_1\ (k_o, x)}{\partial\ x}\right|_{x=0} = 0 = \gamma\ (k_o - 2)\quad \leadsto k_o = 2 \tag{2.17}$$

in the case of h = 0 (compare fig. 1), 2), 3) and 5). We may therefore call k_o the phasetransition point with respect to the polarisation structure of the groups describable by f_{st}(x ; k, h = 0). We may also define - in analogy to physical systems - an order parameter which distinguishes between the two phases of the group system. An appropriate order parameter is given by the square of the stable stationary meanvalue following from the stationary solution of (3.8a)

$$\bar{x}^2 \quad (\infty) = 0 \text{ for } \quad \kappa < \kappa_0 \tag{2.18}$$
$$\bar{x}^2 \quad (\infty) = x_1^2 > 0 \text{ for } \kappa > \kappa_0 \quad ; \quad K_1(\kappa, x_1) = 0$$

In fig. 1) to 5) we show some results of the calculations: Fig. 1) to fig. 4) show the stationary distribution function $f_{st}(x)$, the time development of the meanvalues $\bar{x}(\tau)$ in the dimensionless time coordinate $\tau = \gamma t$ and the drift coefficient $K_1(x)$ for different values of adaptationparameter κ and preference parameter h.

Fig. 1) $\kappa = 0$ means independent decision behaviour of members of the groups; $h = 0$ means no preference between the attitudes 1 and 2. The most probable groups are those with $n_1 = n_2$ or $x = 0$. The meanvalues $\bar{x}(\tau)$ quickly relax to $\bar{x}(\infty) = 0$ independently of the initial value.

Fig. 2) $\kappa = \kappa_0 = 2$; $h = 0$ gives the group statistics at phase transition point of the adaptation parameter κ. The broad distribution function indicates the large critical fluctuations of the single groups around the stationary meanvalue $\bar{x}(\infty) = 0$. The $\bar{x}(\tau)$ show the critical slowing down of the relaxation to $\bar{x}(\infty) = 0$, due to the fact that the driftcoefficient $K_1(x)$ vanishes in higher order at $x = 0$.

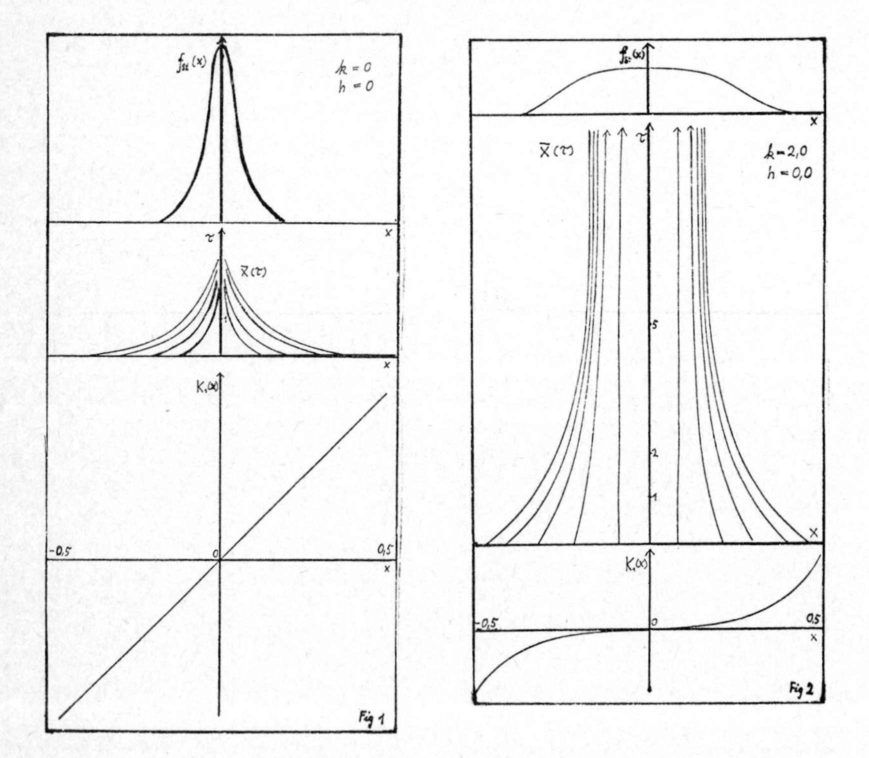

Fig. 3) $\kappa = 2,5$; $h = 0$ shows the behaviour of groups beyond phasetransition for strong mutual adaptation of the members of a group. The most probable groups now have $n_1 \gg n_2$ or $n_2 \gg n_1$ (polarisation structure). The meanvalues $\bar{x}(\tau)$ now tend to $\pm x_1$, with $K_1(\kappa, x_1) = 0$ according to their initial values.

Fig. 4) $\kappa = 2,5$; $h = 0,01$ shows the same case with a small preference parameter leading to an asymmetric distribution function $f_{st}(x; \kappa\ h)$ and an asymmetric relaxation of the $\bar{x}(\tau)$.

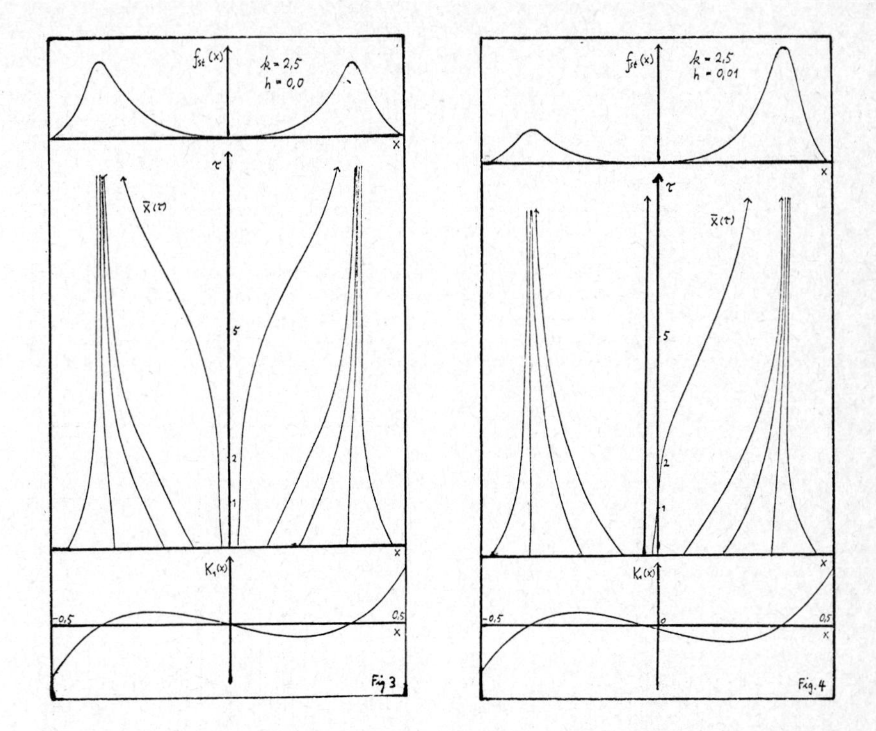

Fig. 5) shows the longrange change of parameter k from $k = 0$ to $k = 2,5$ the corresponding change of the distribution function and the development of a typical sample group with critical fluctuations at $k \approx k_c$.

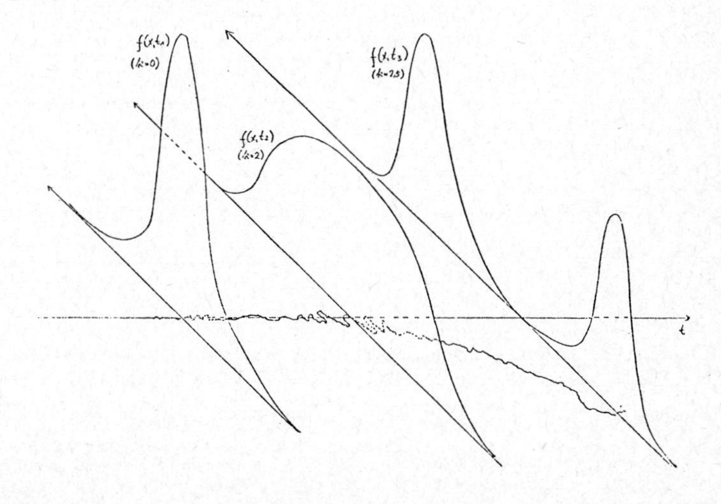

We conclude this section with some general remarks: The fact that our example showed the characteristics of a phase transition is not accidential. In statistical physics we often have a competition between independent fluctuations and cooperative interactions among the elementary units of a macroscopic system. In a critical region a small change of external parameters of the system may lead to a sudden change of its macroscopic structure, because one of the competing effects is made to prevail. This macroscopic change is called a phase transition. In sociology the fundamental transition probabilities also include two competing behaviours: The wish for selfdetermination and individual independence and on the other hand the readyness to cooperation and adaptation. Correspondingly, on the macroscopic level phase-transition phenomena with properties similar to those in physics (critical fluctuations etc.) are also to be expected in society.

III. Coupled Equations for "Situation" and "Attitudes"

We begin this section with some criticism about the insufficiencies of our treatment in section II. There we considered the situation, in which the society lives, as a constant, while in reality we have a dynamical mutual interaction between society and its situation. Now we wish to include at least the "artificial" component of the situation, which by definition underlies the control and influence of man, into our treatment. Further we want to take into account also memory-effects, i.e. retarded interactions, which were not contained in the Fokker-Planck equation treatment.

At the beginning we have to introduce some new concepts allowing to idealize and to quantify the interaction between society and situation. The first and by far not trivial problem is the *quantification* of the *situation:* We have to distinguish between the *"objective situation"* in its complex structure and the *"interpreted situation"*. In contrast to the objective situation the interpreted situation only contains relatively few independent aspects which are, in the opinion of men, relevant for the state of society. Such material or spiritual aspects of the situation are, for instance, the standard life, the structure of economy, the legislation, the political climate, the religious life. It seems now difficult but not impossible to introduce quantitative measures for the state of an aspect, i.e. a component of the interpreted situation, within some range of possibilities. Thus we may try to quantify the standard of life between wealth and poverty, the economy between capitalistic and socialistic structures, the legislation between liberal and restrictive standpoints, the political climate between freedom of decision and totalitarian pressure and the religious life between secularisation and practisation of belief by introducing normalized parameters $y_1 \ldots y_A$ for the A chosen independent aspects $a_1 \ldots a_1$ which may vary in the range $-1 \leq y^A \leq +1$.
The interpreted situation is then described by a vector $\vec{y} = \{y_1 \ldots y_A\}$ in the "situationspace" spanned by the chosen independent aspectcoordinates. The values of the components of $\vec{y}$ are found by interpreting, estimating and weighing the items of the objective situation with respect to their relevance, meaning and relative position in the frame of these aspects. Thus the objective situation is projected into a vector of the interpretational situationspace. (For a first crude approximation it will even be sufficient to introduce only one parameter y varying from "progressive" to "conservative", to characterize the global political situation). Our quantification-procedure of course, implies the supposition that the situationspace has a metric structure in the sense that the evaluation of the objective situation allows for the definition of a sequence of gradually differing degrees ranging from one extreme state to the opposite extreme state for every aspect. In some sense the aspects of the situation play the role of macroscopic observables by which the highly complex objective situation is coarse grained. (In the present consideration we neglect the important problem that different ideologies may give a completely different interpretation and evaluation to the same objective situation and that even the conceptual network of different ideologies may be incommensurable. We assume as a first approximation the possibility of an undistorted estimation of the objective situation by "common sense" with respect to the mentioned aspects).

The next important problem is the relation between the situation space and the attitudes (α) of people: Different attitudes (α) more or less correspond to different designs of and opinions about "ideal" societies characterized by "ideal" situation vectors $\vec{y}^{(\alpha)} = \{y_1^{(\alpha)}, \ldots y_A^{(\alpha)}\}$. If an attitude (α) has more or less closed ideology as its intellectual background, it tends to stress certain onesided positions and to neglect complementary positions. Correspondingly, the "ideal" situations $\vec{y}^{(\alpha)}$ belonging to such ideologic attitudes will mainly lie near the edge of the accessible situationspace. (As the accessible range of the interpretational situationspace itself is a function of the development of the objective situation, e.g. by new technical, medical abilities of man, the relative position of a situationvector $\vec{y}^{(\alpha)}$ belonging to a *rigid* attitude (α) may shift in course of time; for instance, a formerly wellbalanced attitude may become obsolete and extreme in the light of new objective developments).

Let us now try to set up the general form of the coupled equations of motion for the development of the groups of society with different attitudes and of the "situation" represented by a situationvector $\vec{y}$ (t). In this first approach we simplify the description by only considering equations for meanvalues, neglecting the statistical fluctuations which we included in section II. We obtain two coupled sets of equations: The first set describes the development of groups of people with different opinions, behaviours and activities, i.e. different attitudes, because of their interaction with other groups *and* with the changing situation. The second set describes the development of the situation because of the competing activities of the different groups of society trying to influence the objective situation and thereby also the aspects of the interpreted situation, i.e. the situationvector $\vec{y}(t)$.

The first set was already derived in section II (comp. eq. (2.6) and we repeat it here (omitting the bar in $x_{k\alpha}$):

$$\frac{dx_{k,\varepsilon}}{dt} = \sum_{\substack{\alpha=1 \\ (\alpha \neq \varepsilon)}}^{\Omega} \Big(x_{k,\alpha} \, p^{(k)}_{\alpha\to\varepsilon}(\{x_{j,\gamma}\};\{y_e\}) - x_{k,\varepsilon} \, p^{(k)}_{\varepsilon\to\alpha}(\{x_{j,\gamma}\};\{y_e\}) \Big) \tag{3.1}$$

The decisive difference between (3.1) and (2.6) is, however, that the transitionprobabilities $p^{(k)}_{\alpha\to\varepsilon}(\{x_{j,\gamma}\};\{y_e\})$ are now in general functions not only of the group configuration $\{x_{j,\gamma}\}$ but also of the situation described by the situationvector $\vec{y} = \{y_e\}$. In particular, the $p^{(k)}_{\alpha\to\varepsilon}$ may now describe the reaction of the attitudes of people to a developing situation. These reactions may be retarded, may depend on the derivatives of $\vec{y}(t)$ and may be highly nonlinear. Thus, eq. (3.1) may have a very complicated structure. The setting up of the functional form of the $p^{(k)}_{\alpha\to\varepsilon}(\{x_{j,\gamma}\};\{y_e\})$ is a problem of political psychology, for which we will only discuss simple plausible model assumptions.

The second set gives the equation of motion for the situationvector $\vec{y}(t)$. Clearly, those subgroups of the (socially homogeneous) groups j (j=1...Z) which have the attitude (α) try to influence the objective situation - and as a consequence also the interpreted situation $\vec{y}(t)$ - in such a direction, that the "ideal" situation $\vec{y}^{(\alpha)}$ belonging to attitude (α) is approached. The timederivative $\left(\frac{d\vec{y}}{dt}\right)_\alpha$ due to the activity of people with attitude (α) thus will have the form

$$\left(\frac{d\vec{y}}{dt}\right)_\alpha = -K_\alpha \cdot (\vec{y} - \vec{y}^{(\alpha)}) \tag{3.2}$$

where K_α is an activity parameter describing the efficient influence of all people with attitude (α) in the sense of a change of the situation towards $\vec{y}^{(\alpha)}$. The ansatz

$$K_\alpha = \sum_{j=1}^{Z} K_{\alpha j} \tag{3.3}$$

with $K_{\alpha j} = n_{\alpha j} \, a_{\alpha j}(\{x_{k,\beta}\};\{y_e\})$; $n_{\alpha j} = x_{\alpha j} n_j$

takes account of the composition of K_α of the specific activities $K_{\alpha j}$ of the subgroups (j,α); these are proportional to the number $n_{\alpha j}$ of people in these groups and to individual activity parameters $a_{\alpha j}$ which in general may depend on $\{x_{k,\beta}\}$ and on $\{y_e\}$. (For instance, the activity $a_{\alpha j}$ may depend on the distance $|\vec{y} - \vec{y}^{(\alpha)}|$ of $\vec{y}$ from the ideal situation $\vec{y}^{(\alpha)}$ in the sense of decreasing activity if the ideal situation $\vec{y}^{(\alpha)}$ is approached). The full timederivative of $\vec{y}$ is now given by the superposition of the contributions (3.2) of the competing groups whose activities aim at different "ideal situations":

$$\frac{d\vec{y}}{dt} = \sum_{\alpha=1}^{\Omega}\left(\frac{d\vec{y}}{dt}\right)_{\alpha} = -\sum_{\alpha=1}^{\Omega} \kappa_{\alpha}\,(\vec{y} - \vec{y}^{(\alpha)}) \tag{3.4}$$

$$= -\sum_{\alpha=1}^{\Omega}\sum_{j=1}^{Z} n_j\, x_{\alpha j}\, a_{\alpha j}(\{x_{k,\beta}\};\{y_e\})\,(\vec{y} - \vec{y}^{(\alpha)}).$$

The coupled set of equations (3.1), (3.4) describes the motion of the groups of a society and of the situation.

We note that the structure of the functions $p^{(k)}_{\alpha\to\beta}(\{x_{j,\gamma}\};\{y_e\})$ and $a_{\alpha j}(\{x_{k,\beta}\};\{y_e\})$ appearing in (3.1) and (3.4) determine the dynamics of the society + situation on a coarse grained level and thus should also provide quantitative criteria for such general questions, whether, for instance, the society is in a pluralistic and liberal state L or in a restrictive and totalitarian state T. Indeed we expect
(1) an increasingly strong dependence of the decision probabilities $p^{(k)}_{\alpha\to\beta}(\{x_{j,\gamma}\},\{y_e\})$ on the group configuration $\{x_{j\gamma}\}$ for a transition from state L to T, corresponding to the transition from an independent to an adaptive decision behaviour of the groups of society,
(2) a change of the functional dependence of the activity parameters $a_{\alpha j}$ on $\{x_{j\alpha}\}$ and $\{y_e\}$ for a transition from state L to T: In the liberal case L the $a_{\alpha j}$ will depend smoothly on $\{x_{j\gamma}\}$; $\{y_e\}$ providing for a competitive but partly neutralizing pluralistic manifold of activities. In the totalitarian case T the $a_{\alpha j}$ will depend strongly on $\{x_{je}\}$; $\{y_e\}$, for instance, in the sense of suppression of competing activities by the successful group with an attitude (α), thus trying to stabilize the situation $\vec{y}$ in the vicinity of $\vec{y}^{(\alpha)}$ (⟶ "Machtergreifung").

Now we will set up a *"minimal nontrivial model"* as application of the general considerations, and solve it explicitely. We assume only one homogeneous group, i.e. one kind of members of the society, with two attitudes $Z = 1$; $\Omega = 2$. Further we project the objective situation into a onedimensional situationspace, where the parameter y $(-1 \leq y \leq +1)$ may e.g. describe political situations from extreme left (-1) to extreme right (+1). Let us identify the attitudes $\alpha = 1$ and 2 with the "left" or "right" opinion. Correspondingly, the "ideal" situations $y^{(1)}$ and $y^{(2)}$ lie in the vicinity of (-1) and (+1) respectively. For simplicity we will assume symmetry in all general relations between "left" and "right". Thus we should have $y^{(1)} = -y^{(2)}$ and may, for instance, put $y^{(1)} = -1$; $y^{(2)} = +1$. Introducing

$$x_{\alpha} = \frac{n_{\alpha}}{n_1 + n_2}\;;\; x = (x_2 - x_1)\;;\; (x_1 + x_2) = 1,$$

the equations (3.1) and (3.4) take the form

$$\frac{dx}{dt} = -(p_{12} + p_{21})\,x + (p_{12} - p_{21}) \tag{3.5}$$

$$\frac{dy}{dt} = \left(\frac{dy}{dt}\right)_1 + \left(\frac{dy}{dt}\right)_2 = -\kappa^{(1)}(y - y^{(1)}) - \kappa^{(2)}(y - y^{(2)}) \tag{3.6}$$

$$= -(\kappa^{(1)} + \kappa^{(2)})\,y + (\kappa^{(2)} - \kappa^{(1)})\,|y^{(2)}|$$

with $\kappa^{(\alpha)} = n_{\alpha} a_{\alpha} = x_{\alpha}\, n\, a_{\alpha}$; $(n = n_1 + n_2)$

These equations are, however, nonlinear and retarded, according to the choice of the functions $p_{\alpha\beta}$ (x,y) and a_{α} (x,y). We now make plausible simple assumptions about $p_{\alpha\beta}$ and a_{α}. If, in first approximation, the activities a_{α} are constant, we obtain

$$n a_{\alpha} = \kappa\;;\; (\kappa^{(1)} + \kappa^{(2)}) = \kappa\;;\; (\kappa^{(2)} - \kappa^{(1)}) = x\kappa \tag{3.7}$$

On the other hand, the transition probabilities $p_{\alpha\beta}$ should contain cooperative adaptation effects between members of the society (which we studied already in section II) and the reaction on the situation. Both are included in an ansatz of the form

$$p_{12} = \alpha_{12}(y) + \beta_{12}(y)x + \gamma_{12}(y)x^2$$
$$p_{21} = \alpha_{21}(y) + \beta_{21}(y)x + \gamma_{21}(y)x^2 \qquad (3.8)$$

where in the symmetrical case

$$p_{21}(-x,-y) = p_{12}(x,y) \qquad (3.9)$$

$$\rightsquigarrow \alpha_{21}(-y) = \alpha_{12}(y)\ ;\ \beta_{21}(-y) = -\beta_{12}(y)\ ;\ \gamma_{21}(y) = \gamma_{12}(y).$$

We may choose $\beta_{12} = -\beta_{21} = \beta$ and $\gamma_{12} = \gamma_{21} = \gamma$ independent of y; then only the coefficients $\alpha_{12}(y), \alpha_{21}(y)$ contain the reaction of people to the situation. Political psychology and experience teaches that this reaction is relatively inert in a region of moderate situations but may become strong and highly nonlinear if the situation exceeds some tolerable threshold; the reaction then also continues for some relaxation time, which means, that p_{12}, p_{21} are retarded functions of y(t). Idealizing this counter-reaction of the attitudes, if the situation exceeds the threshold value y_c we may put

$$\alpha_{21}(y) = \alpha_0 + \vartheta\left(\mathrm{Max}(y(t') - y_c)\Big|_{t'=t-\tau}^{t'=t}\right) \cdot \alpha \qquad (3.10)$$

$$\alpha_{12}(y) = \alpha_0 + \vartheta\left(\mathrm{Max}(-y(t') - y_c)\Big|_{t'=t-\tau}^{t'=t}\right) \cdot \alpha$$

which means, that $\alpha_{21}(y)$ takes the value $(\alpha_0 + \alpha)$ instead of α_0 at time t, if $y(t')$ exceeded the critical value $y_c > 0$ at any time t' in the interval $t-\tau < t' \leq t$. Inserting (3.7)...(3.10) into (3.5), (3.6) we obtain the final form of the equations of motion

$$\frac{dx}{dt} = 2\alpha_{as}(y) + 2(\beta - \alpha_s(y))x - 2\gamma x^3 \qquad (3.5')$$

$$\frac{dy}{dt} = -\kappa y + \kappa x \cdot |y^{(2)}|\ ;\ y^{(2)} \approx 1. \qquad (3.6')$$

with $\alpha_0 = 0;\ 2\alpha_{as}(y) = \alpha_{12}(y) - \alpha_{21}(y);\ 2\alpha_s(y) = \alpha_{12}(y) + \alpha_{21}(y).$

We repeat the meaning of the parameters of the system:

(3.11)

β = adaptation parameter
γ = saturation of adaptation
y_c = critical situation
α = reactionstrength to critical situation
τ = relaxation time of critical reaction
κ = rate of change of situation.

The evaluation of eq. (3.5'), (3.6') for different values of the parameters (3.11) shows that in spite of its oversimplification the model contains different cases of stable or unstable development of society which seem to be of principle interest. For finite adaptation constants β, γ we obtain - as already discussed in section II, including the full statistics - a symmetry-breaking of attitudes leading by cooperative effects to their development towards a "left" or "right" stationary value, in dependence of the initial condition. Correspondingly, the situation is drawn into the same directions by the activity of people. Now we have to distinguish between different cases:

a) The attitudes x(t) and afterwards the situation y(t) approach a *"moderate"* stationary value x_{st}, y_{st} which lies *below* the critical value y_c. Then no critical reaction appears and the system will *smoothly* approach the *stationary values* x_{st}, y_{st} (comp. fig. 6,7)

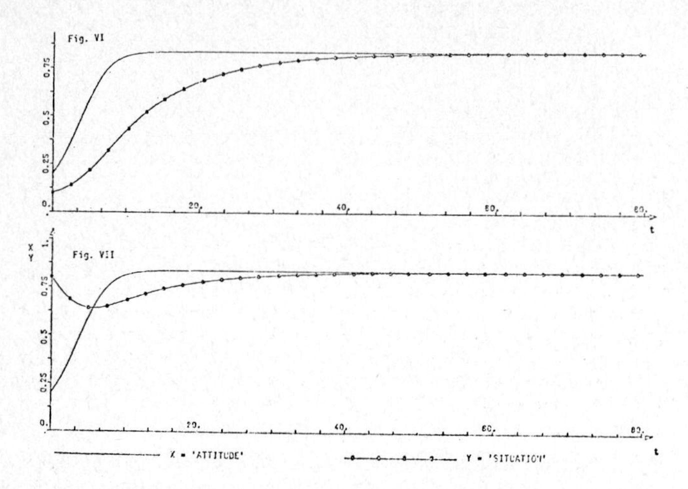

b) If the attitude x(t) and afterwards the situation y(t) tends to an *"extreme"* stationary value x_{st}, y_{st} *above* a critical value y_C, a critical reaction (change of attitudes) will arise as soon as y(t) reaches Y_C. (This reaction is incorporated in our model by the choice of the transition probabilities (3.8) with (3.10)). This leads to a *revolutionary unstable development*. According to the strength and duration of the critical reaction we have to distinguish between the *"small and the big revolutionary cycle"*. In the case of the small cycle b 1) the strength and duration of the critical reaction is not sufficient to bring the global attitude x(t) beyond the neutral point x = 0. After decay of the critical reaction therefore the collective adaptation mechanism again revive tending to the old stationary (extreme) values x_{st}, y_{st} where the cycle may begin anew (comp. fig. 8,9 + 12)

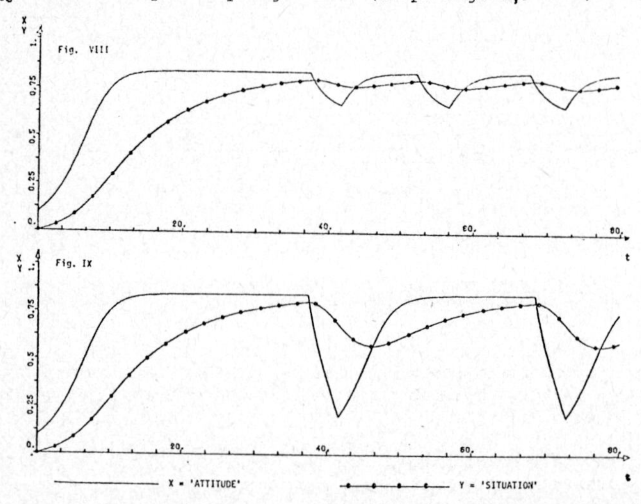

In the case of the big cycle b 2) intensity and duration of the critical reaction are sufficient to lead x(t) beyond the neutral point so that afterwards the collective forces let the attitudes approach to the opposite stationary value $-x_{st}$ followed by the situation $y(t) \rightarrow -y_{st}$. Here, as in our symmetrical case, the same unstable critical counter-reaction may happen (comp. fig. 10, 11 + 13)

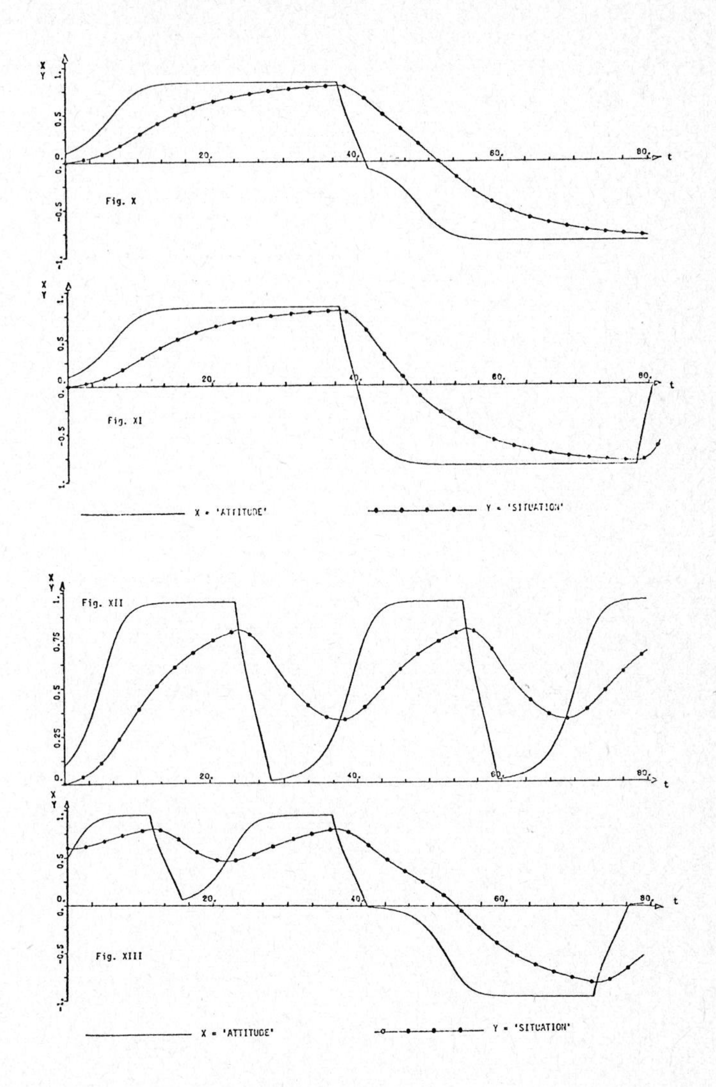

The cases b 1) and b 2) may be considered as a first - still very crude - quantitative approach to the understanding of unstable developments as in Czecho-Slovakia b 1) or in Chile b 2).

We conclude this section with some general remarks:
1.) It seems that a refined treatment of the dynamics of a society including its situation in terms of equations of the form (3.1), (3.4) at least leads to a quanti-

tative classification of different stable or unstable types of development. The Poincaré theory of stationary points and limiting cycles of nonlinear differential equations will be helpful here, though our equations in general are of a more general type.

2.) A further refinement should again include statistics. For instance, the physical theories of metastable states and of the related nucleation problem will have strong analogies in society.

3.) Of course, in a better approximation, the equations should also include, so far as possible, the superposition of perennial irreversible effects. These are due a) to the shift of the accessible situation by irreversible developments in science and technology and b) to the effect of "learning from history" by which the functions $p^{(k)}_{\alpha\to\beta}(\{x_{j\gamma}\};\{y_l\})$ and $a_{\alpha j}(\{x_{k,\gamma}\};\{y_l\})$ in principle may depend on all former states of the system.

Cooperative Phenomena, H. Haken, ed.

COOPERATIVE PHENOMENA IN TELEPHONY

Gerhard Bretschneider

Siemens AG, Forschungslaboratorien

München, Federal Republic of Germany

Let us first take a short look at the present situation of telephony in the world. There are now a total of nearly 300 million telephone sets connected to the public networks, most of which are able - in principle - to originate calls to any other telephone set in the world. The investment value for the whole telephone network is estimated to be as high as 300 billion $. Sometimes the telephone network is called the largest special realtime computer in the world, as it digests the numbering sequences dialed in by the subscribers and connects them as required, doing this for all the subscribers simultaneously. How does the telephone network carry out this job? Let us have a look at Fig. 1. Here you see in the top part a very simplified example showing what a network for only ten subscribers would look like if all the subscribers were linked together directly. You would then need 45 connecting lines. For 100 subscribers this figure would rise to 4950. It is much cheaper to install a central office to make the connections. Then you would only need 10 connecting lines for 10 subscribers or 100 lines for 100 subscribers (see Fig. 1, middle section).

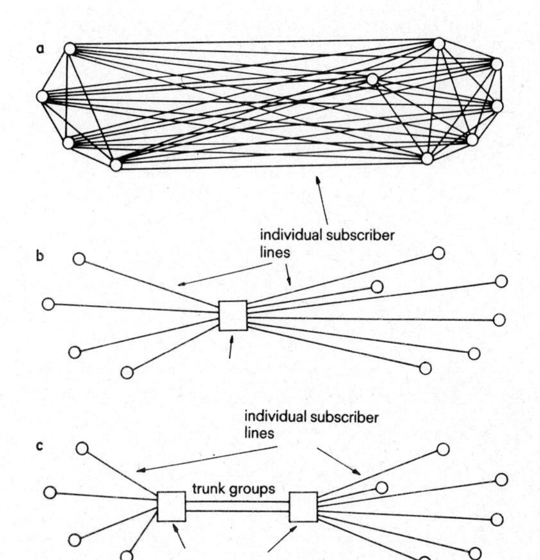

Fig. 1 : Structuring principles of a telephone network

For a larger number of subscribers, especially if more widely distributed, it would be even cheaper to insert more than one central office (see Fig. 1, bottom). In this case a call from a subscriber in the left-hand group to one in the right-hand group must use a line from a trunk connecting the two offices. In this case we meet for the first time a problem of traffic theory: how many trunk lines should be provided between the two offices? This of course depends on the amount of traffic flowing between the two offices. But what is the amount and what are the properties of such traffic?

For an answer to these questions please look at Fig. 2.

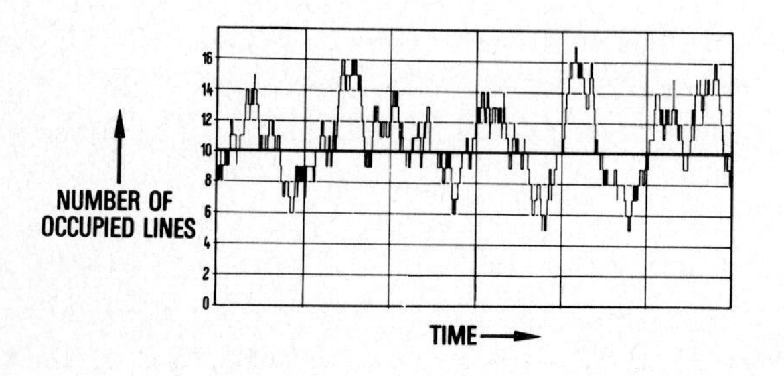

Fig. 2 : Fluctuations of a telephone traffic

In this figure you see what happens if you let the total telephone traffic of say 200 subscribers flow on an unlimited line group. In this figure the vertical scale gives the number of simultaneously occupied lines, and the horizontal scale indicates the time. You see that the mean number of occupied lines is approximately 10, but there are some intervals during which up to 16 lines are needed and there are two short time intervals where even 17 lines would be necessary. Even this number of lines would not be sufficient for accepting all the incoming calls of this subscriber group if we extended the observation time. The number of occupied lines is indeed a statistically fluctuating value, but as extremely high values would only be needed rarely, it is necessary for economic reasons to limit the number of lines to be provided.

So you have to decide what to do with a call which arrives when all the lines are busy. In principle there are three possibilities:

one can refuse the call by giving it a busy tone

one can delay the call until a line becomes free and

one can reroute the call.

All three methods are used in telephone systems, but the possibility of delaying is used - in modern telephone systems - only for internal purposes and cannot be realized by the subscribers.

In reality the telephone networks are much more complicated than shown in Fig. 1. Not only the network of lines between the central offices has to be considered but also the even more complicated network within the offices. In order to handle the problems of congestion in a scientific way they must be broken down and idealized.

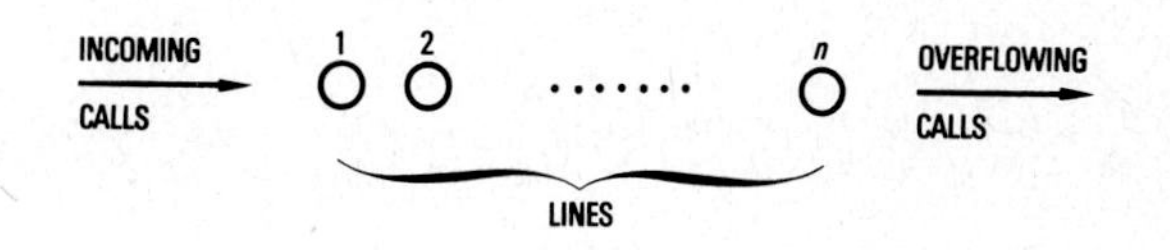

Fig. 3 : Full available trunk group

Consider a finite group of n lines and suppose that it receives a continous flow of calls originating from outside the system. Each arriving call hunts for a idle line and occupies it at once if there is one. Otherwise it is immediately discarded. The interarrival times between the incoming calls, which are the times between two consecutive incoming calls, are statistically distributed independently of each other and - the input process is then called a stochastic process - with the interarrival function

$$P_A\ (>t) = F(t)\ .$$

This means, that the probability of an arbitrary interarrival time being greater than t is given by F(t), with $F(t + \Delta t) \leq F(t)$ and $1 \geq F(t) \geq 0$ for each t. The occupation time for the calls is also a stochastic process with the distribution function

$$P_o\ (>t) = G(t)\ ,$$

the same for each call and independent of the input process.

You can then ask such questions as :

- What is the probability P(s;t) of finding exactly s lines occupied at time t or
- What is the limiting probability of finding all the n lines occupied,

$$P(n) = \lim_{t \to \infty} P(n;\ t)\ .$$

The answers obviously may depend on the number of lines n, on the properties of the functions F(t) and G(t) for the interarrival and occupation times and on further properties. A problem we want to study now.

The general solution to this problem is very complicated. Let us therefore first assume the functions F(t) and G(t) to be negative exponential functions. That is

$$F(t) = e^{-\lambda t}$$

$$G(t) = e^{-\mu t}\ ,$$

with the free parameters λ and μ.

The negative exponential function has one striking property. Suppose that we are looking at the system at time t_o and that we know that the last call arrival was at time t_o-t. What will be the probability that for a further time interal Δt there no call will arrive? This conditional probability is given by

$$P(>t+\Delta t \mid >t) = \frac{F(t+\Delta t)}{F(t)} = \frac{e^{-\lambda(t+\Delta t)}}{e^{-\lambda t}} = e^{-\lambda \Delta t} = F(\Delta t)\ .$$

In the case of the negative exponential function this conditional probability is thus independent of t, that means independent of the time of the last call arrival. In the case of the negative exponential function the past history has no influence on the future behaviour of a process.

For the given telephone traffic problem not only the interarrival times but also the call durations were assumed to be described by a negative exponential distribution. Therefore for this process as a whole in each time instant the "past" is extinguished and "future" depends only on the "present". This might seem to be a severe restriction but it is well justified for certain traffic problems and leads to easily manageable explicit solutions.

Let us now look once more at the negative exponential function. The probability that no call will arrive in a short time interval Δt can be written in the following form

$$P_A(>\Delta t) = e^{-\lambda \Delta t} = 1 - \lambda \Delta t + O\left((\Delta t)^2 \right)$$

where the expression $O((\Delta t)^2)$ means that the further terms are of the order $(\Delta t)^2$. The probability that a call will arrive in Δt is then given by $\Delta \lambda t + O((\Delta t)^2)$.

Similar expressions are true in the case of the occupation time. If more than one event is to be observed, if for instance the probability that neither a call will arrive in the time interval Δt nor one of s existing calls will terminate is to be calculated, you obtain

$$e^{-\lambda \Delta t} (e^{-\mu \Delta t})^s = 1 - (\lambda + s\mu) \Delta t + O((\Delta t)^2) .$$

Now we can proceed to derive a general differential equation system of a well known type. Suppose that the probability of finding s occupied lines at time t is given by $P(s; t)$. We can then derive the following equation system for the probabilites $P(s;t+\Delta t)$ at a short time interval Δt later

$$\begin{aligned} P(s;t+\Delta t) &= P(s-1;t)\lambda \Delta t + P(s;t)(1-\lambda \Delta t - s \cdot \mu \Delta t) \\ &\quad + P(s+1;t)(s+1)\mu \Delta t + O((\Delta t)^2) \end{aligned} \qquad (1)$$

for $0 \leq s \leq n$, and with $P(-1;t) = P(n+1;t) = 0$.

This equation system expresses the sum of the probabilities of four mutually exclusive events by which the system can arrive to state s at time $t + \Delta t$, namely:

1. The first term is the pr. that at time t the system was in state s-1, with pr. P (s-1,t), and a transition to s occurred, with pr. $\lambda\Delta t$.

2. The second term is the pr. that at time t the system was in state s and during Δt no change occurred, with pr.

$$(1 - \lambda\Delta t - s \cdot \mu\Delta t) \ .$$

3. The third term is the pr. that at time t the system was in state s + 1 and a transition to s occurred, with pr. $(s+1)\mu\Delta t$.

4. The last term takes into account the possibility of more than one transition during $(t, t+\Delta t)$.

The situation can be represented by a graph

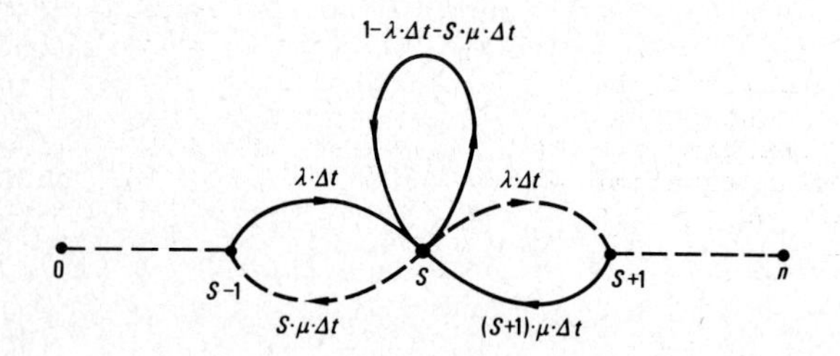

Fig. 4 : Simple "birth and death" - process

The states of the system are the junction points on the graph and the probabilities are written beside the corresponding arrows.

We thus have a special case of a birth and death process, the λ being the birthrates and the μ the deathrates.

If we now transfer the term P(s;t) of the equation (1) to the left side, divide the whole expression by Δt and let Δt approach zero we obtain the following differential equation system, called the Kolmogorov difference - differential equation

$$P'(s;t) = \lambda P(s-1;t) - (\lambda + s\cdot\mu)\,P(s;t) + (s+1)\,\mu P(s+1;t) \qquad (2)$$

$$0 \leqq s \leqq n \quad \text{and} \quad P(-1;t) = P(n+1;t) = 0 \quad .$$

The solution of this system depends on the parameters λ and μ and on the initial values P(s;0) at time zero, as is well known. The P(s;t) are often called in traffic theory the equations of state. This system can be solved explicitly by Laplace transformation. But for telephone traffic problems it is generally sufficient to know only the limiting functions

$$\lim_{t\to\infty} P(s;t) = P(s) \qquad 0 \leqq s \leqq n \quad .$$

The existance of the limit P(s) indicates that the influence of the initial state disappears and the steady state conditions are reached. For the limiting functions we derive from (2) because of P'(s) = 0 the linear equation system

$$0 = \lambda \cdot P(s-1) - (\lambda + s\cdot\mu)P(s) + (s+1)\,\mu\cdot P(s+1) \qquad (3)$$

$$0 \leqq s \leqq n \quad \text{and} \quad P(-1) = P(n+1) = 0 \quad .$$

If we divide these equation by μ and put

$$\frac{\lambda}{\mu} = A$$

we obtain

$$0 = A\cdot P(s-1) - (A+s)\cdot P(s) + (s+1)\,P(s+1 \qquad (4)$$

$$0 \leqq s \leqq n \quad ; \quad P(-1) = P(n+1) = 0 \, .$$

The steady state solution thus depends only on two parameters; the number n of lines of the full available trunk group and the mean traffic volume A offered by the process.

The equation system (4) can be easily solved recursively, resulting in

$$P(s) = \frac{A^s}{s!} P(O)$$

and by use of the additional equation

$$\sum_{i=o}^{n} P(i) = 1$$

we finally obtain

$$P(s) = \frac{\frac{A^s}{s!}}{\sum_{i=o}^{n} \frac{A^i}{i!}} \tag{5}$$

For the probability that a call will be blocked we derive

$$E_n(A) = \frac{A \cdot P(n)}{A} = \frac{\frac{A^n}{n}}{\sum_{i=o}^{n} \frac{A^i}{i!}} \tag{6}$$

This is the famous Erlang loss formula first published in 1917 by A.K. Erlang. In honour of him the unit of the traffic amount is called Erlang. Therefore a traffic needing in the mean 4 lines has an intensity of 4 Erlangs.

The conditions for the validity of the Erlang loss formula are:

- the probability that a call is offered at a certain interval is independent of time and of the number of occupied lines i.e. "unlimited number of equivalent independent traffic sources"
- all the lines of the given bundle are available to arriving calls i.e. "full availability of the bundle"
- a free line will be occupied immediately by an incoming call i.e. "no additional orientation time"
- if a call is offered at a moment when all the lines are busy it is discarded and will not be repeated i.e. "no waiting, no repetition for calls finding no free line"

. the probability of calls ending is independent. For our derivation we had to suppose beyond this, that the probability of a call ending is independent of the time it lasts. But it can be shown that this condition is not necessary for the validity of the Erlang loss formula.

The Erlang loss formula is a sufficient approximation when the number of the equivalent traffic sources is finite but large. It serves therefore as a basis for traffic calculations for most of the telephone administrations in the world.

In the same way we can derive an equation system for the limiting probabilities of state in the case of a pure waiting system with negative exponential input and call durations.

$$\begin{aligned} 0 &= A \cdot P(s-1) - (A+s)\,P(s) + (s+1)\,P(s+1) \\ &\text{for} \quad s < n \;; \quad P(-1) = 0 \quad \text{and} \\ 0 &= A \cdot P(s-1) - (A+n)\,P(s) + nP(s+1) \\ &\text{for} \quad s \geq n \end{aligned} \tag{7}$$

The necessary condition for the existence of this system is now given by

$$\frac{A}{n} = \rho < 1$$

From the system (7) we can derive the probability of waiting for a incoming call. This probability is given by

$$Q(>0) = \sum_{s=n}^{\infty} P(s) = \frac{P(n)}{1 - \frac{A}{n}}$$

This finally gives

$$Q(>0) = \frac{\frac{A^n}{n!} \frac{n}{n-A}}{\sum_{i=o}^{n-1} \frac{A^i}{i!} + \frac{A^n}{n!} \frac{n}{n-A}}$$

This formula, often characterized by $E_{2,n}(A)$, was also found by Erlang in 1917. It is always larger than the loss probability $E_{1,n}(A)$ for the same values of the parameters A and n.
For a waiting system not only the probability of waiting $Q(>0) = E_{2,n}(A)$ is of interest but also the probability of a waiting time larger than a given value t, that means $Q(>t)$. This probability

depends on the so-called queue discipline i.e. the order in which the waiting calls are served.

There are different possibilities, i.e.

1. The first call to arrive will be served first "First in - First out".

2. The call to be served will be chosen at random from the waiting calls. "Random service".

3. The call to be served will be selected according to various priorities imposed on the calls. "Priority service".

4. The last call to arrive will be served first. "Last in - First out".

I'll only mention the formula for the "First in - First out" case. Then we have

$$P(>t) = P(>0) \cdot e^{-\mu(n-A)t} \quad .$$

The mean waiting time M, when calls not waiting are included, is found directly to be

$$M = P(>0)\,\frac{1}{\mu(n-A)} = P(>0)\,\frac{t_m}{(n-A)} \quad .$$

The mean waiting time M_w for the waiting calls alone is then

$$M_w = \frac{t_m}{(n-A)} \quad .$$

This is an interesting relation. It follows that the minimum of the mean waiting time is

$$\frac{t_m}{n} = \frac{\text{mean holding time}}{\text{number of lines}}$$

A call which has to wait must wait in the mean at least for the mean congestion time, no matter how small the offered traffic A and therefore the probability P(>0) of waiting may be.

A discussion of results for other queue disciplines will be omitted here as they are much more complicated and only of interest to specialists.

But let us now go back to the assumptions made for the derivation of both the Erlang formulas. These assumptions were the starting point of telephone traffic theory, more than fifty years ago. A lot of work has since been done by mathematicians and engineers all over the world. Theory progressed in different directions by weakening or altering the initial assumptions. Let me discuss here some of these developments.

You remember the special birth and death process we found when treating the loss system. It is easy to generalize this process to produce a general birth and death process by choosing the following assumptions:

- the system has a finite or countable infinite number of states 0 to .... n ...

- it changes only through transitions from states to their next neighbors, i.e. from state s to s-1 and s+1 only

- if at any time t the system is in state s then the probability that during the time interval $(t, t+\Delta t)$ a transition from s to s+1 (if $s<n$) occurs is

$$\lambda_s \, \Delta t + O\left((t)^2 \right)$$

  where λ_s is a non-negative constant depending on s

- if at any time t the probability that during $(t,t+\Delta t)$ a transistion from s to s-1 occurs (if $s>0$) is:

$$\mu_s \, \Delta t + O\left((\Delta t)^2 \right)$$

  where μ_s is a non-negative constant depending on s

- the probability for more than one transition during $(t,t+\Delta t)$ is

$$O\left((\Delta t)^2 \right) \quad .$$

The change from s to s+1, the birth, may be interpreted as the origination of a new call, whereas the change from s to s-1, the death may be interpreted as the termination of one of the calls already in progress.

It is easy to see that these assumptions include for instance the case of a finite number of traffic sources. We can derive again the Kolmogorov difference-differential system, now reading

$$P'(s;t) = \lambda_{s-1} \, P(s-1;t) - (\lambda_s + \mu_s) \, P(s;t) + \mu_{s+1} \, P(s+1;t)$$

$$0 \leqq s \leqq n \qquad P(-1;t) = P(n+1;t) = 0 \quad ,$$

and for the steady state probabilities a corresponding linear equation system giving finally for the steady state probability of s occupied lines

$$P(s) = \frac{\lambda_o \lambda_1 \cdots \lambda_{s-1}}{\mu_1 \mu_2 \cdots \mu_s} \quad P(o)$$

where P(o) is to be determined by

$$\sum_{i=o}^{n} P(i) = 1 \qquad .$$

In this case we still kept the "extinguishing - the-past - " property of the negative exponential function for the arrival and the termination of calls, resulting in a Markov process. Now we may look at the possibility of dropping this assumption. Assume a process where the interarrival times between incoming calls are still statistically distributed and independent of each other, but the interarrival function

$$P_A(>t) = F(t)$$

now not being the negative exponential function. Then at an arbitrary time instant the probability of the future behavior of the process is no longer independent of its history. But such an input process has still certain points on the time scale where history is extinguished. There are the instants where calls are arriving. At these "equilibrium points" the stochastic process starts afresh according to its interarrival function. Such stochastic processes, where at certain instances the process starts again, are called renewal processes.

They have been dealt with in a multitude of papers on probability theory. Compound processes where all the components are general renewal processes usually are no longer renewal processes themselves as the renewal points of the different processes don't coincide. But there is an important exception, when only one of these processes is a general renewal process, the others being exponentially distributed. In this case renewal points for the compound process exist and for calculations it is possible to take only these points into consideration. Look for instance at a loss system with a fully available bundle of lines where the input process is a general renewal process but the holding times are described by a negative exponential distribution. Although the general congestion process Z(t) in continous time is <u>not</u> Markovian it is possible to extract from it a Markov chain in <u>discrete</u> time, which sufficiently describes the behavior of the process. This method is known as the embedded Markov-chain (D.G. Kendall, 1953) .

In the present case the set Z of renewal points consists of the instants when calls arrive to the system. In fact, Z_ν - the state at the ν-th point - depends only on $Z_{\nu-1}$, because interarrival periods are independent and for the negative exponentially distributed holding times all the points of the time scale are renewal points. The process
$\{Z_\nu; \nu = 1,2,....\}$ is thus the Markov chain with stationary transition probabilities denoted by

$$P_{ij} = p(Z_{\nu+1} = j \mid Z_\nu = i)$$

To calculate the P_{ij} it is necessary to bridge the gap between two consecutive renewal points. For these conditional probabilities the following expressions can be derived

$$P_{ij} = \begin{cases} 0 & \text{for } i < j - 1 \\ \binom{i+1}{j} \int_0^\infty e^{-\mu t j} (1-e^{-\mu t})^{i+1-j} \, dU(t) & \text{for } j - 1 \leq i \leq n - 1 \\ P_{n-1,j} & \text{for } i = n \end{cases}$$

U(t) being the renewal function for the arriving calls and $1/\mu$ being the mean holding time.
The final probabilities for an incoming call to find j lines busy can then be calculated from the linear equation system

$$\sum_{i=0}^{n} P_{ij} P(i) = P(j)$$

with

$$\sum_{i=0}^{n} P(i) = 1$$

The explicit solution of this system may be omitted here as it is rather voluminous and needs a somewhat lengthy derivation.
In certain cases it is even possible to extract a Markov chain from a process without any renewal points. Look for instance at a call process where the arriving calls are flowing at a fully available bundle of n lines, the interarrival times of the calls are distributed on a negative exponential basis with the mean $1/\lambda$, the call durations are constant and given by $1/\mu$ and where blocked calls can wait.

This process as a whole obviously has no renewal points. But now extract from this process a sequence of time instants which are equally spaced in time with distances $1/\mu$ and ask only for the number of calls present. Then the state Z_ν is dependent only on $Z_{\nu-1}$, as all the calls in progress at $Z_{\nu-1}$ will have terminated. All calls present at Z_ν were either waiting calls in state $Z_{\nu-1}$ or arrived during the time interval

$\{t(Z_{\nu-1}), t(Z_\nu)\}$. In this case the limiting probability of finding exactly i calls present at the points of the Markov chain Z_ν is given by

$$P(i) = \left(\sum_{j=o}^{n} P(j) \right) \frac{A^i}{i!} e^{-A} + \sum_{j=1}^{n} P(n+j) \frac{A^{i-j}}{(i-j)!} e^{-A}$$

when $A = \frac{\lambda}{\mu}$.

This can be derived as follows. The event of i calls present at the end of the interval equal to one holding time $1/\mu$ can arise from the following exclusive events

- no more than n calls were present at the beginning of this interval and exactly j calls arrived during $1/\mu$,
- there were n+1 calls present, i.e. one call waiting at the beginning of the interval, and exactly i-1 calls arrived,
- and so on until finally there were n+i calls present at the beginning of the interval and none arrived.

The derivation of the explicit solution of this system may be omitted again .

These are not yet the most advanced results in this field. The use of further refined mathematical methods resulted in solutions even for the case were the input as well as the occupation processes are of the general renewal type. Beyond this many investigations were made on systems modified in a different way. As examples I mention systems with limited waiting positions and with bulk arrival. If you are interested you may refer to the relevant literature.

All these results are of some theoretical importance for telephone traffic engineering as they constitute the background knowledge for the dimensioning processes. But there is another development of greater importance for the practical work of telephone engineers. In this case the input and the termination process of the calls are kept negative exponential but the full availability assumption of the line bundle is dropped. This means that each incoming call is no longer able to choose one line out of the whole bundle but only one of a subset depending on the incoming position and possibly on the found occupation pattern.

As only a few coefficients of the matrix are different from zero, matrices of this type are called sparse matrices. They may be solved by special methods. Iteration methods are well suited. Personally, I have had good experience with the overrelaxation method. I have solved equation systems of this type with up to one thousand unknown probabilities, but it should be possible to solve equation systems with ten thousand unknowns and more.

Though these are rather large numbers for equation systems they are not at all sufficient for arrangements encountered in telephony. Suppose for instance you have ten linegroups with only three lines each thus having a total of thirty lines the corresponding equation system has the rank of

$$(3+1)^{10} = 2^{20}$$

which is more than one million.
For actually encountered arrangements you would have to solve systems with hundreds or thousands of millions of unknowns. It is therefore understandable that in telephone traffic engineering extensive use is made of approximation methods. I'll mention in this connection the problem of overflow but I do not think these specially tailored methods can be the subject of a presentation in this connection. So let me close with a look of a very general method used for the solution of telephone traffic problems too: the simulationmethod.

Simulation of a traffic process, that is the imitation of all the events on an artificial object, i.e. a computer, can be done in the usual way by dividing the time scale into small pieces and deciding for each piece by suitably selected random numbers whether a call should be initiated, terminated or the like. But it should be noted that in order to determine the sought values of traffic parameters it is not necessary to simulate accurately the given process. It may, on the contrary, be advisable to simulate a process which leads to a set of ordinary difference equations with the same stationary solution as the set of differential equations for the given process. The advantage of this procedure may be the elimination of systematic errors due to the non-ideal simulation of the exponential functions and the speeding up of the convergence. This is indeed possible for telephone traffic. Take for instance $N = p + q \cdot n$ equally probable random numbers, where

p = number of random events that may cause the beginning of a call

q = number of random events per outgoing line that may cause the termination of a call, if any, present at the line

n = number of lines

Then the equivalent to a random traffic with the Erlang value

$$A = p/q$$

is produced, as can be shown.

BIBLIOGRAPHY

Beneš, V.E. — General stochastic processes in the theory of queues. Reading (1963)

Beneš, V.E. — Mathematical theory of connecting networks and telephone traffic. New York, London (1965)

Brockmeyer, E., Halstrøm, H.L., Jensen, A. — The life and works of A.K. Erlang. Kopenhagen (1948)

Cohen — The single server queue. Amsterdam (1969)

Cox, D.R., Smith, W.L. — Queues. London (1961)

Gnedenko, B.W., Kovalenkov, I.N. — Introduction to queuing theory. Jerusalem (1968)

Jaiswal, N.K. — Priority Queues. New York (1968)

Kaufmann, A., Cruon, R. — Les phénomènes d'attente. Paris (1968)

Khintchine, A.Y. — Mathematical methods in the theory of queuing. London (1960)

Le Gall, P. — Les systèmes avec et sans attente et les processus stochastiques. Paris (1962)

Meyer, K.H.F. — Wartesysteme mit variabler Bearbeitungsrate. Berlin (1971)

Morse, P.M. — Queues, Inventories and Maintenance. New York (1963)

Newell, G.F. — Applications of queuing theory. London (1971)

Palm, C. — Intensitätsschwankungen im Fernsprechverkehr. Stockholm, Ericsson Techn. (1943)

Pollaczek, F. — Théorie analytique des problèmes stochastiques relatifs à un groupe de lignes téléphoniques avec dispositif d'attente. Paris (1961)